AP® PHYSICS 2 PREP

9th Edition

The Staff of The Princeton Review

PrincetonReview.com

Penguin
Random
House

The Princeton Review
110 East 42nd St, 7th Floor
New York, NY 10017

Published in the United States by Penguin Random House, LLC, New York.

ISBN: 978-0-593-51682-9
ISSN: 2688-1624

The material in this book is up-to-date at the time of publication. However, changes may have been instituted by the testing body in the test after this book was published.

If there are any important late-breaking developments, changes, or corrections to the materials in this book, we will post that information online in the Student Tools. Register your book and check your Student Tools to see if there are any updates posted there.

Editor: Selena Coppock
Production Editors: Liz Dacey, Heidi Torres
Production Artist: Deborah Weber

Printed in the United States of America.

10 9 8 7 6 5 4 3 2 1

9th Edition

For customer service, please contact **editorialsupport@review.com**, and be sure to include:

- full title of the book
- ISBN
- page number

Acknowledgments

The Princeton Review would like to extend special thanks to Felicia Tam for her contributions to the 9th edition of this book. We are also, as always, very appreciative of the time and attention given to each page by Liz Dacey, Heidi Torres, and Deborah Weber.

Contents

at PrincetonReview.com/prep

As easy as 1·2·3

1

Go to PrincetonReview.com/prep or scan the **QR code** and enter the following ISBN for your book:

9780593516829

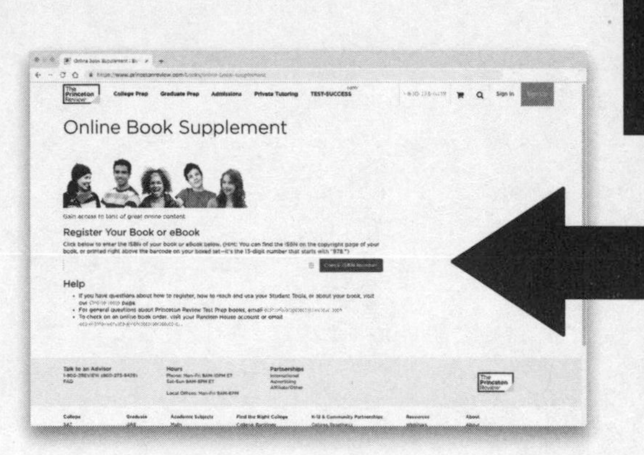

2

Answer a few simple questions to set up an exclusive Princeton Review account. *(If you already have one, you can just log in.)*

3

Enjoy access to your **FREE** content!

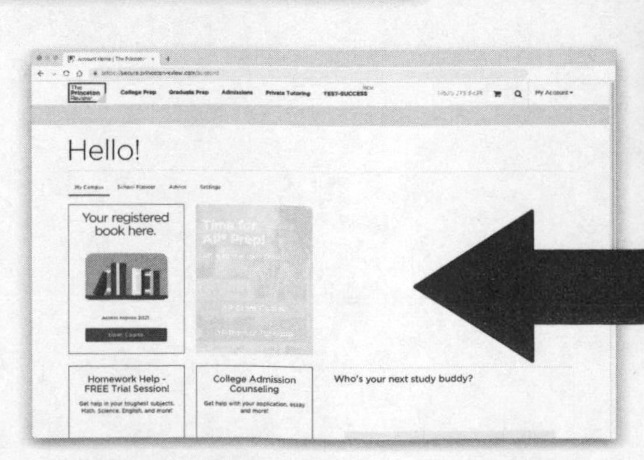

Once you've registered, you can...

- Get our take on any recent or pending updates to the AP Physics 2 Exam
- Access comprehensive study guides and a variety of printable resources, including bubble sheets for the practice tests in the book as well as important equations and formulas
- Take a full-length practice SAT and ACT
- Get valuable advice about the college application process, including tips for writing a great essay and where to apply for financial aid
- Use our searchable rankings of *The Best 389 Colleges* to find out more information about your dream school
- Check to see if there have been any corrections or updates to this edition

Need to report a potential **content** issue?

Contact **EditorialSupport@review.com** and include:

- full title of the book
- ISBN
- page number

Need to report a **technical** issue?

Contact **TPRStudentTech@review.com** and provide:

- your full name
- email address used to register the book
- full book title and ISBN
- Operating system (Mac/PC) and browser (Chrome, Firefox, Safari, etc.)

Look For These Icons Throughout The Book

APPLIED STRATEGIES

ONLINE ARTICLES

PROVEN TECHNIQUES

MORE GREAT BOOKS

TIME-SAVING TIP

Part I
Using This Book to Improve Your AP Score

- Preview: Your Knowledge, Your Expectations
- Your Guide to Using This Book
- How to Begin

PREVIEW: YOUR KNOWLEDGE, YOUR EXPECTATIONS

Your route to a high score on the AP Physics 2 Exam will depend on how you plan to use this book. Please respond to the following questions.

Stay Up to Date!
For late-breaking information about test dates, exam formats, and any other changes pertaining to AP Physics 2, make sure to check the College Board's website at https://apstudents.collegeboard.org/courses/ap-physics-2-algebra-based

1. Rate your level of confidence about your knowledge of the content tested by the AP Physics 2 Exam.

   A. Very confident—I know it all
   B. I'm pretty confident, but there are topics for which I could use help
   C. Not confident—I need quite a bit of support
   D. I'm not sure

2. If you have a goal score in mind, circle your goal score for the AP Physics 2 Exam.

   5 4 3 2 1 I'm not sure yet.

3. What do you expect to learn from this book? Circle all that apply to you.

   A. A general overview of the test and what to expect
   B. Strategies for how to approach the test
   C. The content tested by this exam
   D. I'm not sure yet.

YOUR GUIDE TO USING THIS BOOK

This book is organized to provide as much—or as little—support as you need, so you can use this book in whatever way will be most helpful for improving your score on the AP Physics 2 Exam.

- The remainder of **Part I** provides guidance on how to use this book and helps you determine your strengths and weaknesses.
- **Part II** of this book contains your first practice test, its Diagnostic Answer Key, its answers and explanations, and a scoring guide. (Bubble sheets can be found online in your Student Tools.) Begin your test preparation with this practice test in order to realistically determine:
  - your starting point right now
  - which question types you're ready for and which you might need to practice
  - which content topics you are familiar with and which you will want to carefully review

  Once you have nailed down your strengths and weaknesses with regard to this exam, you can focus your test preparation, build a study plan, and be efficient with your time. Our Diagnostic Answer Key will assist you with this process.

- **Part III** of this book will:
    - provide information about the structure, scoring, and content of the AP Physics 2 Exam
    - help you to make a study plan
    - point you toward additional resources
- **Part IV** of this book will explore the following strategies:
    - how to approach multiple-choice questions
    - how to approach free-response questions
    - how to manage your time to maximize the number of points available to you
- **Part V** of this book covers the content you need for the AP Physics 2 Exam.
- **Part VI** of this book contains Practice Test 2, a Diagnostic Answer Key, detailed answer explanations, and a scoring guide. (Bubble sheets can be found online in your Student Tools.) If you skipped Practice Test 1, we recommend that you do both (with at least a day or two between them) so that you can compare your progress between the two.

  Additionally, taking both tests will help to identify any external issues: if you get a certain type of question wrong both times, you probably need to review it. If you get it wrong only once, you may have run out of time or been distracted by something. In either case, completing both practice tests will allow you to focus on the factors that caused the discrepancy in scores and to be as prepared as possible on the day of the test.

Don't Forget!
When you register your book at PrincetonReview.com/prep following the instructions on pages x–xi, you'll gain access to a wealth of other helpful Student Tools, including study guides and printable bubble sheets.

You may choose to use some parts of this book over others, or you may work through the entire book. Your approach will depend on your needs and how much time you have. Let's now look at how to make this determination.

HOW TO BEGIN

1. **Take Practice Test 1**

   Before you can decide how to use this book, you need to take a practice test. Doing so will give you insight into your strengths and weaknesses, and the test will also help you make an effective study plan. If you're feeling test-phobic, remember that a practice test is a tool for diagnosing yourself—it's not how well you do that matters but how you use information gleaned from your performance to guide your preparation.

   So, before you read further, take AP Physics 2 Practice Test 1 starting on page 9 of this book. Be sure to do so in one sitting, following the instructions that appear before the test.

2. Check Your Answers

Using the Diagnostic Answer Key on page 30, follow our three-step process to identify your strengths and weaknesses with regard to the tested topics. This will help you determine which content review chapters to prioritize when studying this book. Don't worry about the explanations for now, and don't worry about why you missed questions. We'll get to that soon.

3. Reflect on the Test

After you take your first test, respond to the following questions:

- How much time did you spend on the multiple-choice questions?
- How much time did you spend on each free-response question?
- How many multiple-choice questions did you miss?
- Do you feel you had the knowledge to address the subject matter of the free-response questions?

4. Read Part III of this Book and Complete the Self-Evaluation

Part III will provide information on how the test is structured and scored. It will also set out areas of content that are tested.

As you read Part III, reevaluate your answers to the questions above. At the end of Part III, you will revisit the questions above and refine your answers to them. You will then be able to make a study plan, based on your needs and time available, that will allow you to use this book most effectively.

5. Engage with Parts IV and V as Needed

Notice the word *engage*. You'll get more out of this book if you use it intentionally than if you read it passively, hoping for an improved score through osmosis.

The strategy chapters in Part IV will help you think about your approach to the question types on this exam. This part opens with a reminder to think about how you approach questions now and then closes with a reflection section asking you to think about how/whether you will change your approach in the future.

The content chapters in Part V are designed to provide a review of the content tested on the AP Physics 2 Exam, including the level of detail you need to know and how the content is tested. You will have the opportunity to assess your understanding of the content of each chapter through test-appropriate questions.

6. **Take Practice Test 2 and Assess Your Performance**

    Once you feel you have developed the strategies you need and gained the knowledge you lacked, you should take Practice Test 2. You should do so in one sitting, following the instructions at the beginning of the test.

    When you are done, check your answers to the multiple-choice sections. See if a teacher will read your free-response answers and provide feedback.

    Once you have taken the test, reflect on what areas you still need to work on, and revisit the chapters in this book that address those deficiencies. Through this type of reflection and engagement, you will continue to improve.

7. **Keep Working**

    After you have revisited certain chapters in this book, continue the process of testing, reflecting, and engaging with Practice Test 2. Consider what additional work you need to do and how you will change your strategic approach to different parts of the test. You can continue to explore areas that can stand to be improved and engage in those areas right up to the day of the test. As we will discuss in Part III, there are other resources available to you, including a wealth of information online at AP Students.

Got a Question?
For answers to test-prep questions for all your tests and additional test-taking tips, subscribe to our YouTube channel at www.YouTube.com/ThePrincetonReview

Part II
Practice Test 1

Practice Test 1

AP® Physics 2 Exam

SECTION I: Multiple-Choice Questions

DO NOT OPEN THIS BOOKLET UNTIL YOU ARE TOLD TO DO SO.

At a Glance

Total Time
90 minutes
Number of Questions
50
Percent of Total Grade
50%
Writing Instrument
Pen required

Instructions

Section I of this examination contains 50 multiple-choice questions. Fill in only the ovals for numbers 1 through 50 on your answer sheet.

CALCULATORS MAY BE USED IN BOTH SECTIONS OF THE EXAMINATION.

Indicate all of your answers to the multiple-choice questions on the answer sheet. No credit will be given for anything written in this exam booklet, but you may use the booklet for notes or scratch work. After you have decided which of the suggested answers is best, completely fill in the corresponding oval on the answer sheet. Give only one answer to each question. If you change an answer, be sure that the previous mark is erased completely. Here is a sample question and answer.

Sample Question	Sample Answer
Chicago is a (A) state (B) city (C) country (D) continent	Ⓐ ● Ⓒ Ⓓ

Use your time effectively, working as quickly as you can without losing accuracy. Do not spend too much time on any one question. Go on to other questions and come back to the ones you have not answered if you have time. It is not expected that everyone will know the answers to all the multiple-choice questions.

About Guessing

Many candidates wonder whether or not to guess the answers to questions about which they are not certain. Multiple-choice scores are based on the number of questions answered correctly. Points are not deducted for incorrect answers, and no points are awarded for unanswered questions. Because points are not deducted for incorrect answers, you are encouraged to answer all multiple-choice questions. On any questions you do not know the answer to, you should eliminate as many choices as you can, and then select the best answer among the remaining choices.

GO ON TO THE NEXT PAGE.

ADVANCED PLACEMENT PHYSICS 2 TABLE OF INFORMATION

CONSTANTS AND CONVERSION FACTORS	
Proton mass, $m_p = 1.67 \times 10^{-27}$ kg	Electron charge magnitude, $e = 1.60 \times 10^{-19}$ C
Neutron mass, $m_n = 1.67 \times 10^{-27}$ kg	1 electron volt, $1 \text{ eV} = 1.60 \times 10^{-19}$ J
Electron mass, $m_e = 9.11 \times 10^{-31}$ kg	Speed of light, $c = 3.00 \times 10^8$ m/s
Avogadro's number, $N_A = 6.02 \times 10^{23}$ mol^{-1}	Universal gravitational constant, $G = 6.67 \times 10^{-11}$ m^3/kg$\cdot$s^2
Universal gas constant, $R = 8.31$ J/(mol$\cdot$K)	Acceleration due to gravity at Earth's surface, $g = 9.8$ m/s^2
Boltzmann's constant, $k_B = 1.38 \times 10\ ^{23}$ J/K	

1 unified atomic mass unit,	$1 \text{ u} = 1.66 \times 10^{-27}$ kg $= 931$ MeV/c^2
Planck's constant,	$h = 6.63 \times 10^{-34}$ J$\cdot$s $= 4.14 \times 10^{-15}$ eV$\cdot$s
	$hc = 1.99 \times 10^{-25}$ J$\cdot$m $= 1.24 \times 10^3$ eV$\cdot$nm
Vacuum permittivity,	$\varepsilon_0 = 8.85 \times 10^{-12}$ C^2/N$\cdot$m^2
Coulomb's law constant,	$k = 1/4\pi\varepsilon_0 = 9.0 \times 10^9$ N$\cdot$m^2/C^2
Vacuum permeability,	$\mu_0 = 4\pi \times 10^{-7}$ (T$\cdot$m)/A
Magnetic constant,	$k' = \mu_0/4\pi = 1 \times 10^{-7}$ (T$\cdot$m)/A
1 atmosphere pressure,	$1 \text{ atm} = 1.0 \times 10^5$ N/m^2 $= 1.0 \times 10^5$ Pa

UNIT SYMBOLS								
	meter,	m	mole,	mol	watt,	W	farad,	F
	kilogram,	kg	hertz,	Hz	coulomb,	C	tesla,	T
	second,	s	newton,	N	volt,	V	degree Celsius,	°C
	ampere,	A	pascal,	Pa	ohm,	Ω	electron volt,	eV
	kelvin,	K	joule,	J	henry,	H		

PREFIXES		
Factor	Prefix	Symbol
10^{12}	tera	T
10^9	giga	G
10^6	mega	M
10^3	kilo	k
10^{-2}	centi	c
10^{-3}	milli	m
10^{-6}	micro	μ
10^{-9}	nano	n
10^{-12}	pico	p

VALUES OF TRIGONOMETRIC FUNCTIONS FOR COMMON ANGLES							
θ	0°	30°	37°	45°	53°	60°	90°
$\sin\theta$	0	1/2	3/5	$\sqrt{2}/2$	4/5	$\sqrt{3}/2$	1
$\cos\theta$	1	$\sqrt{3}/2$	4/5	$\sqrt{2}/2$	3/5	1/2	0
$\tan\theta$	0	$\sqrt{3}/3$	3/4	1	4/3	$\sqrt{3}$	∞

The following conventions are used in this exam.

I. The frame of reference of any problem is assumed to be inertial unless otherwise stated.
II. In all situations, positive work is defined as work done on a system.
III. The direction of current is conventional current: the direction in which positive charge would drift.
IV. Assume all batteries and meters are ideal unless otherwise stated.
V. Assume edge effects for the electric field of a parallel plate capacitor unless otherwise stated.
VI. For any isolated electrically charged object, the electric potential is defined as zero at infinite distance from the charged object.

GO ON TO THE NEXT PAGE.

ADVANCED PLACEMENT PHYSICS 2 EQUATIONS

MECHANICS

$v_x = v_{x0} + a_x t$

$x = x_0 + v_{x0} t + \frac{1}{2} a_x t^2$

$v_x^2 = v_{x0}^2 + 2a_x(x - x_0)$

$\vec{a} = \frac{\sum \vec{F}}{m} = \frac{\vec{F}_{net}}{m}$

$|\vec{F}_f| \le \mu |\vec{F}_n|$

$a_c = \frac{v^2}{r}$

$\vec{p} = m\vec{v}$

$\Delta\vec{p} = \vec{F}\,\Delta t$

$K = \frac{1}{2}mv^2$

$\Delta E = W = F_{\parallel} d = Fd\cos\theta$

$P = \frac{\Delta E}{\Delta t}$

$\theta = \theta_0 + \omega_0 t + \frac{1}{2}\alpha t^2$

$\omega = \omega_0 + \alpha t$

$x = A\cos(\omega t) = A\cos(2\pi f t)$

$x_{cm} = \frac{\sum m_i x_i}{\sum m_i}$

$\vec{\alpha} = \frac{\sum \vec{\tau}}{I} = \frac{\vec{\tau}_{net}}{I}$

$\tau = r_{\perp} F = rF\sin\theta$

$L = I\omega$

$\Delta L = \tau\,\Delta t$

$K = \frac{1}{2}I\omega^2$

$|\vec{F}_s| = k|\vec{x}|$

$U_s = \frac{1}{2}kx^2$

$\Delta U_g = mg\,\Delta y$

$T = \frac{2\pi}{\omega} = \frac{1}{f}$

$T_s = 2\pi\sqrt{\frac{m}{k}}$

$T_p = 2\pi\sqrt{\frac{\ell}{g}}$

$|\vec{F}_g| = G\frac{m_1 m_2}{r^2}$

$\vec{g} = \frac{\vec{F}_g}{m}$

$U_G = -\frac{Gm_1 m_2}{r}$

a = acceleration
A = amplitude
d = distance
E = energy
F = force
f = frequency
I = rotational inertia
K = kinetic energy
k = spring constant
L = angular momentum
ℓ = length
m = mass
P = power
p = momentum
r = radius or separation
T = period
t = time
U = potential energy
v = speed
W = work done on a system
x = position
y = height
α = angular acceleration
μ = coefficient of friction
θ = angle
τ = torque
ω = angular speed

ELECTRICITY AND MAGNETISM

$|\vec{F}_E| = \frac{1}{4\pi\varepsilon_0}\frac{|q_1 q_2|}{r^2}$

$\vec{E} = \frac{\vec{F}_E}{q}$

$|\vec{E}| = \frac{1}{4\pi\varepsilon_0}\frac{|q|}{r^2}$

$\Delta U_E = q\Delta V$

$V = \frac{1}{4\pi\varepsilon_0}\frac{q}{r}$

$|\vec{E}| = \left|\frac{\Delta V}{\Delta r}\right|$

$\Delta V = \frac{Q}{C}$

$C = \kappa\varepsilon_0\frac{A}{d}$

$E = \frac{Q}{\varepsilon_0 A}$

$U_C = \frac{1}{2}Q\Delta V = \frac{1}{2}C(\Delta V)^2$

$I = \frac{\Delta Q}{\Delta t}$

$R = \frac{\rho\ell}{A}$

$P = I\,\Delta V$

$I = \frac{\Delta V}{R}$

$R_s = \sum_i R_i$

$\frac{1}{R_p} = \sum_i \frac{1}{R_i}$

$C_p = \sum_i C_i$

$\frac{1}{C_s} = \sum_i \frac{1}{C_i}$

$B = \frac{\mu_0}{2\pi}\frac{I}{r}$

A = area
B = magnetic field
C = capacitance
d = distance
E = electric field
$\mathcal{E}$ = emf
F = force
I = current
ℓ = length
P = power
Q = charge
q = point charge
R = resistance
r = separation
t = time
U = potential (stored) energy
V = electric potential
v = speed
κ = dielectric constant
ρ = resistivity
θ = angle
Φ = flux

$\vec{F}_M = q\vec{v} \times \vec{B}$

$|\vec{F}_M| = |q\vec{v}||\sin\theta||\vec{B}|$

$\vec{F}_M = I\vec{\ell} \times \vec{B}$

$|\vec{F}_M| = |I\vec{\ell}||\sin\theta||\vec{B}|$

$\Phi_B = \vec{B} \cdot \vec{A}$

$\Phi_B = |\vec{B}|\cos\theta|\vec{A}|$

$\mathcal{E} = -\frac{\Delta\Phi_B}{\Delta t}$

$\mathcal{E} = B\ell v$

GO ON TO THE NEXT PAGE.

ADVANCED PLACEMENT PHYSICS 2 EQUATIONS

FLUID MECHANICS AND THERMAL PHYSICS

$\rho = \frac{m}{V}$

$P = \frac{F}{A}$

$P = P_0 + \rho g h$

$F_b = \rho V g$

$A_1 v_1 = A_2 v_2$

$P_1 + \rho g y_1 + \frac{1}{2}\rho v_1^2 = P_2 + \rho g y_2 + \frac{1}{2}\rho v_2^2$

$\frac{Q}{\Delta t} = \frac{kA\,\Delta T}{L}$

$PV = nRT = Nk_B T$

$K = \frac{3}{2}k_B T$

$W = -P\Delta V$

$\Delta U = Q + W$

A = area
F = force
h = depth
k = thermal conductivity
K = kinetic energy
L = thickness
m = mass
n = number of moles
N = number of molecules
P = pressure
Q = energy transferred to a system by heating
T = temperature
t = time
U = internal energy
V = volume
v = speed
W = work done on a system
y = height
ρ = density

MODERN PHYSICS

$E = hf$

$K_{\text{max}} = hf - \phi$

$\lambda = \frac{h}{p}$

$E = mc^2$

E = energy
f = frequency
K = kinetic energy
m = mass
p = momentum
λ = wavelength
ϕ = work function

WAVES AND OPTICS

$\lambda = \frac{v}{f}$

$n = \frac{c}{v}$

$n_1 \sin\theta_1 = n_2 \sin\theta_2$

$\frac{1}{s_i} + \frac{1}{s_o} = \frac{1}{f}$

$|M| = \left|\frac{h_i}{h_o}\right| = \left|\frac{s_i}{s_o}\right|$

$\Delta L = m\lambda$

$d\sin\theta = m\lambda$

d = separation
f = frequency or focal length
h = height
L = distance
M = magnification
m = an integer
n = index of refraction
s = distance
v = speed
λ = wavelength
θ = angle

GEOMETRY AND TRIGONOMETRY

Rectangle
$A = bh$

Triangle
$A = \frac{1}{2}bh$

Circle
$A = \pi r^2$
$C = 2\pi r$

Rectangular solid
$V = \ell w h$

Cylinder
$V = \pi r^2 \ell$
$S = 2\pi r\ell + 2\pi r^2$

Sphere
$V = \frac{4}{3}\pi r^3$
$S = 4\pi r^2$

A = area
C = circumference
V = volume
S = surface area
b = base
h = height
ℓ = length
w = width
r = radius

Right triangle
$c^2 = a^2 + b^2$

$\sin\theta = \frac{a}{c}$

$\cos\theta = \frac{b}{c}$

$\tan\theta = \frac{a}{b}$

c
a
θ
90°
b

GO ON TO THE NEXT PAGE.

PHYSICS 2
SECTION I
Time—90 minutes
50 Questions

Note: To simplify calculations, you may use $g = 10$ m/s^2 in all problems.

Directions: Each of the questions or incomplete statements below is followed by four suggested answers or completions. Select the one that is best in each case and mark it on your sheet.

1. A positively charged particle enters a region between the plates of a parallel-plate capacitor. The particle is moving initially parallel to the plates. What is the correct description of the trajectory of the particle in the region between the plates?

   (A) The particle will continue straight.
   (B) The particle will move toward the negative plate along a straight-line path.
   (C) The particle will move toward the negative plate along a curved path.
   (D) The particle will move toward the positive plate along a curved path.

2. Which of the following arguments does NOT support the assertion that a parallel-plate capacitor held with one plate parallel to the floor is an electric analog of the gravitational field near the surface of the Earth?

   (A) The voltage is constant along horizontal lines.
   (B) The field changes linearly with the distance from the lower plate.
   (C) Both the gravitational field lines and electric field lines are vertical.
   (D) Both the electrical and gravitational force are governed by inverse square laws.

3. An electromagnetic wave moves along the positive *z*-axis. The wave is polarized so that the magnetic field is confined in the *y*-axis. Where is there a nonzero electric field from this wave?

   (A) The electric field exists everywhere.
   (B) The electric field is only along the *x*-axis.
   (C) The electric field is only along the *y*-axis.
   (D) The electric field is only along the *z*-axis.

4. Three neutral conducting spheres sit on insulating bases. The spheres are separated by a very large distance. The sphere in the center is given a positive charge. The spheres are brought close together, as shown above, but not allowed to come into contact with one another. Which is the correct description of the net charge on each sphere?

   (A) All three spheres are positively charged.
   (B) The center sphere is positively charged and the two outer spheres are negatively charged.
   (C) The center sphere is positively charged and the other two are neutral.
   (D) All three spheres are neutral.

GO ON TO THE NEXT PAGE.

Questions 5–7 refer to the following diagram.

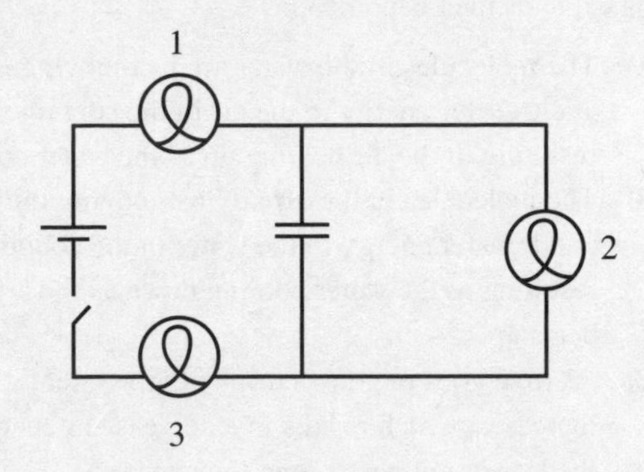

5. The circuit above contains a battery, a switch, three identical light bulbs, and a capacitor. The capacitor is initially uncharged. The instant the switch is closed, which of the bulbs are lit?

    (A) All three bulbs are unlit.
    (B) Only bulb 1 is lit.
    (C) Only bulb 2 is lit.
    (D) Bulb 1 and bulb 3 are lit.

6. After the capacitor is fully charged, which bulbs are lit?

    (A) All three bulbs are unlit.
    (B) Bulb 1 and bulb 3 are lit.
    (C) Only bulb 2 is lit.
    (D) All three bulbs are lit.

7. When the capacitor is fully charged, which components have equal voltage drops across them?

    (A) The capacitor and the battery
    (B) The capacitor and bulb 2
    (C) The battery and the sum of bulbs 1 and 3
    (D) All three of the above selections have equal voltage drops across them.

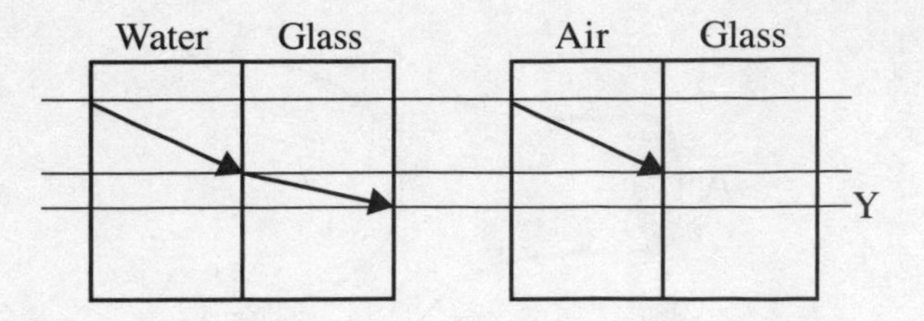

8. A ray of light enters the pair of tanks, as shown above. The light rays in the water and glass tank are shown at the left, with the light exiting the pair of tanks at a point Y. The light ray in the air tank is identical to the ray in the water tank, entering at the same height and same angle. Both sets of tanks use the same glass. How will the ray exiting the second glass tank compare to height Y?

    (A) The ray will not exit the second tank due to total internal reflection.
    (B) The exiting ray will be closer to the top of the tank than Y.
    (C) The exiting ray will be at the same height as Y.
    (D) The exiting ray will be farther from the top of the tank than Y.

9. A mirror may be used to create a real image under which of the following conditions?

    (A) Use a concave mirror and place the object between the mirror and the focal point.
    (B) Use a concave mirror and place the object beyond the focal point.
    (C) Use a convex mirror and place the object between the mirror and the focal point.
    (D) Use a convex mirror and place the object beyond the focal point.

10. The pressure at location A in a pipe is known. The speed of the fluid in the pipe and the pipe diameter are also known at position A. Position B is another location in the pipe. Which pieces of data are required to calculate the speed of the fluid at position B?

    (A) The diameter at B only
    (B) The diameter at B and the height of B relative to A
    (C) The diameter at B, the height of B relative to A, and the pressure at B
    (D) The diameter at B, the pressure at B, the height of B relative to A, and the fluid density

GO ON TO THE NEXT PAGE.

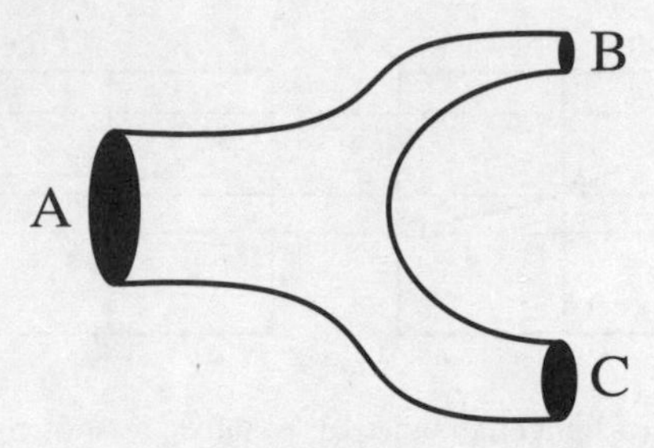

11. The tube shown above carries water. At some point, the tube splits, as shown above. Three points within the tube are labeled A, B, and C. The greatest diameter is at A and the smallest diameter is at B. How do the pressures in the pipes compare at the three points?

(A) $P_A > P_C > P_B$
(B) $P_A = P_B = P_C$
(C) $P_B = P_C > P_A$
(D) $P_B > P_C > P_A$

12. A paper-filled three ring binder sits in a room for a long time. A student touches the metal rings of the binder and the paper. She expects the two objects to feel the same temperature, but she observes that the paper feels warmer than the metal. Which of the following correctly explains her observation?

(A) The objects were in thermal equilibrium, but she must have felt the paper first, as when you feel multiple objects at the same temperature, the ones felt first will feel warmest.
(B) The objects were in thermal equilibrium, but the energy flow occurs more slowly between her fingers and the paper than her fingers and the metal, resulting in the paper feeling warmer.
(C) The objects were in thermal equilibrium and the paper warms up quickly when touched. Her perception is not instantaneous, but is the average temperature over a period of time and higher for the paper.
(D) The paper was warmer than the metal because temperature is related to the kinetic energy of the molecules, and the more massive metal molecules will have less kinetic energy than the paper molecules.

13. A hot cup of water is placed within an insulating container. Later, the water has cooled. Which of the following explains the phenomenon?

(A) The molecules in the water are fast-moving initially and transfer energy in the air in the container, resulting in the air heating up as the water cools.
(B) The molecules in the air are fast-moving initially and transfer energy to the water in the container, resulting in the water cooling down as the air heats up.
(C) As time goes on, the entropy of the system increases, which results in a decrease in energy and the loss of temperature in the cup.
(D) The total heat in the container must remain constant unless there is leakage to the outside surroundings, so the water can only cool down if the surroundings heat up.

14. A very large sheet of metal has a net negative charge. An electron is placed above the center of the sheet. Which of the following correctly describes the force on the sheet of charge from the electron?

(A) The force pulls the sheet upward.
(B) The force pushes the sheet downward.
(C) The electron causes a torque on the sheet, but no net upward or downward force exists.
(D) There is no force on the sheet because the electron is a test charge.

15. An experiment is conducted on a circuit consisting of a battery and several wires. The wires are all made of the same metal. It is found that as the cross-sectional area of the wire increases, the current measured coming out of the battery increases linearly. What can be concluded about the wire from this data?

(A) The battery has an internal resistance, which can be found from the slope of the area vs. current graph.
(B) The battery has an internal resistance, which can be found from the *y*-intercept of the area vs. current graph.
(C) The wires have the same length, which can be found from the slope of the area vs. current graph.
(D) The wires have the same length, which can be found from the *y*-intercept of the area vs. current graph.

GO ON TO THE NEXT PAGE.

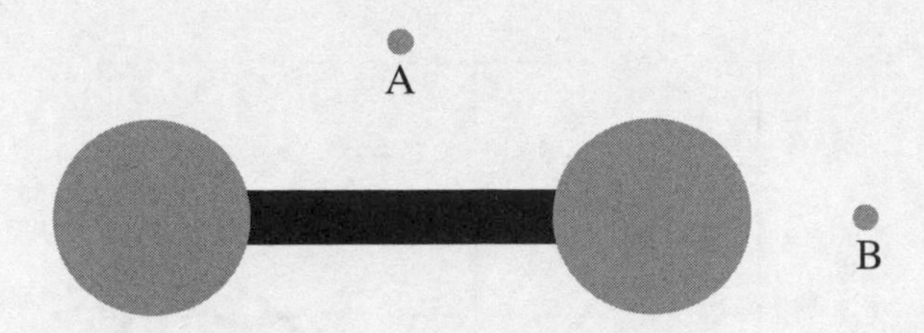

16. An insulating rod separates two conducting spheres as shown above. Point A is midway between the spheres. Point B lies on the axis of the rod. Which of the following arrangements of charges would result in a net torque on the rod?

    (A) Charge both spheres negatively and place a positive charge at point A.
    (B) Charge one sphere negatively and the other positively and place a positive charge at point A.
    (C) Charge both spheres negatively and place a positive charge at point B.
    (D) Charge one sphere negatively and the other positively and place a positive charge at point B.

17. The process of charging by induction requires which physical property of a system?

    (A) Charge polarization
    (B) Uneven charge distribution
    (C) Zero net charge
    (D) Contact between a conductor and an insulator

18. To determine the direction of the force on a charge that is moving near a current-carrying wire, which data do you need?

    (A) Only the direction the charge is moving
    (B) The sign of the charge and its direction of motion
    (C) The sign of the charge, the direction of motion, and the direction of the current
    (D) The sign of the charge, the direction of motion, the direction of the current, and the position of the charge relative to the wire

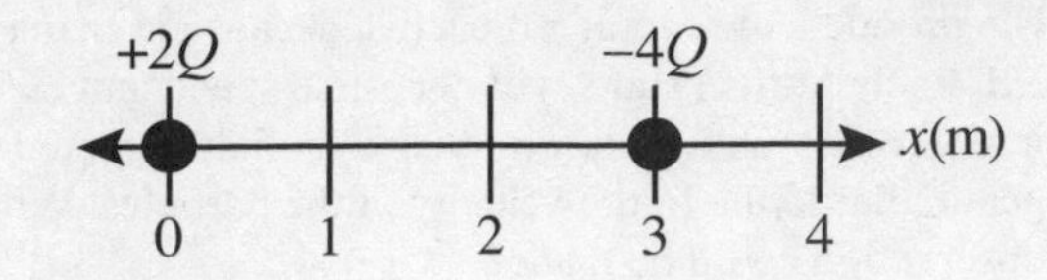

19. A positive charge of $2Q$ and a negative charge of $4Q$ are arranged at positions as shown above. What is the correct ranking of the electric potential at points along the x-axis?

    (A) $V(x = 4\text{ m}) = V(x = 2\text{ m}) > V(x = 1\text{ m})$
    (B) $V(x = 4\text{ m}) > V(x = 1\text{ m}) > V(x = 2\text{ m})$
    (C) $V(x = 1\text{ m}) > V(x = 2\text{ m}) > V(x = 4\text{ m})$
    (D) $V(x = 4\text{ m}) > V(x = 2\text{ m}) > V(x = 1\text{ m})$

20. A large piece of wood has the same weight as a small rock. Both the wood and rock are placed into a pool of liquid. The rock sinks while the wood floats. How do the buoyant forces on the wood and rock compare?

    (A) The buoyant forces are the same on each.
    (B) The buoyant force on the wood is greater than on the rock.
    (C) The buoyant force on the wood is less than on the rock.
    (D) The buoyant forces cannot be compared without the density of the liquid.

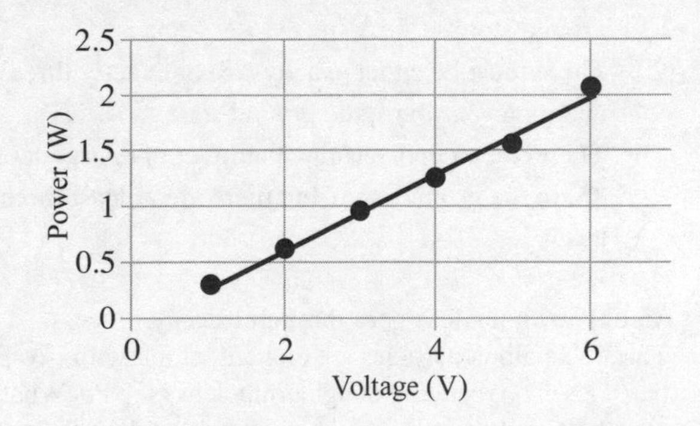

21. An experiment is conducted to determine the power output from a circuit as various voltages are supplied to the circuit. The circuit is set up so that it draws a constant current. What is the value of the current in the circuit that produced the graph above?

    (A) 0.00 A
    (B) 0.25 A
    (C) 1.0 A
    (D) 6.0 A

GO ON TO THE NEXT PAGE.

22. The products of several radioactive decays are being studied. Each particle starts with the same speed and enters into a region with a uniform magnetic field directed perpendicular to the initial velocity of the particles. Which observation could be made?

(A) A positron and an electron both turn in the same direction, but the electron turns with a larger radius.
(B) An alpha particle and an electron both turn in the same direction, but the alpha particle turns with a larger radius.
(C) An alpha particle and an electron both turn in opposite directions, but the alpha particle turns with a larger radius.
(D) A positron and an electron both turn in opposite directions, but the electron turns with a larger radius.

23. A hypothetical atom contains an undetermined number of energy levels above the ground state. A gas of this atom is entirely in the ground state. Light with a broad spectrum is shined upon the gas and the spectrum of the light is recorded after shining through the gas. Three wavelengths of light are observed to have diminished intensity. What can be concluded from this?

(A) There must be exactly two energy levels above the ground state.
(B) There must be exactly three energy levels above the ground state.
(C) There must be either exactly two or exactly three energy levels above the ground state.
(D) There are an undetermined number of energy levels above the ground state, but there are at least three levels.

24. An unknown nucleus goes through a decay process. Three ${}^{4}_{2}$He alpha particles are emitted, along with two β^- particles. The resulting daughter nucleus is ${}^{206}_{82}$Pb. What was the atomic number of the unknown nucleus that started the process?

(A) 84
(B) 86
(C) 88
(D) 90

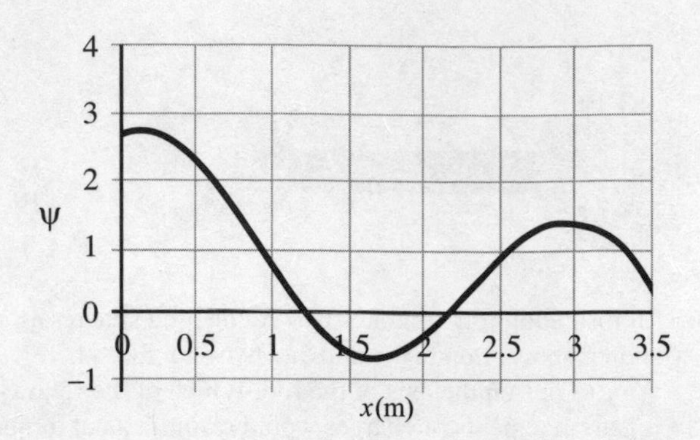

25. The wave function of a quantum object vs. position is graphed above. Which of the following correctly ranks the probabilities of observing the particle at the listed positions?

(A) $P(x = 0.5\text{ m}) > P(x = 2.25\text{ m}) > P(x = 1.75\text{ m})$
(B) $P(x = 0.5\text{ m}) > P(x = 1.75\text{ m}) > P(x = 1.25\text{ m})$
(C) $P(x = 3.0\text{ m}) > P(x = 1.75\text{ m}) > P(x = 2.0\text{ m})$
(D) $P(x = 3.5\text{ m}) > P(x = 3.0\text{ m}) > P(x = 1.75\text{ m})$

26. Radiothorium-228 decays into Radon-220 through two alpha decays, as shown below.

$$^{228}_{90}\text{Th} \rightarrow {}^{220}_{86}\text{Rn} + {}^{4}_{2}\text{He} + {}^{4}_{2}\text{He}$$

Which equation correctly describes the energy released during this process?

(A) $(m_{Th} - m_{Rn})c^2$
(B) $(m_{Th} + 2m_{He} - m_{Rn})c^2$
(C) $(m_{Th} - 2m_{He} - m_{Rn})c^2$
(D) $(m_{Rn} + 2m_{He} - m_{Th})c^2$

GO ON TO THE NEXT PAGE.

Questions 27 and 28 refer to the following graph.

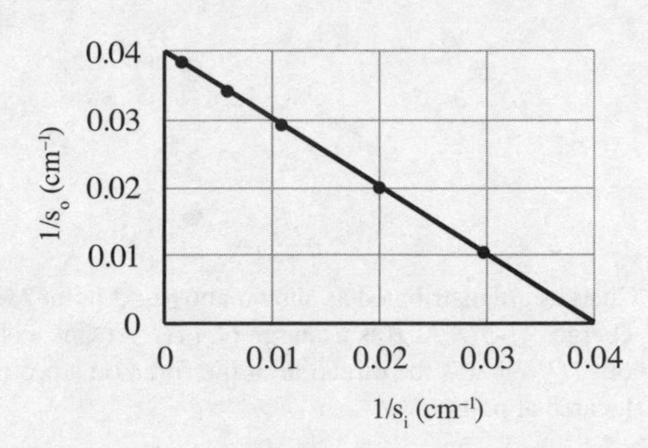

A thin lens is used to make an image for several different object distances. The image and object distances are used to make the above graph.

27. What is the approximate focal length of the lens?

(A) 1 cm
(B) 10 cm
(C) 25 cm
(D) 100 cm

28. In order to achieve a magnification of 1, how far from the lens should an object be placed?

(A) 2 cm
(B) 20 cm
(C) 50 cm
(D) 200 cm

29. A container has a round lid, with mass *M*, and makes an airtight seal with the body of the container. Some of the air in the container is pumped out. The container is turned over, but the lid stays shut. Which of the following explanations for this phenomenon is correct?

(A) Pumping the air out of the container decreases the gravitational force on the lid.
(B) Pumping the air out creates an upward force that will always balance with the gravitational force.
(C) The force on the lid downward from the molecules inside is less than the force on the lid from the particles in the air surrounding the container.
(D) The force on the lid downward from the molecules inside is less than the difference between the force on the lid from the particles in the air surrounding the container and the gravitational force on the lid.

30. A mercury thermometer is placed in a glass of ice-cold water. After some time goes by, the mercury thermometer can be read to determine that the temperature of the water is 0°C. Several minutes later, the reading on the thermometer has not changed. What has happened to the molecules of mercury in the thermometer?

(A) Over time, all the molecules slow down, but the rate of slowing decreases as temperature decreases, so it is not noticeable between the two readings.
(B) The mercury molecules stopped moving upon reaching a temperature of 0°C and remained stopped for the second reading.
(C) The mercury molecules reached thermal equilibrium before the first reading, so all their molecules had the same nonzero speed when both readings were taken.
(D) The mercury molecules reached thermal equilibrium before the first reading, but between the readings, some of the molecules sped up and others slowed down, resulting in the same reading.

GO ON TO THE NEXT PAGE.

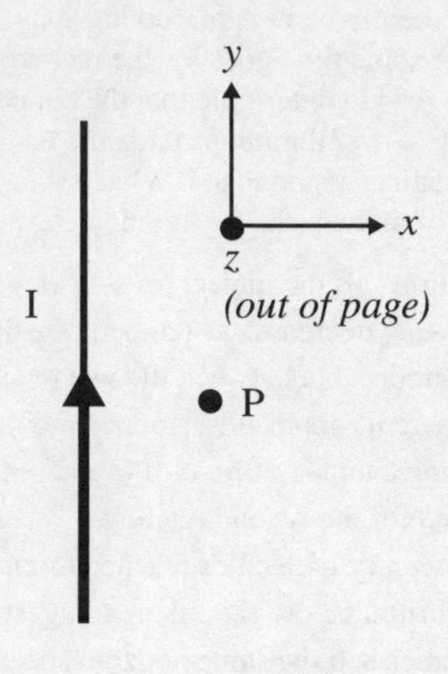

31. A current-carrying wire and coordinate system is shown above. Initially, the wire carries a current I toward the top of the page. The amount of current is steadily decreased until it is 0 A, then steadily increased until it reaches a value of I in the downward direction. This change in current takes time t. A graph is made of the magnetic field versus time at the observation point P for the time duration from $t = 0$ until t is produced. Which statement is true concerning the graph if $+z$ is above the horizontal axis and $-z$ is below the horizontal axis?

(A) The graph has a constant slope.
(B) The graph is piecewise linear with a negative slope for the first half and a positive slope for the second half.
(C) The graph is piecewise linear with a positive slope for the first half and a negative slope for the second half.
(D) The graph is a curve showing an inverse relationship.

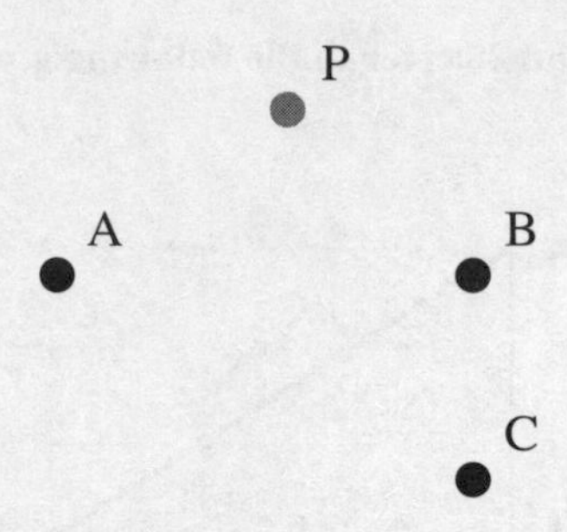

32. Charges are distributed as shown above. At point A is a charge of $+3Q$. At B is a charge of $+1Q$. At C is a charge of $-1Q$. What is the direction of the force on a proton located at point P?

(A) Up and to the left
(B) Down and to the left
(C) Up and to the right
(D) Down and to the right

33. All of the following observations of electric field diagrams are correct EXCEPT

(A) longer arrows correspond to a greater field magnitude
(B) field vectors will be larger at positions closer to sources
(C) arrow lengths will decrease linearly with distance from sources
(D) arrows point in the direction in which a positive charge would experience a force

34. A neutral sphere of metal is placed in a region of uniform electric field. The electric field points toward the top of the page. Which of the following diagrams shows the arrangement of the charges on the surface of the sphere once electrostatic equilibrium is reached?

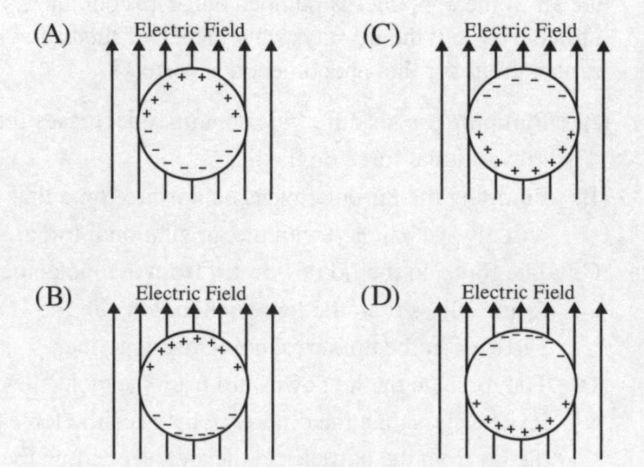

GO ON TO THE NEXT PAGE.

35. An electron enters a region of magnetic field created by a long current-carrying wire. Along which path could the electron travel without being deflected?

    (A) The electron is traveling parallel to the wire and in the same direction that the current is flowing.
    (B) The electron is traveling parallel to the wire and in the opposite direction from the direction that the current is flowing.
    (C) The electron is traveling straight away from the wire.
    (D) The electron is traveling in a circle centered at the wire.

36. Which process will result in charging an uncharged sphere using a charged rod?

    (A) Bringing the rod near the sphere and touching the two objects together, then removing the rod
    (B) Grounding the sphere first, then bringing the charged rod near the sphere and touching the two objects together, then removing the rod
    (C) Bringing the rod near the sphere, but not touching them together, then removing the rod
    (D) Grounding the sphere, bringing the rod near the sphere but not touching them together, then removing the rod

Questions 37 and 38 refer to the following diagram.

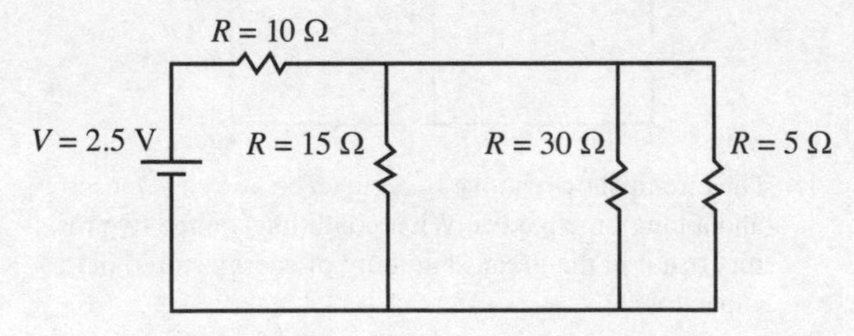

37. The current supported by the battery is most nearly equal to what value?

    (A) 0.042 A
    (B) 0.19 A
    (C) 0.30 A
    (D) 0.68 A

38. A fifth resistor is placed in the circuit. It is parallel beside the 5 Ω resistor. The voltage drop across this resistor is found to be $V = 0.50$ V. What is the resistance of the additional resistor?

    (A) 5 Ω
    (B) 10 Ω
    (C) 15 Ω
    (D) 30 Ω

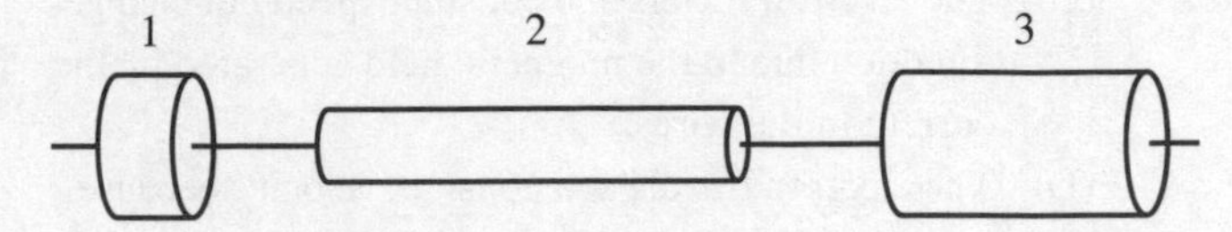

39. Three cylinders of the same metal act as resistors arranged in series, as shown above. Which of the following correctly ranks the voltage drops across the three resistors?

    (A) $V_1 = V_2 = V_3$
    (B) $V_3 > V_1 > V_2$
    (C) $V_2 > V_3 > V_1$
    (D) $V_1 > V_3 > V_2$

40. An electron experiences both electric and magnetic forces. Those forces are balanced. What must be true about the motion of the particle and the alignment of the two fields?

    (A) The fields point along the same axis, and the velocity of the electron is parallel to that axis.
    (B) The fields point along the same axis, and the velocity of the electron is perpendicular to that axis.
    (C) The fields point along the perpendicular axes, and the velocity is parallel to the axis of the electric field.
    (D) The fields point along the perpendicular axes, and the velocity along the third perpendicular axis.

41. A pair of electrons is held a fixed distance apart. Which of the following correctly describes the electrical and gravitational forces between the electrons?

    (A) The electrical force is much greater than the gravitational force.
    (B) The gravitational force is much greater than the electrical force.
    (C) The forces are equal as both the gravitational force and electrical force are governed by inverse square laws.
    (D) The forces cannot be compared without knowing the value of the distance between the electrons.

GO ON TO THE NEXT PAGE.

42. An electron travels down the center of a solenoid, carrying a current I. Which of the following explanations for the motion of the electron is correct?

(A) The electron accelerates due to the electric force from the charges in the current in the wire.
(B) The electron accelerates due to the magnetic field generated by the current in the wires of the solenoid.
(C) The electron travels at a constant speed but changes direction due to the magnetic field generated by the current in the wires.
(D) The electron travels at a constant velocity because there is no net force on the electron.

43. A voltage is sent through a tube containing hydrogen gas. The gas emits light. When the light is sent into a spectrometer, several distinct bright lines are seen. Why does this occur?

(A) The voltage causes the gas to heat up to specific temperatures, which are characterized by the lines.
(B) The voltage causes gas molecules to move at specific speeds, which are characterized by the lines.
(C) The voltage sets up standing waves due to the pressure of the gas in the tube, which can be characterized by the lines.
(D) The voltage excites electrons in the gas into specific energy levels, which can be characterized by the lines.

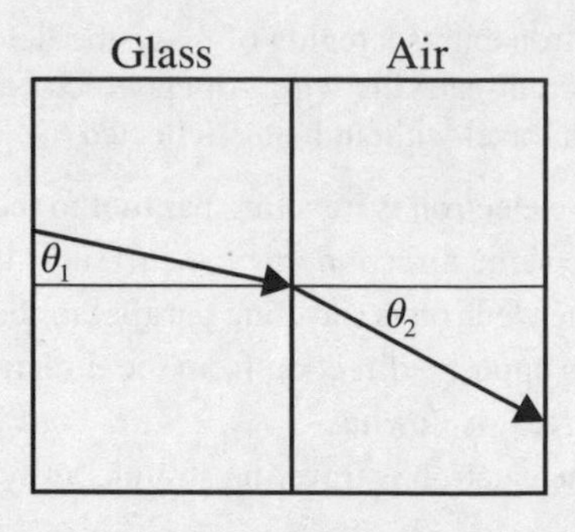

44. An experiment is conducted to determine the critical angle for light going from glass into air, as shown above. A linear plot is made with a vertical axis of sin (θ_1) and a horizontal axis of sin (θ_2). How is the critical angle determined from the graph?

(A) The critical angle cannot be found from the line.
(B) The critical angle can be found from the y-intercept of the line.
(C) The critical angle can be found from the horizontal axis value, which corresponds to the maximum vertical value.
(D) The critical angle can be found from the vertical axis value, which corresponds to the maximum horizontal value.

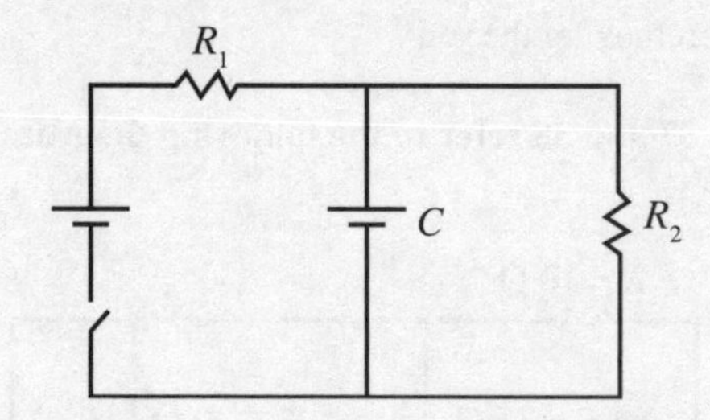

45. The circuit shown above is set up. The switch is closed and a long time passes. What conditions on the two resistors result in the greatest amount of energy stored in the capacitor?

(A) The energy stored in the capacitor will be greatest if $R_1 > R_2$.
(B) The energy stored in the capacitor will be greatest if $R_1 = R_2$.
(C) The energy stored in the capacitor will be greatest if $R_1 < R_2$.
(D) The energy will be the same regardless of the resistor values.

GO ON TO THE NEXT PAGE.

Directions: For questions 46–50 below, two of the suggested answers will be correct. Select the two answers that are best in each case, and then fill in both of the corresponding circles on the answer sheet.

46. A wire is placed vertically and carries a current in the upward direction. A compass is located directly to the north of the wire. Which of the following observations can be made about the compass needle deflection? Select two answers.

(A) The deflection is greater farther from the wire.
(B) The deflection is to the west.
(C) The deflection is greater if the current is increased.
(D) The deflection changes direction as the compass is moved upward.

47. An ideal gas is confined to a leakproof box. What type of processes could occur to cause the gas to absorb heat, but have no net work done on the gas? Select two answers.

(A) Isothermal expansion to a new volume, then isobaric compression to the original volume
(B) A doubling of pressure at constant volume
(C) A doubling of pressure at constant volume, then doubling the volume at constant pressure, then halving the pressure at constant volume, and finally halving the volume at constant pressure to return to the initial state
(D) A doubling of pressure at constant volume, then halving the pressure at constant volume, then doubling the volume at constant pressure, then finally halving the volume at constant pressure to return to the initial state

48. Which of the following results of the photoelectric effect give support for the particle nature of light? Select two answers.

(A) There is a threshold frequency below which no electrons are emitted.
(B) Above the threshold frequency, the number of electrons emitted increases with increasing intensity.
(C) The stopping voltage is related only to the maximum kinetic energy electron, not all of the electrons ejected from the metal.
(D) The stopping voltage increases linearly with frequency above the threshold frequency regardless of the light intensity.

49. A circuit contains four resistors R_1, R_2, R_3, and R_4 and a battery. The first two are in series and the second two are in parallel. How can a fifth resistor, R_5, be added to the circuit to increase the current supported by the battery? Select two answers.

(A) Place R_5 in parallel with R_1.
(B) Place R_5 in series with R_2.
(C) Place R_5 in parallel with R_3.
(D) Place R_5 in series with R_4.

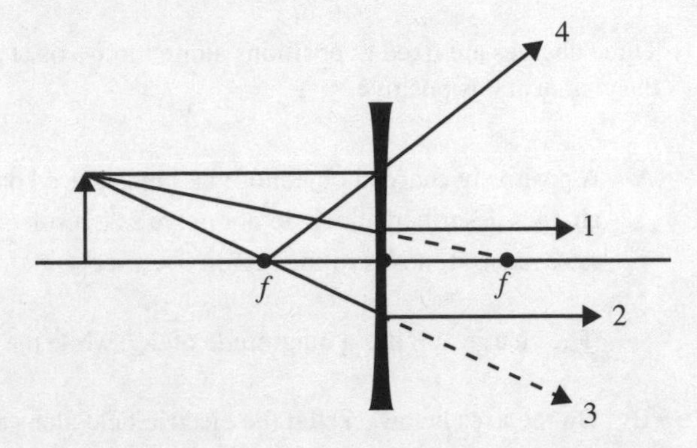

50. For the diverging lens shown above, which principal rays are correctly drawn? Select two answers.

(A) 1
(B) 2
(C) 3
(D) 4

END OF SECTION I

PHYSICS 2
SECTION II
Time—90 minutes
4 Questions

Directions: Questions 1 and 2 are long free-response questions that require about 25 minutes to answer. Questions 3 and 4 are short free-response questions that require about 20 minutes to answer. On test day, you will be asked to show your work for each part in the space provided after that part. For this practice test, you may use scrap paper.

1. Three charges are fixed at positions along the x-axis at positions $-d$, 0, and $+d$. The charges at $-d$ and $+d$ are both negative, and the charge at 0 is positive.

    (A) A positively charged object of mass m is placed on the x-axis between 0 and $+d$, closer to the position $x = 0$. If the three charges described above do not move as a result of this new charged object, describe the motion of the object after it is released as it moves in the region $0 < x < d$.

    The charge at 0 has a magnitude of $2Q$, while the other two charges have a magnitude of Q.

    (B) On the axes below, sketch the electric field along the x-axis in the vicinity of the charges. An electric field to the right should be graphed as positive and a field pointing left should be graphed as negative.

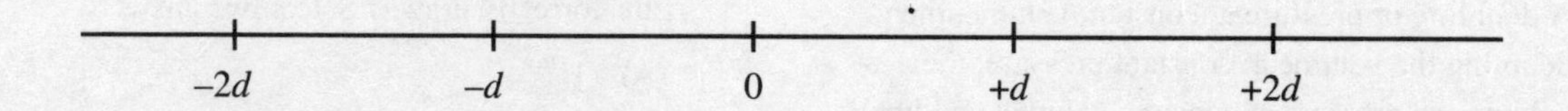

    (C) Write a mathematical function, $E(x)$, that gives the value of the electric field at any position along the x-axis for $0 < x < d$. Give your answer in terms of Q, d, and fundamental constants.

    (D) In order to originally assemble the three charges on the x-axis, some work had to be done. Consider arranging the charges along the x-axis in the following manner: first, bring the $+2Q$ charge to position $x = 0$, then bring the $-Q$ charge to $x = +d$, and finally, bring in the last charge. Bringing the $+2Q$ charge to position 0 required no work. Bringing in the second charge required an amount of work W. Explain whether bringing in the third charge will require more work, less work, or an amount of work equal to W.

GO ON TO THE NEXT PAGE.

2. A student has a convex lens of unknown focal length. He lights a candle in a darkened room and uses the lens and moves a screen until he forms a sharp image. He then records the distance from the candle to the lens and the distance from the lens to the screen. Below is a sketch of his setup and his data.

s_o (cm)	s_i (cm)
15	61.5
30	19.3
50	15.6
80	13.9
90	13.7
110	13.4

Selected values of the object and image distances

s_o^{-1} (cm^{-1})	s_i^{-1} (cm^{-1})
0.067	0.016
0.050	0.034
0.033	0.052
0.020	0.064
0.017	0.065
0.014	0.068

Selected values of the inverse of the object and inverse of the image distances

(A) Explain how a graph of $1/s_o$ vs. $1/s_i$ can be used to find the focal length of the lens.

(B) Create a graph of and find the focal length of the lens used in the experiment.

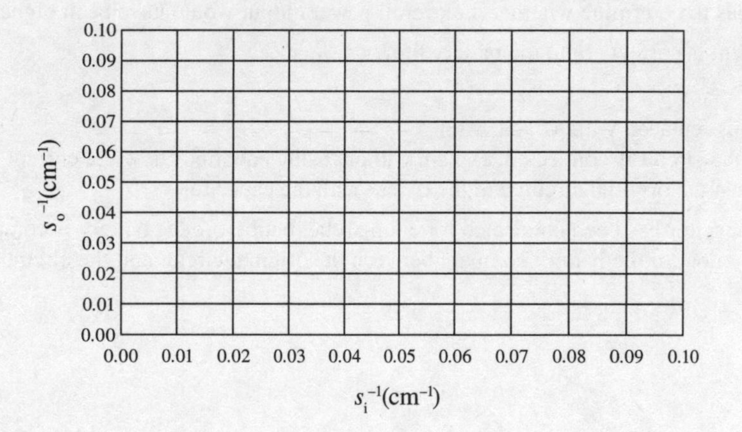

(C) Use ray tracing to make a sketch when the object is 6 cm from the lens.

(D) The top half of the lens is now covered by a sheet of cardboard so that light rays can strike only the bottom half of the lens. Briefly explain what effect this has on the image and how this would affect a ray-tracing diagram.

GO ON TO THE NEXT PAGE.

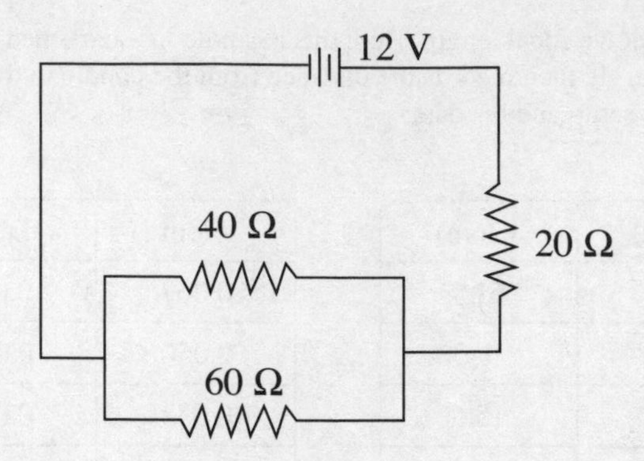

3. The circuit shown is built and the voltage source supplies voltage for 15 minutes before the battery is completely drained. Assume the voltage supplied by the battery is constant at 12 V until the battery is drained, after which the battery supplies 0 V.

(A) What is the equivalent resistance of the circuit?

(B) Two students are discussing the apparatus. Student 1 says, "If the 20 Ω resistor were not present, the overall resistance of the circuit would have been lower and the battery would have lasted longer." Student 2 says, "If the 20 Ω resistor were not present, I think the power output would have been higher and the battery would have drained faster."

i. Use equations to show whether the overall resistance would have been lower without the 20 Ω resistor present.

ii. Use equations to determine whether the overall power output would have been higher.

iii. Which student is correct about the battery life?

(C) The 20 Ω resistor is replaced with a capacitor.

i. As soon as the circuit is connected, explain without using equations how the current supported by the battery compares between the original circuit and the circuit with the capacitor.

ii. After the capacitor has been connected for a long time, but before the battery is completely drained, how does the current drawn out of the battery compare between the original circuit and the circuit with the capacitor?

GO ON TO THE NEXT PAGE.

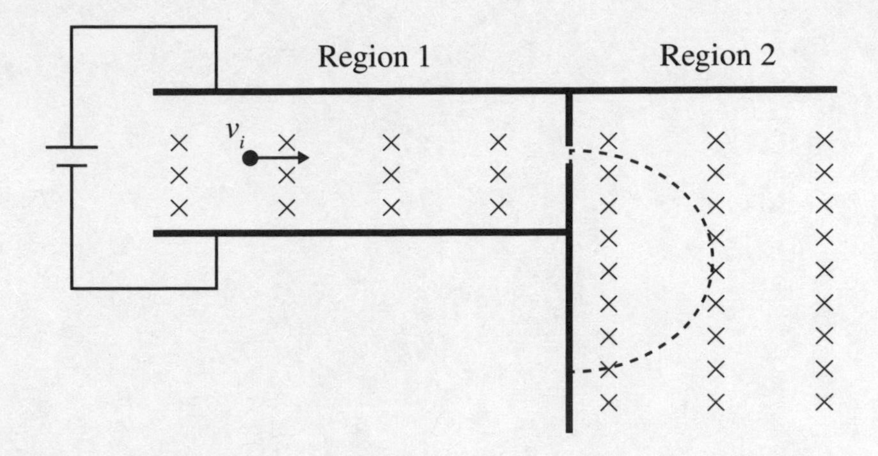

4. In both region 1 and region 2, there is a uniform magnetic field, B, directed into the page. There is a uniform electric field, E, in region 1 established by the battery.

(A) A charged particle moves through region 1 undeflected.

i. Explain how this balance occurs between the electric and magnetic forces.

ii. What would happen to the particle if it was moving at a speed $v_2 > v_i$?

iii. How, if at all, would the setup in region 1 have to change if the particle was of the other sign?

(B) A particle at speed v_i travels through region 1 undeflected. It then enters region 2 and follows the dotted path that is indicated.

i. Is this a positively or negatively charged particle?

ii. What distance from the opening in the boundary between region 1 and region 2 does the particle hit the wall? Answer in terms of m, B, E, v_i, and q.

(C) A similar apparatus is used to determine the percentage of two different isotopes of similarly ionized gases in a gas sample. Explain how such an apparatus can differentiate between the two isotopes.

(D) Another similar apparatus is unable to distinguish between two objects:

The first object has amu 14 and a charge of $-1e$.

The second object has amu 28 and a charge of $-2e$.

Why can this detector not distinguish between these two objects?

STOP

END OF EXAM

Practice Test 1: Diagnostic Answer Key and Explanations

PRACTICE TEST 1: DIAGNOSTIC ANSWER KEY

Let's take a look at how you did on Practice Test 1. Follow the three-step process in the diagnostic answer key below and go read the explanations for any questions you got wrong, or you struggled with but got correct. Once you finish working through the answer key and the explanations, go to the next chapter to make your study plan.

Check your answers and mark any correct answers with a ✔ in the appropriate column.

Section I: Multiple Choice							
Q #	Ans.	✔	Chapter #, Section Title	Q #	Ans.	✔	Chapter #, Section Title
1	C		**5,** The Electric Field **6,** Electric Field and Capacitors	19	C		**6,** Electric Potential
2	D		**6,** Electric Field and Capacitors	20	B		**3,** Buoyancy
3	B		**8,** Magnetic Fields Created by Current-Carrying Wire	21	B		**7,** Power Dissipation
4	C		**5,** Conductors and Insulators	22	C		**8,** The Magnetic Force on a Moving Charge
5	D		**7,** RC Circuits with Capacitors in Steady State	23	D		**10,** The Bohr Model of the Atom
6	D		**7,** RC Circuits with Capacitors in Steady State	24	B		**10,** Nuclear Reactions
7	B		**7,** RC Circuits with Capacitors in Steady State	25	B		**10,** The Wave Function
8	B		**9,** Reflection and Refraction	26	D		**10,** Disintegration Energy
9	B		**9,** Ray Tracing for Mirrors	27	C		**9,** Thin Lenses
10	A		**3,** Bernoulli's Equation: Conservation of Energy in Fluids	28	C		**9,** Thin Lenses
11	A		**3,** The Continuity Equation: Flow Rate and Conservation of Mass **3,** Bernoulli's Equation: Conservation of Energy in Fluids	29	D		**3,** Pressure
12	B		**4,** The Laws of Thermodynamics **4,** Heat Transfer	30	D		**4,** The Laws of Thermodynamics
13	A		**4,** Heat Transfer	31	A		**8,** Magnetic Fields Created by Current-Carrying Wire
14	B		**5,** Coulomb's Law	32	C		**5,** Coulomb's Law
15	C		**7,** Resistors and Resistance **7,** Ohm's Law	33	C		**5,** The Electric Field
16	B		**5,** Coulomb's Law	34	B		**5,** The Electric Field
17	B		**5,** Conductors and Insulators	35	D		**8,** The Magnetic Force on a Moving Charge **8,** Magnetic Fields Created by Current-Carrying Wire
18	D		**8,** The Magnetic Force on a Moving Charge **8,** Magnetic Fields Created by Current-Carrying Wire	36	A		**5,** Conductors and Insulators

Section I: Multiple Choice—Continued							
Q #	Ans.	✔	Chapter #, Section Title	Q #	Ans.	✔	Chapter #, Section Title
37	B		**7,** Analysis of Circuits with Resistors	44	D		**9,** Total Internal Reflection
38	B		**7,** Analysis of Circuits with Resistors	45	C		**7,** RC Circuits with Capacitors in Steady State
39	C		**7,** Resistors and Resistance **7,** Ohm's Law	46	B, C		**8,** Magnetic Fields Created by Current-Carrying Wire
40	D		**5,** The Electric Field **8,** The Magnetic Force on a Moving Charge	47	B, D		**4,** The Laws of Thermodynamics
41	A		**5,** Coulomb's Law	48	A, D		**10,** Photons and the Photoelectric Effect
42	D		**8,** The Magnetic Force on a Moving Charge **8,** Solenoids Create Uniform Fields	49	A, C		**7,** Analysis of Circuits with Resistors
43	D		**10,** The Bohr Model of the Atom	50	A, D		**9,** Ray Tracing for Lenses

Section II: Free Response			
Q #	Ans.	✔	Chapter #, Section Title
1(A)	See Explanation*		**5,** Coulomb's Law
1(B) 1(C)	See Explanation*		**5,** The Electric Field
1(D)	See Explanation*		**6,** Electric Potential Energy
2(A) 2(B)	See Explanation*		**9,** Thin Lenses
2(C) 2(D)	See Explanation*		**9,** Ray Tracing for Lenses
3(A)	See Explanation*		**7,** Combining Resistors and Equivalent Resistance
3(B)	See Explanation*		**7,** Analysis of Circuits with Resistors
3(C)	See Explanation*		**7,** RC Circuits with Capacitors in Steady State
4(A)	See Explanation*		**5,** The Electric Field **8,** The Magnetic Force on a Moving Charge
4(B)	See Explanation*		**8,** The Magnetic Force on a Moving Charge
4(C)	See Explanation*		**8,** The Magnetic Force on a Moving Charge **10,** Nuclear Physics
4(D)	See Explanation*		**8,** The Magnetic Force on a Moving Charge

**Explanations begin on page 33.*

Once you have checked your answers, remember to return to page 4 and respond to the Reflect questions.

Tally your correct answers from Step 1 by chapter. For each chapter, write the number of correct answers in the appropriate box. Then, divide your correct answers by the number of total questions (which we've provided) to get your percent correct.

CHAPTER 3 TEST SELF-EVALUATION

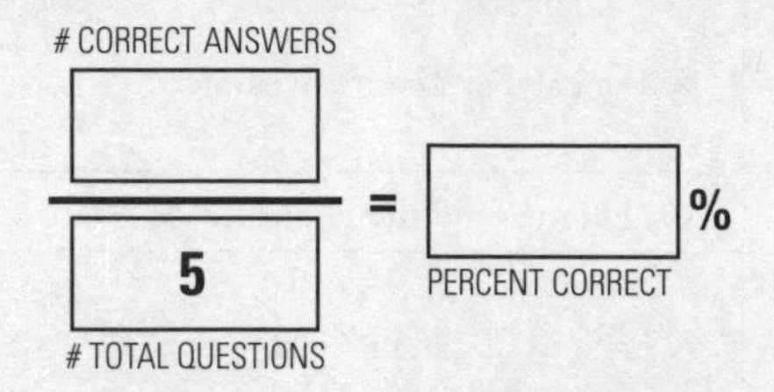

CHAPTER 4 TEST SELF-EVALUATION

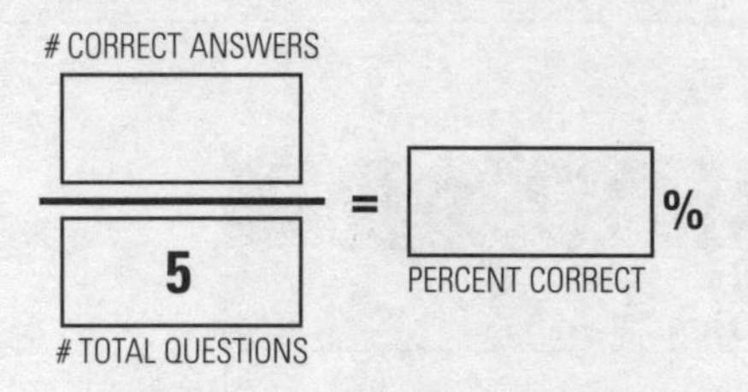

CHAPTER 5 TEST SELF-EVALUATION

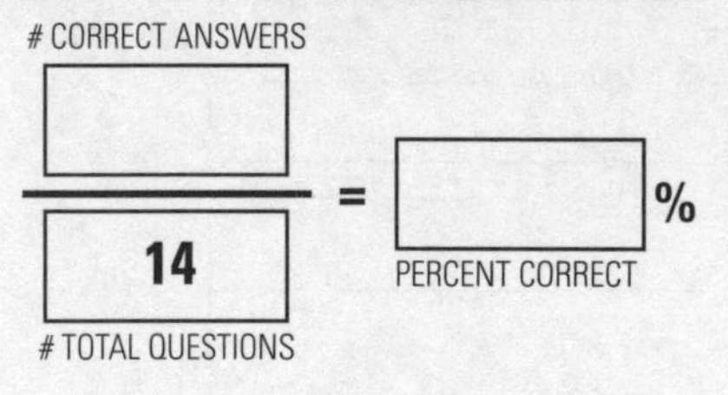

CHAPTER 6 TEST SELF-EVALUATION

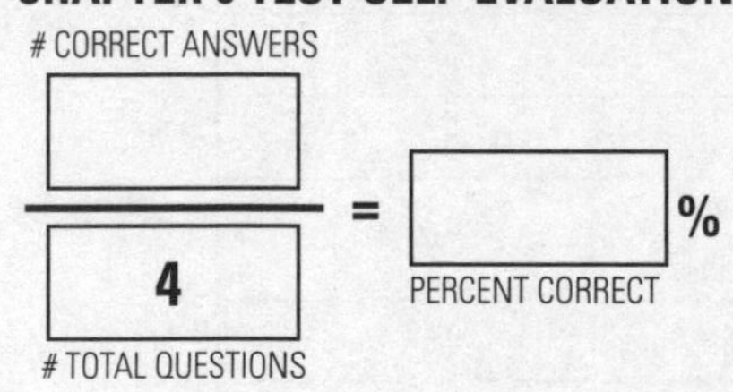

CHAPTER 7 TEST SELF-EVALUATION

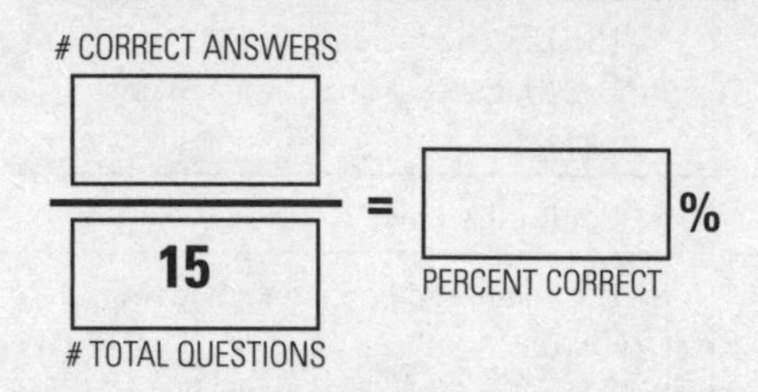

CHAPTER 8 TEST SELF-EVALUATION

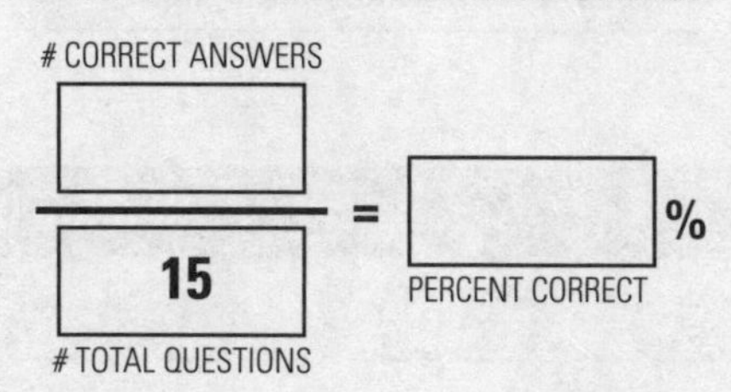

CHAPTER 9 TEST SELF-EVALUATION

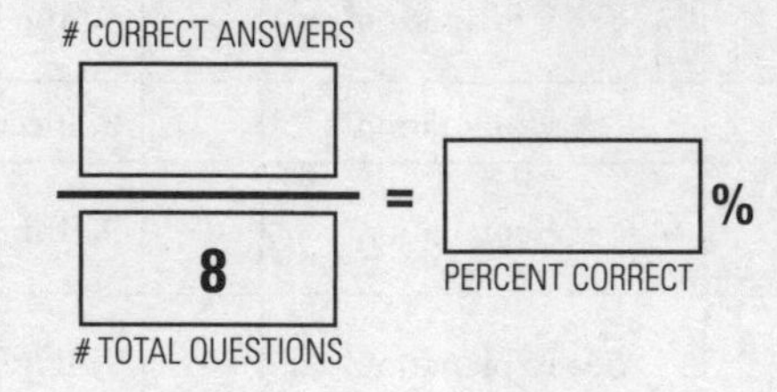

CHAPTER 10 TEST SELF-EVALUATION

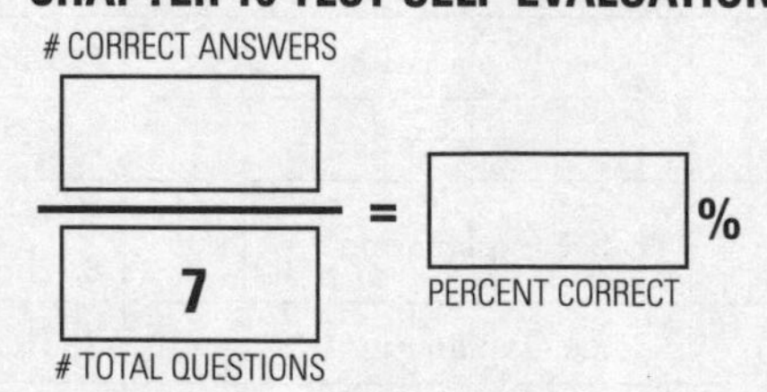

Use the results above to customize your study plan. You may want to start with, or give more attention to, the chapters with the lowest percents correct.

PRACTICE TEST 1: ANSWERS AND EXPLANATIONS

Section I: Multiple Choice

1. **C** There will be a uniform electric field pointing away from the positive plate and toward the negative plate. A positive particle will move in the direction of the field. Because there will be a constant force on the particle, it will have a curved trajectory.

2. **D** All four statements are true. However, when dealing with the gravitational field near the surface of the Earth, the inverse square nature of the gravitational force is not a factor. Furthermore, the electric force within a parallel-plate capacitor is constant due to the constant electric field between the plates.

3. **B** The direction an electromagnetic wave travels must be perpendicular to both the electric and magnetic fields. The electric field will always also be perpendicular to the magnetic field. The wave travels along the z-direction. The magnetic field is along the y-axis. The electric field must be on the x-axis.

4. **C** The spheres do not touch, so there cannot be any charging by contact. While the outside spheres will have non-symmetric charge distributions, each will still have a net charge of 0 C.

5. **D** The uncharged capacitor acts as a wire, shorting bulb 2. A complete loop is made consisting of bulb 1, the capacitor, and bulb 3.

6. **D** When the capacitor is fully charged, there is still a continuous conducting path that includes the battery and all three lightbulbs. Therefore, all three bulbs will be lit when the capacitor is fully charged.

7. **B** There is a closed path from the battery, through bulb 1, bulb 2, and bulb 3, so current will flow across bulbs 1 and 3, causing a voltage drop at each bulb so the voltage across the battery will be greater than the capacitor, making (A) and (D) incorrect. Choice (C) is incorrect because once the capacitor is full, there will still be current flowing in bulb 2, causing a voltage drop across bulb 2. Choice (B) must be correct because the voltage drop across purely parallel components must be equal.

8. **B** The index of refraction of air is less than the index of refraction of water. According to Snell's Law, $n_1\sin(\theta_1) = n_2\sin(\theta_2)$, a smaller n_1 results in a smaller θ_2. The ray in the glass will be at an angle closer to the normal than the ray in the glass from the water/glass apparatus.

9. **B** Convex mirrors will always create virtual images, so eliminate (C) and (D). A concave mirror will create a virtual image when the object is placed within the focal length of the mirror, so eliminate (A). The only way to create a real image with a mirror is to use a concave mirror with the object beyond the focal point.

10. **A** The Continuity Equation, $A_1v_1 = A_2v_2$, allows one to determine the speed at the second location because the volume flow rate, Av, can be determined at location A, so only the area at location B is needed. Therefore, only the diameter at location B is needed.

11. **A** Pressure is highest where flow speed is slowest. At the widest point in the tube, position A, the flow speed will be lowest, and hence, the pressure is highest at A. Because B is narrower than C, the flow speed at B will be greater than at C, making the pressure lowest at point B.

12. **B** According to the Second Law of Thermodynamics, whenever two objects are in the same environment for a long period of time, they will eventually end up at the same temperature. But even if two objects are the same temperature, one of them will feel cooler when it is able to transfer heat more rapidly than the other. Metal conducts heat better than paper, making the metal feel colder than the paper.

13. **A** The Second Law of Thermodynamics says that energy will flow from the higher temperature part of a system into the lower temperature part of the system.

14. **B** Newton's Third Law dictates that the electron must exert a force on the sheet, which is opposite to the force the sheet exerts on the electron. The electron will be repelled by the sheet, so the electron will be pushed up. The sheet must be pushed down.

15. **C** The problem states that the cross-sectional area of the wire is linearly proportional to the current. Ohm's Law, combined with the definition of resistivity, gives $V = I\frac{\rho L}{A}$. Algebra gives $A = \left(\frac{\rho}{V}L\right)I$. Because ρ and V are held constant, the slope of the area versus current graph can be used to find the length of the wires.

16. **B** By the Process of Elimination, placing charges at B will result in the lever arm being parallel to the force and will not cause torque, eliminating (C) and (D). The sign of the charges on the two spheres must be different to have a net torque.

17. **B** Induction requires drawing charge toward one location on a conductor, and then isolating that charge either by using a grounding wire or by dividing the conductor.

18. **D** Force on a moving charge is determined using the right-hand rule. In the right-hand rule, the thumb points with $q\mathbf{v}$, the fingers point with $\mathbf{B}$, and the palm faces in the direction of the magnetic force. To find the direction of the magnetic field from a current-carrying wire, you must know the direction the current flows and the location of the moving charge relative to the wire.

19. **C** In the presence of a positive and a negative point charge, the electric potential will be higher at locations closer to the positive charge and farther from the negative charge.

20. **B** An object will float when the buoyant force balances with the weight of the object. The buoyant force increases as the volume of the object that is submerged increases; therefore, a heavier floating object will have a greater volume submerged than a lighter floating object. An object will sink if, when the entire volume is submerged, the buoyant force is still less than the weight of the object. The buoyant force on the wood equals its weight, while the buoyant force on the rock is less than the weight of the rock.

21. **B** Power dissipated in a circuit is directly proportional to the product of the current and voltage. The slope of the line of a power versus voltage graph gives the current.

22. **C** Charged particles will deflect in a magnetic field. They will move along a circular arc. The direction can be found using the right-hand rule. The right-hand rule depends on the sign of the charge, so like charged objects deflect in the same direction. The object with the greater mass-to-charge ratio will deflect along a larger circle.

23. **D** Three wavelengths of light have been absorbed, indicating that there are three energy differences that correspond to an atomic transition, eliminating (A) and (C). It is possible that there are multiple levels with the same spacing, making (D) correct.

24. **B** Alpha decay removes 4 nucleons, two of which are protons. The atomic number depends only on the number of protons. Three alpha particles cause the atomic number to decrease by 6. Each β^- decay causes the atomic number to increase by 1. Three alpha particles and two β^- particles cause the atomic number to decrease by 6 and then increase by 2. Since the final atomic number is 82, the initial atomic number must have been 86.

25. **B** The square of the wave function of a quantum object can be interpreted as the likelihood of an observation occurring at the stated position.

26. **D** Rest-mass energy of a particle is given by $E = mc^2$. The released energy is given by the mass that remains after the decay less the mass that existed before the decay.

27. **C** The relationship between image distance, object distance, and focal length is given by $\frac{1}{f} = \frac{1}{s_i} + \frac{1}{s_o}$, which is equivalent to $\frac{1}{s_o} = -\frac{1}{s_i} + \frac{1}{f}$. To find the focal length, you need the reciprocal of the y-intercept of a graph of $\frac{1}{s_o}$ versus $\frac{1}{s_i}$. Therefore, $\frac{1}{(0.04\text{ cm}^{-1})} = 25\text{ cm}$.

28. **C** A magnification of 1 occurs when $s_o = s_i$. The image distance and the object distance are equal when $\frac{1}{s_o} = 0.02\ \text{cm}^{-1}$ on the graph, which means $s_o = 50$ cm.

29. **D** Decreasing pressure inside the container decreases the force for the air molecules inside the container compared to the force from the air molecules outside the container. In order for the vertical forces to balance, the total upward force must be equal to the net downward force. The upward force is only from the external air pushing against the lid, while the downward forces are from the air inside the container, the force of gravity on the lid, and a normal force from the sides of the container pushing down on the lid where the seal occurs.

30. **D** When objects reach thermal equilibrium, their average kinetic energy becomes constant. However, there is a constant exchange of energy within the system as some molecules speed up and others slow down.

31. **A** Initially, the magnetic field is strong and points in the –*z* direction. So, the vertical intercept for the graph is a large negative value. As the current steadily decreases, the magnetic field will steadily decrease. The slope of the graph is positive as the magnetic field becomes less negative and approaches a value of 0. The magnetic field changes direction when the current flow changes direction. Furthermore, the magnetic field increases steadily in magnitude as the current flow steadily increases. Therefore, the graph continues above the horizontal axis with the same slope while the magnetic field increases in strength and points in the +*z* direction.

32. **C** From the charge at point A, the force will be up and to the right. From B, the force will be up and to the left. The right component of A will be stronger than the leftward component of B. From the charge at C, the force will point down and to the right. The upward force from either A or B will be greater than the downward force from C.

33. **C** Field vectors represent the force a test charge would experience at a given position relative to a charge distribution. For point charges, field strength will decrease as an inverse-square as positions get farther from the sources.

34. **B** The positive charges will align with the field. Because the edge of the sphere is curved, there will be a greater concentration of charges near the middle of the curve than at the edges.

35. **D** The electron will be deflected if there is a nonzero magnetic force. The magnetic force will be zero when the magnetic field is parallel to the direction of travel. The magnetic field around a current-carrying wire is created in concentric circles.

36. **A** The process of charging by contact involves touching a charged object to an uncharged conductor.

37. **B** To find the equivalent resistance of the circuit, first combine the parallel resistors. $\frac{1}{R_{parallel}} = \frac{1}{15} + \frac{1}{30} + \frac{1}{5} \rightarrow R_{parallel} = 3.3\ \Omega$. The combination of parallel resistors is in series with the other resistor, so $R_{eq} = 10 + 3.3 = 13.3\ \Omega$. The current out of the battery is found from Ohm's Law: $I = \frac{V}{R_{eq}} = \frac{2.5}{13.3} = 0.19$ A.

38. **B** The voltage drop in each of the parallel paths must be the same at $V = 0.50$ V. The Loop Rule indicates that in any closed loop, the voltage drop must equal the voltage of the battery. The $2.5 - 0.5 = 2.0$ V voltage drop must occur across the 10 Ω resistor. Ohm's Law indicates that the current through that resistor must be 0.2 A, which is the total current in the circuit.

    The Junction Rule states that this total current must also be the sum of the current in all four branches. The current in the 15 Ω resistor is $I = \frac{0.50}{15} = 0.033$ A, in the 30 Ω resistor is $I = \frac{0.50}{30} = 0.017$ A, and in the 5 Ω resistor is $I = \frac{0.50}{5} = 0.010$. The remaining current must be $0.20 - 0.033 - 0.017 = 0.05$ A.

    Using Ohm's Law on the unknown resistor, $R = \frac{V}{I} = \frac{0.50}{0.05} = 10\ \Omega$.

39. **C** The resistance of a wire is directly proportional to the length of the wire and inversely proportional to the cross-sectional area. 1 must have the least resistance and 2 must have the most resistance. Because the resistors are in series, they have the same current through them. Ohm's Law states that the greatest voltage drop is across the greatest resistance value.

40. **D** The magnetic field must be perpendicular to the direction of motion in order for there to be a magnetic force. The magnetic force will be perpendicular to both the field and the velocity. The electric force is parallel to the electric field. Therefore, the electric field must be perpendicular to the magnetic field, and both fields must be perpendicular to the velocity.

41. **A** For electrons, the numerator of the Coulomb force is $9 \times 10^9 * (1.6 \times 10^{-19})^2$, while the numerator of the gravitational force is $6.67 \times 10^{-11}\ (9.11 \times 10^{-31})^2$. The denominators are equal for both forces, so the Coulomb force is much greater.

42. **D** Inside a solenoid, an electron will be equally repelled by any charges within the wires, so there will be a zero net electric force. The velocity of the electron traveling along the axis of the solenoid is parallel to the magnetic field, so there will be no magnetic force.

43. **D** Electrons are excited into higher energy levels with the input of energy from the voltage source. As the excited electron falls back to lower energy levels, light is emitted at frequencies that characterize the energy differences between levels.

44. **D** The critical angle occurs at the angle for θ_1 when the value of $\sin(\theta_2) = 1$, which is the maximum value that $\sin(\theta)$ can obtain.

45. **C** The fully charged capacitor will have the same voltage as R_2, since they are in parallel with each other. When the capacitor is fully charged, it acts like an open switch, and the circuit becomes a battery in series with the two resistors. In order to increase the energy stored in the capacitor, a greater voltage needs to be applied across its plates, which means a greater voltage drop needs to be found across R_2. This can happen if less of the battery's voltage is dropped across R_1.

46. **B, C** The amount and direction of compass needle deflection indicates the strength of the magnetic field and its direction. The field from a current-carrying wire is $B = \frac{\mu_0}{2\pi}\frac{I}{r}$, so the field increases as the current increases. The right-hand rule with wires says to determine the direction, the thumb points in the direction of current, and the fingers curl in the direction of the field.

47. **B, D** Work is done on a gas when volume changes at constant pressure, or it can be found from the area under a pressure-volume curve. Choice (B) involves no change in volume. The first two steps in (D) involve no volume change. The last two steps involve the same amount of work first being done by the gas and then the same amount being done on the gas.

48. **A, D** The existence of a frequency below which no light is emitted argues for the particle nature of light. If light were purely a wave, then at sufficiently high intensity, the energy would be available in the wave to transfer to the electron. The linear increase in the stopping potential and frequency, and the independence of the stopping potential on the intensity, also argues for the particle nature of light. A wave explanation would indicate that a higher intensity should give the electrons more energy.

49. **A, C** The overall current will increase when the overall resistance decreases. Placing an additional resistor in parallel with an existing resistor, whether the existing resistor is part of a series combination or part of a parallel combination, will lower the overall resistance of the circuit.

50. **A, D** For a diverging lens, the "focus" is on the far side of the lens. So (1) shows a light ray traveling into the focus and then bending parallel to the optic axis. (4) shows a light ray traveling parallel to the optic axis, and it leaves on a line connecting the other focal point.

Section II: Free Response

1. (A) A positively charged object will be repelled from the positive charge at position 0 and attracted to the negative charge at position $+d$, so it will experience a force to the right during the entirety of its motion. The object will move from its release position directly toward the negative charge at position d with a velocity that increases during the entirety of the motion.

(B) The field will be to the right in the two regions $-\infty < x < -d$ and $0 < x < d$, so this is where the plot will be positive. The field magnitude will go toward infinity as the position approaches the charges at $-d$, 0, and $+d$. The charge at 0 has a larger magnitude, so the positions where the charge has the smallest magnitude should be closer to the charges at $+d$ and $-d$.

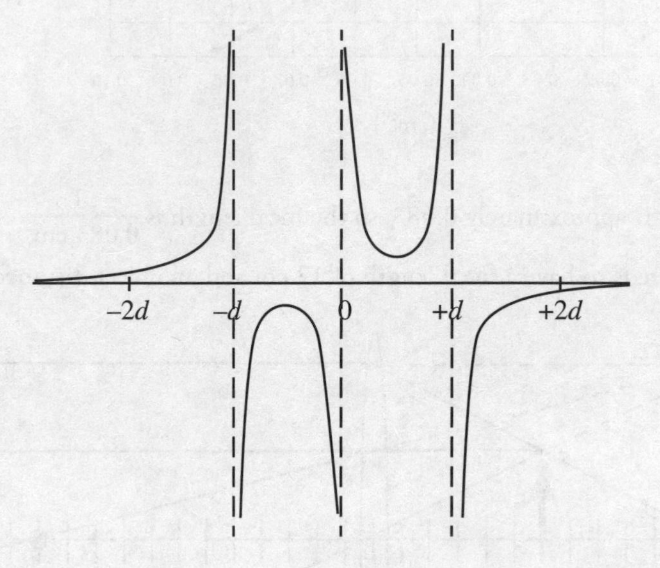

(C) In the region between 0 and d, the net charge will depend on the influence of each of the three charges on the x-axis.

$E_{-d}(x) = \dfrac{kQ}{(d+x)^2}$ is directed to the left.

$E_0(x) = \dfrac{k(2Q)}{(x)^2}$ is directed to the right.

$E_{+d}(x) = \dfrac{kQ}{(d-x)^2}$ is directed to the right.

Thus, $E(x) = \dfrac{kQ}{(d-x)^2} + \dfrac{k(2Q)}{(x)^2} - \dfrac{kQ}{(d+x)^2}$ is directed to the right.

(D) The work required to bring a charge to a position is equal to the electric potential at that position multiplied by the amount of charge that is moved. Once the first two charges have been assembled, the electric potential at position $x = -d$ will be lower than when there was only a positive charge at $x = 0$ because the charge at $x = d$ is negative. Therefore, the work to bring in the third charge will be less than W.

2. (A) $\frac{1}{f} = \frac{1}{s_i} + \frac{1}{s_o}$, so a plot of s_o^{-1} versus s_i^{-1} will have a y-intercept equal to $\frac{1}{f}$.

(B)

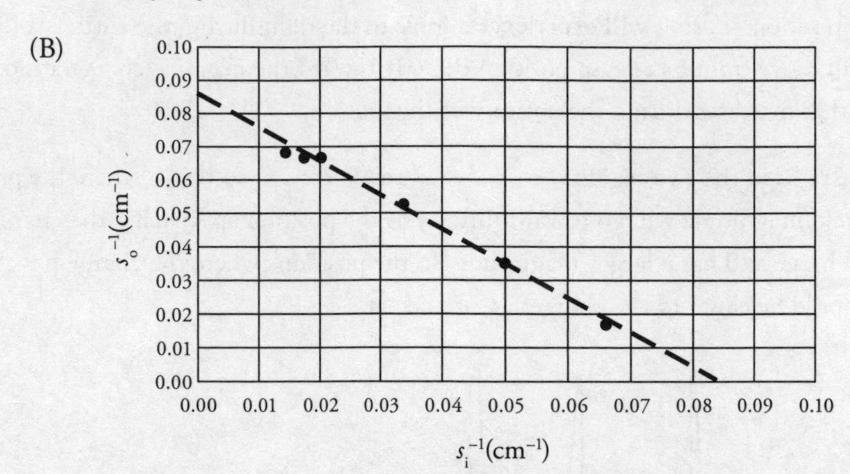

The y-intercept is approximately 0.083, so the focal length is $\frac{1}{0.083\ \text{cm}^{-1}} = 12$ cm.

(C) The diagram needs to have a focal length of 12 cm and an object distance of 6 cm.

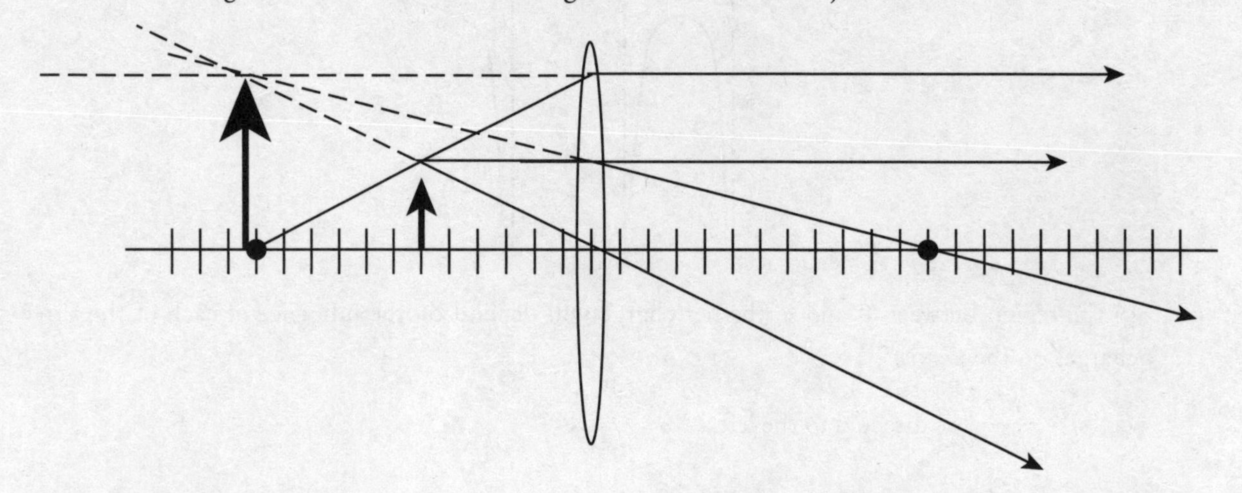

(D) The brightness of the image would decrease, but the location of the image would not change. The ray diagram would require using rays that are not the principal rays, as all three principal rays are blocked. However, all rays coming from the object converge at the image position, so the image position would not change.

3. (A) The resistors in parallel are combined first.

$$R_{eq}^{-1} = 40^{-1} + 60^{-1}$$

$$R_{eq} = 24$$

The total equivalent resistance is then $R_{total} = 24 + 20 = 44\ \Omega$.

(B) (i) Without the resistor, there would be only the parallel resistor combination.

$$R_{eq}^{-1} = 40^{-1} + 60^{-1}$$

$$R_{eq} = 24\ \Omega$$

(ii) $P = IV = \left(\frac{V}{R}\right)V$, so a lower resistance results in a higher power output for the same battery.

(iii) A higher power output means that it will take less time for the energy to be used up, so student 2 is correct.

(C) (i) As soon as the circuit is connected with the capacitor, the capacitor acts as if it had no resistance. This is exactly like (B)(i). The lower resistance would cause a greater current to be supported by the battery.

(ii) After the capacitor is fully charged, no more current can flow in the branch of the circuit with the capacitor. The capacitor is in series with the battery, so there will be no current supported by the battery.

4. (A) (i) The electric force is parallel to the electric field in region 1. This points vertically, either up or down, depending on the sign of the charge. The magnetic force is perpendicular to both the velocity and the field, so the magnetic force is also in the vertical dimension. The forces balance when the electric force equals the magnetic force in magnitude.

$$F_E = qE = qvB$$

The value of the charge cancels, so any particle with a speed $v = E/B$ will experience balanced forces.

(ii) A particle moving faster than E/B will experience a greater magnetic force than electric force. It will be deflected and will not go through the hole between region 1 and region 2.

(iii) There would be no change in outcome if the sign of the particle were reversed, but the directions of both the electric and magnetic forces would reverse. They should still balance for $v = \frac{E}{B}$.

(B) (i) As soon as the particle enters region 2, it has a velocity to the right, a B field into the page, and experiences a downward directed force. From the right-hand rule, the particle must be negatively charged.

(ii) The magnetic force acts as the centripetal force for the particle in region 2.

$$qv_iB = \frac{mv_i^2}{R}$$

$$qB = \frac{mv_i}{R}$$

$$R = \frac{mv_i}{qB}$$

The particle hits the wall at twice the radius, $2mv_i/qB$.

Note: The answer can also be written substituting $v = \frac{E}{B}$.

(C) Both isotopes have the same charge. Any particle entering region 2 through the hole has the same speed. The isotopes have different masses. Based on the answer to (B)(ii), an isotope with a larger mass will hit the wall at a greater distance from the hole than an isotope with a smaller mass.

(D) When the objects have different charges and masses, the radius from (B)(ii) depends on the ratio of mass to charge. Because the ratio for the given example is the same, $-1e$: 14 amu, the apparatus will not distinguish between them.

HOW TO SCORE PRACTICE TEST 1

Section I: Multiple Choice

________ × 1.50 = ________

Number Correct (out of 50) — Weighted Section I Score (Do not round)

Section II: Free Response

Question 1: ________ (out of 12) × 1.7045 = ________ (Do not round)

Question 2: ________ (out of 12) × 1.7045 = ________ (Do not round)

Question 3: ________ (out of 10) × 1.7045 = ________ (Do not round)

Question 4: ________ (out of 10) × 1.7045 = ________ (Do not round)

Sum = ________

Weighted Section II Score (Do not round)

AP Score Conversion Chart Physics 2	
Composite Score Range	AP Score
107–150	5
90–106	4
73–89	3
56–72	2
0–55	1

Composite Score

________ + ________ = ________

Weighted Section I Score + Weighted Section II Score = Composite Score (Round to nearest whole number)

Part III
About the AP Physics 2 Exam

- The Structure of the AP Physics 2 Exam
- How AP Exams Are Used
- Other Resources
- Designing Your Study Plan

THE STRUCTURE OF THE AP PHYSICS 2 EXAM

The AP Physics 2 Exam consists of two sections: a multiple-choice section and a free-response section. Each question in the multiple-choice section is followed by four possible responses. In the single-select subsection, only one of these possible answers is correct. In the multi-select subsection, two of these possible answers are correct. It is your job to choose the right answer(s). Each correct answer is worth one point.

Section	Timing	Scoring	Question Type	Number of Questions
I: Multiple Choice	90 minutes	50% of exam score	Single-select (discrete questions and questions in sets with one correct answer)	45
			Multi-select (discrete questions with two correct answers)	5
				Total—50

Section	Timing	Scoring	Question Type	Number of Questions
II: Free Response	90 minutes	50% of exam score	Experimental Design	1
			Qualitative/ Quantitative Translation	1
			Short Answer	2
				Total—4

Be Calculating

Students may use a four-function, scientific, or graphing calculator on the entire exam. However, be aware that some models have unapproved features (like keyboards, styluses, or wireless capability) and are not permitted, so please check the College Board website. If you're still unsure, you can bring two calculators so that if one is rejected (or not functioning), you do not have to rely on a school-provided backup that you may be unfamiliar with.

The free-response section requires you to write out your solutions, showing your work. The total amount of time for this section is 90 minutes, so you have an average of 22.5 minutes per question. Unlike the multiple-choice section, which is scored by computer, the free-response section is graded by high school and college teachers. They have guidelines for awarding partial credit, so you don't need to correctly answer every part to get points. You are allowed to use a calculator on the entire AP Physics 2 Exam (programmable or graphing calculators are okay, but ones with a typewriter-style keyboard are not), and a table of equations is provided for your use. The two sections—multiple-choice and free-response—are weighted equally, so each is worth 50 percent of your grade.

Grades on the AP Physics 2 Exam are reported as a number: either 1, 2, 3, 4, or 5. The following table provides descriptions of these scores along with the distribution of how students scored for the 2022 exam administrations.

Score	2022 Percentage	Credit Recommendation	College Grade Equivalent
5	16.3%	Extremely Well Qualified	A
4	18.1%	Well qualified	A–, B+, B
3	35.3%	Qualified	B–, C+, C
2	24.1%	Possibly Qualified	–
1	6.3%	No Recommendation	–

Scores from the May 2022 test administration. Data taken from the College Board website.

Colleges are generally looking for a 4 or 5, but some may grant credit for a 3. How well do you have to do to earn such a grade? Each test is curved, and specific cutoffs for each grade vary a little from year to year, but here's a rough idea of how many points you must earn—as a percentage of the maximum possible raw score—to achieve each of the grades 2 through 5:

AP Exam Grade	Percentage Needed
5	≥ 75%
4	≥ 60%
3	≥ 45%
2	≥ 25%

AP Physics 2: Algebra-Based Course Content

You may be using this book as a supplementary text as you take an AP Physics 2 course at your high school, or you may be using it on your own. The College Board is very detailed in what it requires your AP teacher to cover in their AP Physics 2 course. The following summary is the College Board's explanation of what you should know.

> Students explore principles of fluids, thermodynamics, electricity, magnetism, optics, and topics in modern physics. The course is based on seven Big Ideas, which encompass core scientific principles, theories, and processes that cut across traditional boundaries and provide a broad way of thinking about the physical world. The course also focuses on the seven Science Practices, which are the practices that allow for scientific reasoning.

The course content is broken down into seven units, which all have several topics. In Part V of this book, the course content is organized by topic. This is similar to the structure of the majority of physics textbooks. The Big Ideas and Science Practices are spiraled throughout the content areas.

The following are the seven Big Ideas:

1. Objects and systems have properties such as mass and charge. Systems may have internal structure.
2. Fields existing in space can be used to explain interactions.
3. The interactions of an object with other objects can be described by forces.
4. Interactions between systems can result in changes in those systems.
5. Changes that occur as a result of interactions are constrained by conservation laws.
6. Waves can transfer energy and momentum from one location to another without the permanent transfer of mass and serve as a mathematical model for the description of other phenomena.
7. The mathematics of probability can be used to describe the behavior of complex systems and to interpret the behavior of quantum mechanical systems.

The following are seven Science Practices:

1. The student can use representations and models to communicate scientific phenomena and solve scientific problems.
2. The student can use mathematics appropriately.
3. The student can engage in scientific questioning to extend thinking or to guide investigations within the context of the AP course.
4. The student can plan and implement data collection strategies in relation to a particular scientific question.
5. The student can perform data analysis and evaluation of evidence.
6. The student can work with scientific explanations and theories.
7. The student is able to connect and relate knowledge across various scales, concepts, and representations in and across domains.

The Big Ideas and Science Practices are interlaced throughout the Course Content Units.

The following are the Course Content Units:

1. Fluids
2. Thermodynamics
3. Electric Force, Field, and Potential
4. Electric Circuits
5. Magnetism and Electromagnetic Induction
6. Geometric and Physical Optics
7. Quantum, Atomic, and Nuclear Physics

One example of how these are related to one another would be a question in the Course Content Unit of Atomic Physics that relates to the Science Practice of performing data analysis and integrates the Big Idea that interactions are constrained by conservation laws.

The spiraling of the Big Ideas occurs as the same idea is encountered in multiple course units. For example, almost all course units rely on the concept of a system and an understanding that a system has internal properties that influence the behavior of the system. The concept of a fluid system, to which the mathematical treatments for fluid dynamics apply, is very different from the concept of a magnetic system, which will respond in various ways to an external magnetic field. Therefore, many students and teachers find the content unit structure more intuitive.

The Course Content Units are assigned approximate weightings for the multiple-choice portion of the AP Physics 2 Exam.

Content Unit	Approximate AP Exam Weighting Percentages
Fluids	10–12%
Thermodynamics	12–18%
Electric Force, Field, and Potential	18–22%
Electric Circuits	10–14%
Magnetism and Electromagnetic Induction	10–12%
Geometric and Physical Optics	12–14%
Quantum, Atomic, and Nuclear Physics	10–12%

This book is aligned with the College Board AP Physics 2 Course and Exam Description: Effective Fall 2020. Sidebars have been placed throughout Part V to indicate the course content units and topics that relate to particular portions of the review.

HOW AP EXAMS ARE USED

Different colleges use AP Exams in different ways, so it is important that you go to a particular college's website to determine how it uses AP Exams. The three items below represent the main ways in which AP Exam scores can be used:

- **College Credit.** Some colleges will give you college credit if you score well on an AP Exam. These credits count toward your graduation requirements, meaning that you can take fewer courses while in college. Given the cost of college, this could be quite a benefit, indeed.
- **Satisfy Requirements.** Some colleges will allow you to "place out" of certain requirements if you do well on an AP Exam, even if they do not give you actual college credits. For example, you might not need to take an introductory-level course, or perhaps you might not need to take a class in a certain discipline at all.
- **Admissions Plus.** Even if your AP Exam will not result in college credit or even allow you to place out of certain courses, most colleges will respect your decision to push yourself by taking an AP Course or even an AP Exam outside of a course. A high score on an AP Exam shows a command of more difficult content than is taught in many high school courses, and colleges may take that into account during the admissions process.

Stay Tuned for Updates
In 2022, the College Board announced that some topics in AP Physics 1 and AP Physics 2 (fluids, waves) will realign in the next year or two. At the time of this printing, the realignment seems to have been pushed off to the 2023–2024 school year. To learn more about specific exam information, visit the College Board's website for breaking news.

Looking for More Help with Your APs?
We now offer specialized AP tutoring and course packages that guarantee a 4 or 5 on the AP exam. To see which courses are offered and available, visit PrincetonReview.com/college/ap-test-prep

OTHER RESOURCES

There are many resources available to help you improve your score on the AP Physics 2 Exam, not the least of which are your **teachers**. If you are taking an AP class, you may be able to get extra attention from your teacher, such as obtaining feedback on your essays. If you are not in an AP course, reach out to someone who teaches AP Physics 2 and ask if that teacher will review your essays or otherwise help you with content.

Another wonderful resource is **AP Students**, the official site of the AP Exams. The scope of the information at this site is quite broad and includes:

- the course description, which provides details on what content is covered and sample questions
- free-response prompts from previous years and exam tips
- access to AP Classroom if you are enrolled in a course (teacher assistance required)

The AP Students web page is apstudents.collegeboard.org.

Finally, The Princeton Review offers tutoring for the AP Physics 2 Exam. Our expert instructors can help you refine your strategic approach and add to your content knowledge. For more information, call 1-800-2REVIEW.

Prep Like a Pro
Need some help devising a plan of action for your studying? Check out our free AP Physics 2 Exam study guide on your AP Student Tools. See the "Get More (Free) Content" page for details about accessing your online tools.

DESIGNING YOUR STUDY PLAN

As part of the Introduction, you identified some areas of potential improvement. Now delve further into your performance on Test 1, with the goal of developing a study plan appropriate to your needs and time commitment.

Read the answers and explanations associated with the multiple-choice questions (starting at page 33). After you have done so, respond to the following questions:

- Review the Practice Test 1 Diagnostic Answer Key, and make a list of the chapter topics in Step 1. Next to each topic, indicate your rank of the topic as follows: "1" means "I need a lot of work on this," "2" means "I need to beef up my knowledge," and "3" means "I know this topic well."
- How many days/weeks/months away is your exam?
- What time of day is your best, most focused study time?
- How much time per day/week/month will you devote to preparing for your exam?
- When will you do this preparation? (Be as specific as possible: Mondays and Wednesdays from 3:00 p.m. to 4:00 p.m., for example.)
- Based on the answers above, will you focus on strategy (Part IV) or content (Part V) or both?
- What are your overall goals in using this book?

Part IV
Test-Taking Strategies for the AP Physics 2 Exam

- Preview
1. How to Approach Multiple-Choice Questions
2. How to Approach Free-Response Questions

PREVIEW

Review your Practice Test 1 results and then respond to the following questions:

- How many multiple-choice questions did you miss even though you knew the answer?
- On how many multiple-choice questions did you guess randomly?
- How many multiple-choice questions did you miss after eliminating some answers and guessing based on the remaining answers?
- Did you find any of the free-response questions easier or harder than the others—and, if so, why?

HOW TO USE THE CHAPTERS IN THIS PART

For the following Strategy chapters, think about what you are doing now before you read the chapters. As you read and engage in the directed practice, be sure to appreciate the ways you can change your approach.

Chapter 1
How to Approach Multiple-Choice Questions

APPROACHING THE MULTIPLE-CHOICE SECTION

All the multiple-choice questions will have a similar format: each will be followed by four answer choices. At times, it may seem that there could be more than one possible correct answer. In fact, answers resulting from common mistakes in calculation are often included in the four answer choices to trap you.

For the single-select problems, which represent 45 out of the 50 multiple-choice problems, there is only one correct answer! For the last 5 problems of the multiple-choice section, there will be two correct answers per question. Keep in mind that this section is identified before you start it.

Use the Answer Sheet

For the multiple-choice section, you write the answers not in the test booklet but on a separate answer sheet (very similar to the ones we've supplied for you online in your Student Tools—download and print the bubble sheet.) Four oval-shaped answer bubbles follow the question number, one for each possible answer. Don't forget to fill in all your answers on the answer sheet. Don't just mark them in the test booklet. Marks in the test booklet will not be graded. Also, make sure that your filled-in answers correspond to the correct question numbers! Check your answer sheet after every five answers to make sure you haven't skipped any bubbles by mistake.

The Two-Pass System

The AP Physics 2 Exam covers a broad range of topics. There's no way, even with our extensive review, that you will know everything about every topic in algebra-based physics. So, what should you do?

Use the two-pass system to save time!

Adopt a two-pass system. The two-pass system entails going through the test and answering the easy questions first. Save the more time-consuming questions for later. (Don't worry—you'll have time to do them later!) First, read the question and decide if it is a "now" or "later" question. If you decide this is a "now" question, answer it in the test booklet. If it is a "later" question, come back to it. Once you have finished all the "now" questions on a double page, transfer the answers to your bubble sheet. Flip the page and repeat the process. If this seems like too much extra work, read on to learn why this practice makes sense.

Once you've finished all the "now" questions, move on to the "later" questions. Start with the easier questions first. These are the ones that require calculations or that require you to eliminate the answer choices (in essence, the correct answer does not jump out at you immediately). Transfer your answers to your bubble sheet as soon as you answer these "later" questions.

Watch Out for Those Bubbles!

Because you're skipping problems, you need to keep careful track of the bubbles on your answer sheet. As we just mentioned, one way to accomplish this is by answering all the questions on facing pages and then transferring your choices to the answer sheet. If you prefer to enter them one by one, make sure you double-check the number beside the ovals before filling them in. We'd hate to see you lose points because you forgot to skip a bubble!

Know How to Linearize Graphs to Equations

Graphs, data tables, and mathematical relations (equation of fit lines) are all separate ways to represent the outcome of experiments, and the AP Physics 2 Exam will expect you to know how to use all of these representations. For the multiple-choice portion of the AP Physics 2 Exam (as well as the free-response portion), you are supplied with an equation sheet. You should know what graphs of these equations look like (a graphing calculator may be helpful for this), but most often, you will be asked which variables need to be graphed to produce a direct proportion. For Coulomb's Law, $F_{el} = \frac{1}{4\pi\varepsilon_0}\frac{Q_1Q_2}{r^2}$, there are several data that would produce a linear graph. If F_{el} and Q_1 are data, then a graph of F_{el} versus Q_1 will produce a linear plot with slope of $\frac{1}{4\pi\varepsilon_0}\frac{Q_2}{r^2}$, but if F_{el} and r are data, then a graph of F_{el} versus $\frac{1}{r^2}$ produces a linear plot with slope of $\frac{1}{4\pi\varepsilon_0}Q_1Q_2$. Producing a linear graph for data or from an equation is called **linearization**.

Process of Elimination (POE)

On most tests, you need to know your material backward and forward in order to get the right answer. In other words, if you don't know the concept beforehand, you probably won't answer the question correctly. This is particularly true of fill-in-the-blank and essay questions. We're taught to think that the only way to get a question right is by knowing the answer. However, that's not the case on Section I of the AP Physics 2 Exam. You can get a perfect score on this portion of the test without knowing a single right answer—provided you know all the wrong answers!

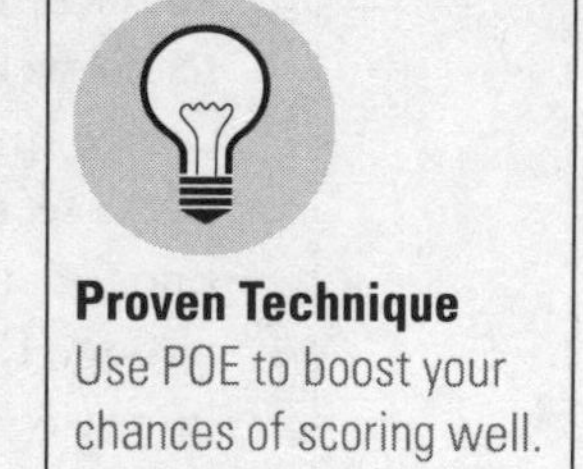

Proven Technique
Use POE to boost your chances of scoring well.

What are we talking about? This is perhaps the most important technique to use on the multiple-choice section of the exam. Let's take a look at an example.

Apply the Strategy
Which answer choices can you eliminate first?

41. A gas undergoes an expansion-compression cycle. If, plotted on a P–V diagram, the cycle is counterclockwise and the work is 300 J in magnitude, what was the heat transfer during this cycle?

    (A) 300 J into the system
    (B) 300 J out of the system
    (C) 600 J into the system
    (D) 600 J out of the system

Now, if this were a fill-in-the-blank-style question, you might be in a heap of trouble. But let's take a look at what we've got. If we can eliminate a few of the answer choices, we can get closer to making an educated guess.

For a cycle, $\Delta U = 0$, we know the magnitude of the work and heat transfer have to be the same. This eliminates (C) and (D). Since the cycle is counterclockwise on a P–V diagram, the work has to be positive, so the heat transfer is negative. A negative heat transfer is heat leaving the system. Choice (B) is correct.

For a cycle, $\Delta U = 0$, we know the magnitude of the work and heat transfer have to be the same. This eliminates (C) and (D). Since the cycle is counterclockwise on a P–V diagram, the work has to be positive, so the heat transfer is negative. A negative heat transfer is heat leaving the system. Choice (B) is correct.

We think we've illustrated our point: Process of Elimination is the best way to approach the multiple-choice questions. Even when you don't know the answer right off the bat, you'll surely know that two or three of the answer choices are not correct. What then?

Aggressive Guessing

You are scored only on the number of questions you get right, so guessing can't hurt you. But can it help you? It sure can. Let's say you randomly guess the same letter on four questions. Odds are you'll get one right, so you've already increased your score by one point. Now, let's add POE into the equation. If you can eliminate as many as two answer choices from each question, your chances of getting them right increase, and so does your overall score. Remember, don't leave any bubbles blank on test day!

General Advice

Answering 50 multiple-choice questions in 90 minutes can be challenging. Make sure to pace yourself accordingly, and remember that you do not need to answer every question correctly to do well. Exploit the multiple-choice structure of this section. For 45 of your 50 multiple-choice questions, there are three wrong answers and only a single correct one (for the multi-select section, the last 5 questions, there are two wrong and two correct). So, even if you don't know exactly which one is the right answer, you can eliminate some that you know for sure are wrong. Then you can make an educated guess from among the answers that are left and greatly increase your odds of getting that question correct.

Problems with graphs and diagrams are usually the fastest to solve, and problems with an explanation for each answer usually take the longest to work through. Do not spend too much time on any one problem, or you may not get to easier problems further into the test.

The practice exams in this book are written to give you an idea of the format of the test, to illustrate the difficulty of the questions, and to allow you to practice pacing yourself. As closely as possible, take the practice tests under the same conditions you will encounter during the real exam.

Reflect

Respond to the following questions:

- How long will you spend on multiple-choice questions?
- How will you change your approach to multiple-choice questions?
- What is your multiple-choice guessing strategy?
- Will you seek further help, outside of this book (such as a teacher, tutor, or AP Students), on how to approach the questions that you will see on the AP Physics 2 Exam?

Chapter 2
How to Approach Free-Response Questions

THE FREE-RESPONSE SECTION: WHAT TO EXPECT

On the free-response section, be sure to show the graders what you're thinking. Write clearly—that is *very* important—and show your steps. If you make a mistake in one part and carry an incorrect result to a later part of the question, you can still earn valuable points if your method is correct. But the graders cannot give you credit for work they can't follow or can't read. And, where appropriate, be sure to include units in your final answers.

Taking time to read the questions—making sure you understand them—will save you time in the long run.

The most important advice we can give you for the free-response section of the AP Physics 2 Exam is to read the questions carefully and answer according to exactly what the questions are asking you to do. Credit for the answers depends not only on the quality of the solutions but also on how they are explained. On the AP Physics 2 Exam, words such as "justify," "explain," "calculate," "what is," "determine," and "derive" have specific meanings, and the graders are looking for very precise approaches in your explanations in order to give maximum credit.

Questions that ask you to "justify" are looking for you to show an understanding in words of the principles underlying physical phenomena and to perform the mathematical operations needed to arrive at the correct answer. The words "justify" and "explain" require that you support your answers with text, equations, calculations, diagrams, or graphs. In some cases, the text or equations must elucidate physics fundamentals or laws; in other cases, they will serve to analyze the behavior of different values or different types of variables in the equation.

The word "calculate" requires you to show numerical or algebraic work to arrive at the final answer. In contrast, "what is" and "determine" questions signify that full credit may be given without showing mathematical work. Just remember, showing work that leads to the correct answer is always a good idea when possible, especially since showing work may still earn you partial credit even if the answer is not correct.

"Derive" questions are looking for a more specific approach, which entails beginning the solution with one or more fundamental equations and then arriving at the final answer through the proper use of mathematics, usually involving some algebra.

APPROACHING THE FREE-RESPONSE SECTION

Section II is worth 50 percent of your grade on the AP Physics 2 Exam. This section is composed of four free-response questions. You're given a total of 90 minutes for this section. There are three types of problems on the exam: one experimental design question worth 12 points, one quantitative/qualitative translation question worth 12 points, and two short-answer questions worth 10 points each. One of the short-answer problems will require a paragraph-length argument. At the top of each question, next to the question number, is the number of points for the problem as well as the recommended amount of time to spend on that problem. The suggested times are 25 minutes each for 12-point problems and 20 minutes each for 10-point problems. Pacing is a personal decision, though, so feel free to build your own timeline for the free-response section.

Clearly Explain and Justify Your Answers

Remember that your answers to the free-response questions are graded by readers and not by computers. Communication is a very important part of AP Physics 2. Compose your answers in precise sentences. Just getting the correct numerical answer is not enough. The majority of questions will require a written explanation or justification, either in lieu of a numerical answer or in addition to a calculation. You will need to explain your reasoning behind the technique that you selected and communicate your answer in the context of the problem. For many problems, points are given "for indicating" some physical reality of the situation, so either a written justification or a calculated solution is acceptable.

Do not expect the graders to read between the lines. Your grader will be an experienced AP Physics 2 teacher, so you will not need to explain every detail of each step in your thought process, but you need to present your solution in a systematic manner using solid logic and appropriate language. The first step in almost any written response should cite a specific physical principle. And remember: although you won't earn points for neatness, the graders can't give you a grade if they can't read and understand your solution!

Use Only the Space You Need

Do not try to fill up the space provided for each question. The space given is usually more than enough. The people who design the tests realize that some students write in big letters and some students make mistakes and need extra space for corrections. So, if you have a complete solution, don't worry about the extra space. Writing more will not earn you extra credit. In fact, many students tend to go overboard and shoot themselves in the foot by making a mistake after they've already written the right answer.

Room to Write

On the actual test, you will be given space along with the bubble sheet to record your answers for each free-response question. You should use scrap paper for the free-responses on the practice tests in this book. After you've gotten the hang of the timing, be aware of how much space each response is taking up, in case you need to write in smaller print or use fewer words on the test.

Use the Numbering to Guide Your Answer!

Questions will typically have several subparts. When the subparts are related directly to one another, they will be numbered with letters and lowercase Roman numerals (for example, *a.i* and *a.ii*). Subparts that are not directly related to one another, but that are still related to the main stem of the problem, are numbered with different letters (for example, *a* and *b*). It is common for a question to have a mixture of both types of subparts. (For example, a problem could be numbered *a.i, a.ii, b, c.i, c.ii.* Subparts *a.i* and *a.ii* would be related to each other, while part *b* would be related back to the main stem of the problem but not to either of the *a* parts. Then both of the *c* parts would be related to each other, but not to the *a* or *b* parts.) Try to answer all of the subparts. If one part is giving you trouble, look at the next subpart to see if you can find a way to approach it. Since the answer to *b* does not usually depend on part *a*, you can take a fresh approach to answering when the numbering of the subparts changes.

Read and Respond to the Entire Question

Many times, a single subpart of a question will ask you to explain multiple things. In order to earn full points, you must address every item the question asks. Such answers typically require no more than a single sentence per item to fully explain. For example, a subpart in a circuit question may ask how, if a particular resistor were removed, the total current, the total resistance, and the total voltage of the circuit would change. You could not receive full credit unless you addressed all three of those quantities in your answer.

Use Common Sense

Always use your common sense in answering questions. If your solutions to part (*a.i*) are incorrect, but are substituted correctly into part (*a.ii*), you will likely not be penalized for getting an incorrect answer to part (*a.ii*). However, if your answer to (*a.ii*) is physically impossible, such as getting a speed of light in vacuum greater than 3.00×10^8 m/s or finding a number of electrons of 3.5, you will not receive credit even if your answer accurately relies on the answer to a previous part. To receive the highest possible score, make your errors easily recognizable and make sure all of your answers are physically reasonable.

Think Like a Grader

When answering questions, try to think about what kind of answer the grader is expecting. Look at past free-response questions and grading rubrics on the College Board website. These examples will give you some idea of how the answers should be phrased. The graders are told to keep in mind that there are two aspects to the scoring of free-response answers: showing statistical knowledge and communicating that knowledge. Again, responses should be written as clearly as possible in complete sentences. You don't need to show all the steps of a calculation, but you must explain how you got your answer and why you chose the technique you used.

Think Before You Write

Abraham Lincoln once said that if he had eight hours to chop down a tree, he would spend six of them sharpening his axe. Be sure to spend some time thinking about what the question is, what answers are being asked for, what answers might make sense, and what your intuition is before starting to write. These questions aren't meant to trick you, so all the information you need is given. If you think you don't have the right information, you may have misunderstood the question. In some calculations, it is easy to get confused, so think about whether your answers make sense in terms of what the question is asking. If you have some idea of what the answer should look like before starting to write, then you will avoid getting sidetracked and wasting time on dead-ends.

Types of Free-Response Questions and How to Answer Them

The AP Physics 2 Exam will assess your ability to perform an **experimental analysis**. This will be one of the two 12-point questions. During the experimental design question, you will be expected to do the following things:

- Draw a clearly labeled diagram for the experimental procedure that you will follow.
- Distinguish between quantities that are directly measured and those that will need to be calculated.
- Describe an experimental procedure using an appropriate amount of detail. A typical appropriate level of detail will include between 4 and 6 steps, describe the arrangements of materials and the measurements that must be taken, and explain which equipment is used to make each measurement described.
    - Steps such as "gather material," "record data," and "analyze data" should be omitted as they are implicit in performing any experimental analysis.
    - Descriptions that involve too much detail should also be avoided.
- Using the graph paper supplied to you, draw or describe a graph from data collected in the experiment (based on a data table supplied for a similar experiment).
    - In drawing or describing a graph, you need to label the vertical and horizontal axes, including labeling values and units.
    - If you need to produce a graph, you must scale the axes so the data is shown on the graph provided.
        - Your data points must take up at least half of the space on the available graph in order to receive full credit.
    - Draw or describe a best-fit line for linear data. If drawing the fit line:
        - The best-fit line should contain approximately an equal number of points above the line as below the line.
        - You should avoid drawing the best-fit line through the first and last data points.
        - You may create a best-fit line using a linear regression on a graphing calculator, but you must explicitly write that you have done this in your answer.
- Describe how to use the slope of the best-fit line and relate it to a physical quantity.
    - Using any method other than finding the slope of the best-fit line will result in lost points.
- You may be asked to discuss which measurement or variable in a procedure contributes most to overall uncertainty in the final result and on conclusions drawn from a given data set.
- You should be able to review and critique an experimental design or procedure and decide whether the conclusions can be justified based on the procedure and the evidence presented.

The AP Physics 2 Exam will have a question about **translating quantitative measurements into qualitative explanations,** or vice versa. This will be the other 12-point problem. This problem will require performing some calculations, so being familiar with the equations on the equation sheet will help you answer this question. You will be asked to relate your calculations to a non-calculated qualitative description of the physical phenomenon. In optics, this could be a question about how the magnification calculated is shown in a ray diagram. In electrostatics, it could be a description of how an electric potential map relates to the calculated electric field.

The AP Physics 2 Exam will have a **short-answer problem that includes the instruction to answer "in a coherent, paragraph-length response."** For full credit on this question, it is required that the argument be presented primarily in prose. If the argument relies too heavily on diagrams, graphs, equations, or calculations, you will not receive full credit. You need to clearly demonstrate your line of reasoning. According to the College Board, your answer "should be a coherent, organized, and sequential description of the analysis of a situation." To construct such an argument, you should begin by citing applicable physical principles (this might include using conservation of energy and charge for a circuit description or Bernoulli's Principle for a fluid dynamics question). You then need to sequentially describe the situation as physics dictates it will evolve. Your response should be at least 4 sentences long. Your sentences should not rely heavily on the use of equations or on descriptions of diagrams.

The AP Physics 2 Exam will have a **short-answer problem that has you analyze an argument among a pair of fictitious students.** You may be asked specifically in separate sub-parts what parts of each student's argument are correct (as parts *i* and *ii*) and what parts of their argument are incorrect (as parts *iii* and *iv*). The question may also be a single subpart asking you to address the claims of each student and explain why the student is correct or incorrect. Regardless of how the question is phrased, your response needs to include a sentence restating the part of the argument made by the student that is correct or incorrect and a sentence explaining why the argument is valid or invalid.

Reflect

Respond to the following questions:

- How much time will you spend on the short free-response questions? What about the long free-response questions?
- What will you do before you begin writing your free-response answers?
- Will you seek further help, outside of this book (such as a teacher, tutor, or AP Students), on how to approach the questions that you will see on the AP Physics 2 Exam?

Part V
Content Review for the AP Physics 2 Exam

HOW TO USE THE CHAPTERS IN THIS PART

You may need to review the following content chapters more than once. Your goal is to obtain command of the content, and a single read of a chapter may not be sufficient. At the end of each chapter, reflect on the following questions to help you determine whether or not you truly have mastered the content of that chapter:

- For which content topics discussed in this chapter do you feel you have achieved sufficient command to answer multiple-choice questions correctly?
- For which content topics discussed in this chapter do you feel you have achieved sufficient command to confidently answer free-response questions?
- For which content topics discussed in this chapter do you feel you need more work before you can answer multiple-choice questions correctly?
- For which content topics discussed in this chapter do you feel you need more work before you can confidently answer free-response questions?
- What parts of this chapter are you going to re-review?
- Will you seek further help outside of this book (such as a teacher, tutor, or AP Students) on any of the content in this chapter—and, if so, on what content?

One key difference between our book and the official AP Physics 2 Equation Sheet is that our book employs a **bold font** to indicate a vector quantity and a non-bold font for magnitude only, whereas the College Board uses arrow hats to denote vector quantities and the absolute value symbol for magnitudes.

Equation Sheet Sidebars

Throughout the content review chapters in Part V, you'll see the words "Equation Sheet" in the sidebars. This is meant to draw your attention to the physics equations featured on the College Board's equation sheet that will be provided to you during your AP Physics 2 Exam. So these are equations that are extremely important for you to be familiar with.

Chapter 3
Fluids

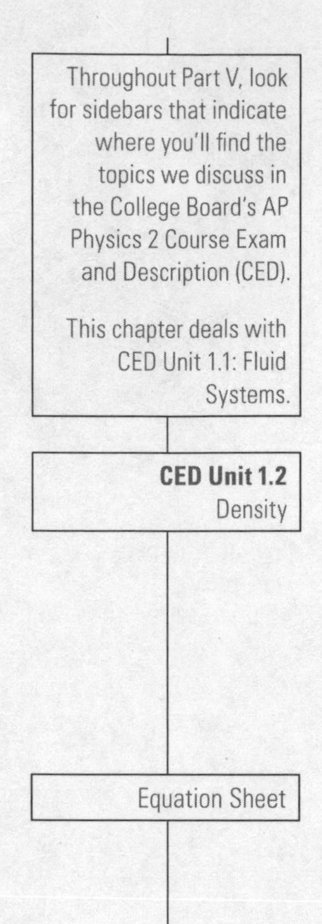

INTRODUCTION

In this chapter, we'll discuss some of the fundamental concepts dealing with substances that can flow, which are known as **fluids**. The term *fluid* refers to both liquids and gases. While there are distinctions between liquids and gases, this chapter focuses on the similarities of all fluids.

DENSITY

Although the concept of *mass* is central to your study of mechanics (because of the all-important equation $\mathbf{F}_{net} = m\mathbf{a}$), it is the substance's *density* that turns out to be more useful in fluid mechanics.

By definition, the density of a substance is its mass per unit volume, and it's typically denoted by the letter ρ (the Greek letter *rho*):

$$\text{density} = \frac{\text{mass}}{\text{volume}}$$

$$\rho = \frac{m}{V}$$

Note that this equation immediately implies that the mass, m, of an object, is equal to the product of that density, ρ, of that object and V, the volume of that object, whenever the density of the object is constant.

For example, if 10^{-3} m^3 of oil has a mass of 0.8 kg, then the density of this oil is

$$\rho = \frac{m}{V} = \frac{0.8 \text{ kg}}{10^{-3} \text{ m}^3} = 800 \frac{\text{kg}}{\text{m}^3}$$

While it is not required knowledge for an AP Physics 2 Exam, it is useful to know that the density of water (at "standard temperature and pressure," or STP) is very close to 1,000 kg/m^3. The density of fluids and gases changes at varying temperatures and pressures, hence the inclusion of the phrase "at standard temperature and pressure." STP is defined at a temperate of 0°C and a pressure of 1×10^5 Pa, which is standard atmospheric pressure at the surface of the Earth listed on the equation sheet. With this density of water in mind, you can see that the oil calculated in the example above has a lower density than water, and therefore the oil would float on top of the water, as we will see in the next section.

PRESSURE

CED Unit 1.3
Fluids: Pressure and Forces

If we place an object in a fluid, the fluid exerts a contact force on the object. How that force is distributed over any small area of the object's surface defines the **pressure**:

$$\text{pressure} = \frac{\text{force}_\perp}{\text{area}}$$
$$P = \frac{F_\perp}{A}$$

Equation Sheet

The subscript $\perp$ (which means *perpendicular*) is meant to emphasize that the pressure is defined to be the magnitude of the force that acts perpendicular to the surface, divided by the area. Because force is measured in newtons (N) and area is expressed in square meters (m^2), the SI unit for pressure is the newton per square meter. This unit is given its own name: the **pascal**, abbreviated Pa:

$$\text{SI unit of pressure: } 1 \text{ pascal} = 1 \text{ Pa} = 1\,\frac{\text{N}}{\text{m}^2}$$

One pascal is a very tiny amount of pressure; for example, a nickel on a table exerts about 140 Pa of pressure, just due to its weight alone. For this reason, you'll often see pressures expressed in kPa (kilopascals, where 1 kPa = 10^3 Pa) or even in MPa (megapascals, where 1 MPa = 10^6 Pa). Another common unit for pressure is the atmosphere (atm). At sea level, atmospheric pressure, P_{atm}, is about 100,000 Pa; this is 1 **atmosphere**.

Example 1 A vertical column made of cement has a base area of 0.5 m^2. If its height is 2 m, and the density of cement is 3,000 kg/m^3, how much pressure does this column exert on the ground?

Solution. The force the column exerts on the ground is equal to its weight, *mg*, so we'll find the pressure it exerts by dividing this by the base area, *A*. The mass of the column is equal to ρV, which we calculate as follows:

$$m = \rho V = \rho A h = (3\times10^3\ \frac{\text{kg}}{\text{m}^3})(0.5\text{ m}^2)(2\text{ m}) = 3\times10^3\text{ kg}$$

Therefore,

$$P = \frac{F}{A} = \frac{mg}{A} = \frac{(3\times10^3\text{ kg})(9.8\frac{\text{N}}{\text{kg}})}{0.5\text{ m}^2} = 5.88\times10^4\text{ Pa}$$

The units given for g, the acceleration due to gravity at the Earth's surface, are m/s^2 when calculated from the time-rate of change in velocity. These units are equivalent to N/kg when the acceleration is instead calculated using $F = ma$. In studying fluids, N/kg are typically more useful units.

HYDROSTATIC PRESSURE

Imagine that we have a tank with a lid on top, filled with some liquid. Suspended from this lid is a string, attached to a thin sheet of metal that hangs horizontally. The figures below show two views of this tank:

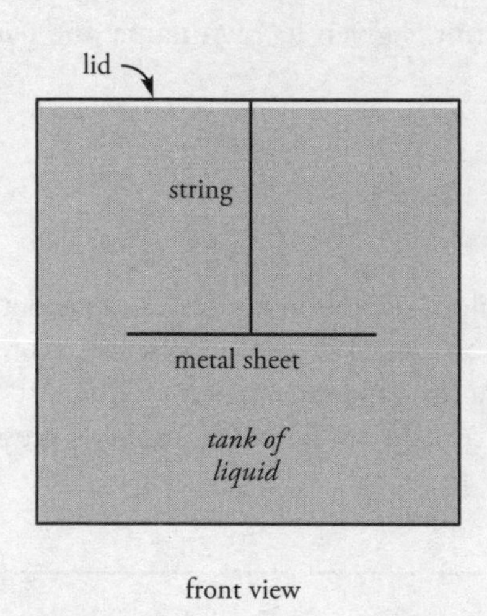

front view

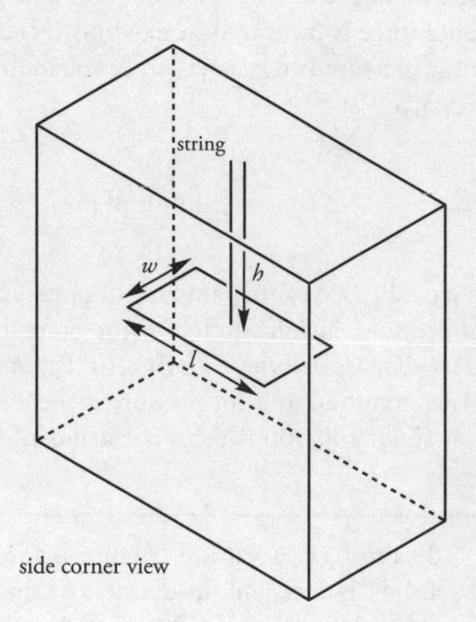

side corner view

The weight of the liquid produces a force that pushes down on the metal sheet. If the sheet has length l and width w, and is at depth h below the surface of the liquid, then the weight of the liquid on top of the sheet is

$$F_g = mg = \rho Vg = \rho(lwh)g$$

where ρ is the liquid's density. If we divide this weight by the area of the sheet ($A = lw$), we get the pressure due to the liquid:

$$P_{\text{liquid}} = \frac{\text{force}}{\text{area}} = \frac{F_{g\,\text{liquid}}}{A} = \frac{\rho(lwh)g}{lw} = \rho gh$$

Since the liquid is at rest, this is known as **hydrostatic pressure**.

Note that the hydrostatic pressure due to the liquid, $P_{\text{liquid}} = \rho gh$, depends only on the density of the liquid and the depth below the surface; in fact, it's proportional to both of these quantities. One important consequence of this is that the shape of the container doesn't matter. For example, if all the containers in the figure below are filled with the same liquid, then the pressure is the same at every point along the horizontal dashed line (and within a container). This is because every point on this line is at the same depth, *h*, below the surface of the liquid.

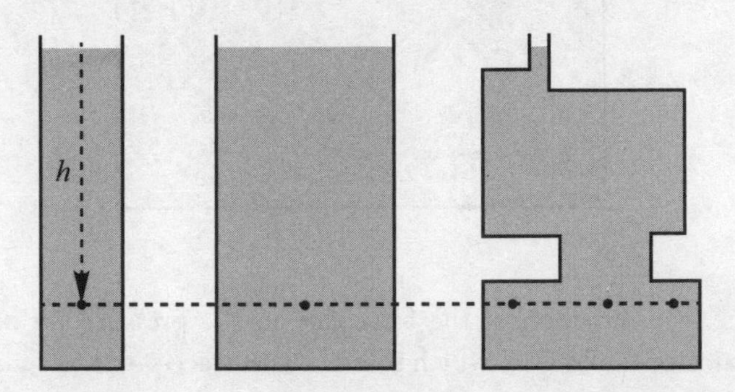

If the liquid in the tank were open to the atmosphere, then the total (or absolute) pressure at depth *h* would be equal to the pressure pushing down on the surface—the atmospheric pressure, P_{atm}—plus the pressure due to the liquid alone. In general, the pressure on the surface of the liquid is some pressure P_0 and the total pressure is

$$\text{total (absolute) pressure: } P_{\text{total}} = P_0 + P_{\text{liquid}} = P_0 + \rho gh$$

When the pressure at the top of the liquid is atmospheric pressure (as in the scenario above), then our equation for total pressure reduces to

$$P_{\text{total}} = P_{\text{atm}} + \rho gh$$

Because pressure is the *magnitude* of the force per area, pressure is a scalar. It has no direction. The direction of the force due to the pressure on any small surface is perpendicular to that surface. For example, in the figure on the next page, the pressure at Point A is the same as the pressure at Point B, because they're at the same depth.

It's All About Depth
The pressure exerted on an object by a fluid is proportional only to the density of the fluid and the depth of the object, but is independent of the object's mass. An elephant and a feather at the same depth in the ocean will feel the same pressure from the water.

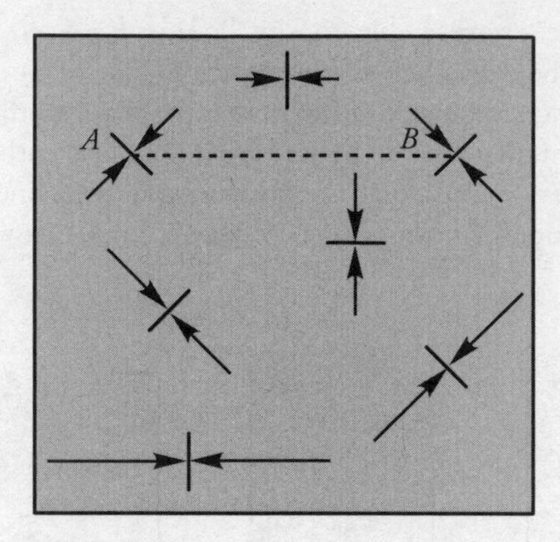

But, as you can see, the direction of the force due to the pressure varies depending on the orientation of the surface—and even which side of the surface—the force is pushing on.

CED Unit 1.4
Fluids and Free-Body Diagrams

CED Unit 1.5
Buoyancy

BUOYANCY

Let's place a block in our tank of fluid. Because the pressure on each side of the block depends on its average depth, we see that there's more pressure on the bottom of the block than there is on the top. Because the block is rectangular and the top and bottom have the same area, there's a greater force pushing up on the block than there is pushing down on it. The forces due to the pressure on the other four sides cancel out (because they are at the same depth), so the net force on the block is upward.

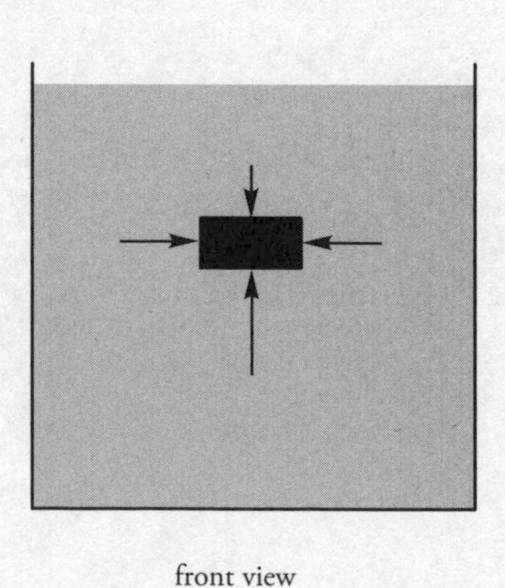

front view

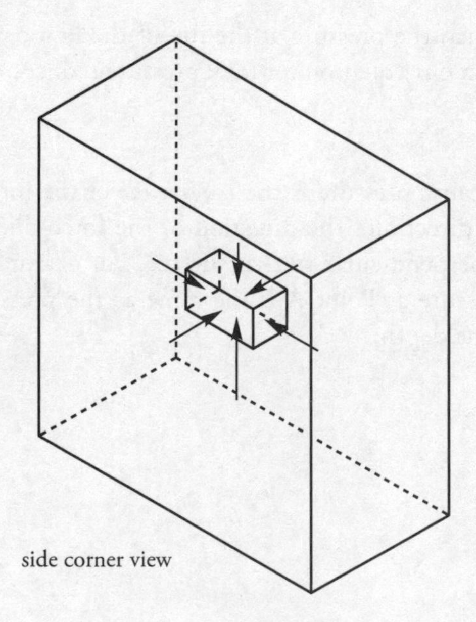

side corner view

This net upward force is called the **buoyant force** (or just **buoyancy** for short), denoted $\mathbf{F}_{buoy}$. We calculate the magnitude of the buoyant force using **Archimedes' principle**; in words, Archimedes' principle says

> The strength of the buoyant force on the object is equal to the weight of the fluid displaced by the object.

When an object is partially or completely submerged in a fluid, the volume of the object submerged, which we call V_{sub}, is the volume of the fluid displaced. By multiplying this volume by the density of the fluid, we get the mass of the fluid displaced; then, multiplying this mass by g gives us the weight of the fluid displaced. So, here's Archimedes' principle as a mathematical equation:

$$\text{Buoyant force: } F_{buoy} = \rho_{fluid} V_{sub}\, g$$

When an object floats, its submerged volume is just enough to make the buoyant force it feels balance its weight. So, if an object's density is ρ_{object} and its (total) volume is V, its weight will be $mg = \rho_{object} Vg$. The buoyant force it feels is $\rho_{fluid} V_{sub}\, g$. Setting these equal to each other, we find that

$$\frac{V_{sub}}{V} = \frac{\rho_{object}}{\rho_{fluid}}$$

So, if $\rho_{object} < \rho_{fluid}$, then the object will float, and the fraction of its volume submerged is the same as the ratio of its density to the fluid's density. For example, if the object's density is 2/3 the density of the fluid, then the object will float, and 2/3 of the object will be submerged.

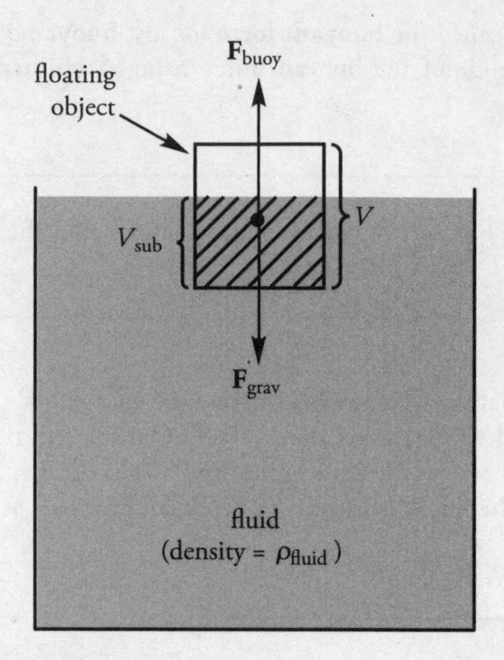

If an object is denser than the fluid, it will sink. In this case, even if the entire object is submerged (in an attempt to maximize V_{sub} and the buoyant force), its weight is still greater than the buoyant force, and down it goes. And if an object just happens to have the same density as the fluid, it will be happy hovering (in static equilibrium) anywhere underneath the fluid.

Let's summarize, and go through the steps of an object floating and then an object as it sinks and hits the bottom of a tank. In the diagram below, we have an object that is floating in a liquid.

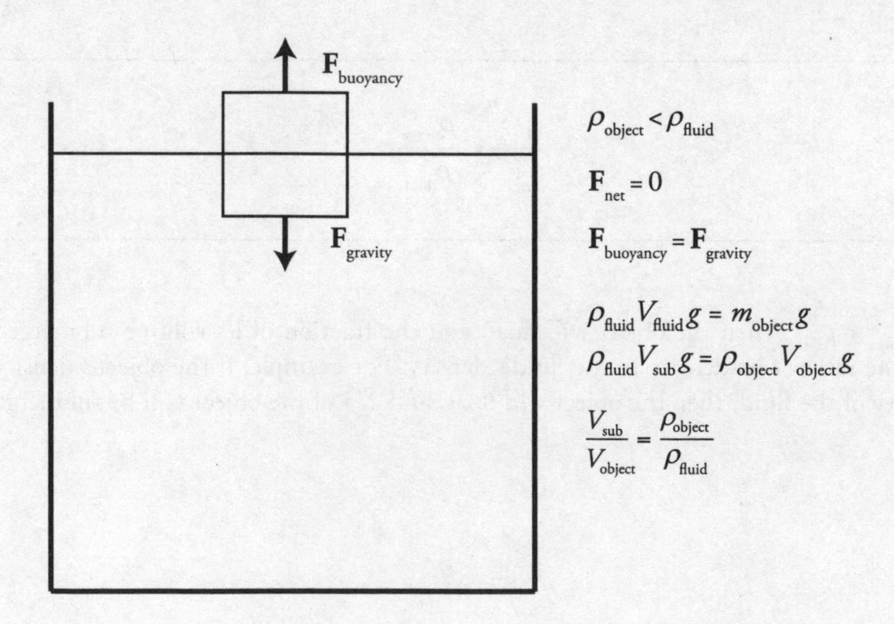

Below is a diagram of an object with a density greater than the fluid density as it sinks to the bottom of the tank.

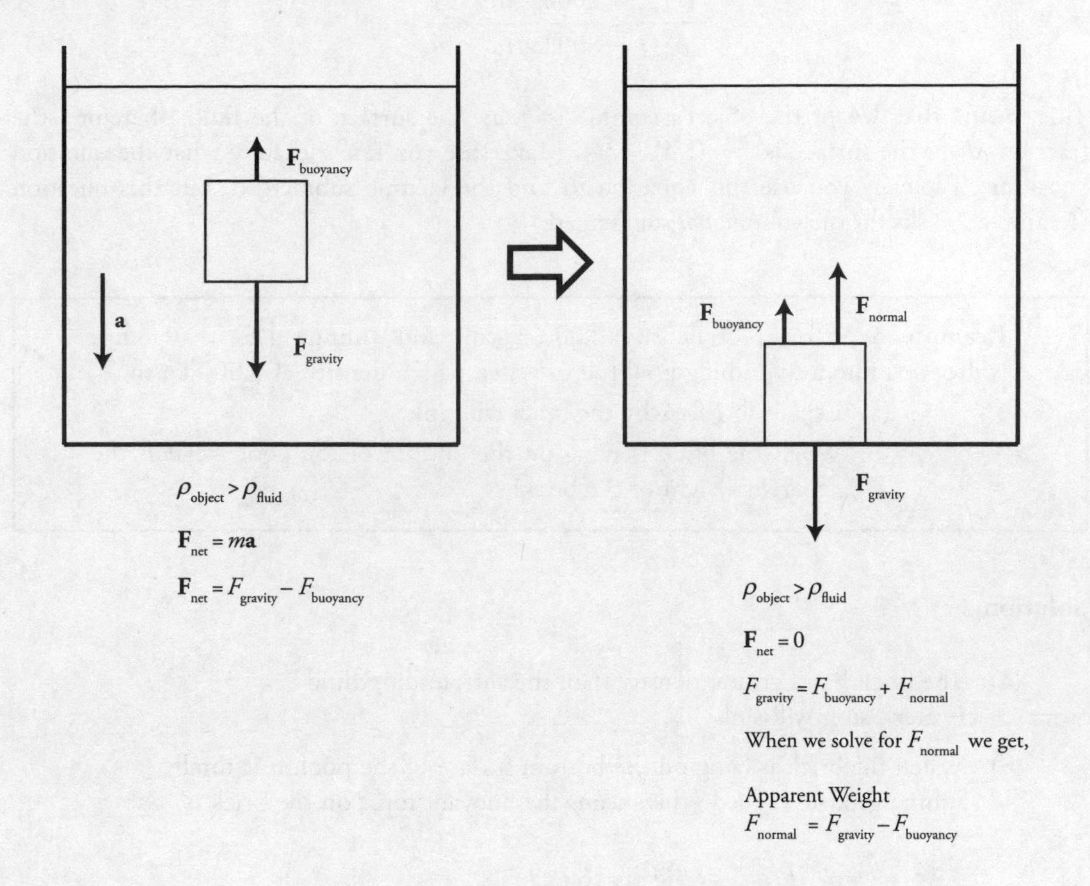

$\rho_{object} > \rho_{fluid}$

$\mathbf{F}_{net} = m\mathbf{a}$

$\mathbf{F}_{net} = F_{gravity} - F_{buoyancy}$

$\rho_{object} > \rho_{fluid}$

$\mathbf{F}_{net} = 0$

$F_{gravity} = F_{buoyancy} + F_{normal}$

When we solve for F_{normal} we get,

Apparent Weight

$F_{normal} = F_{gravity} - F_{buoyancy}$

Because the buoyant force is always directed upward, and the force of gravity is defined to point downward, these forces are always directed in opposite directions. The result of the buoyant force opposing gravity is that any object in a fluid appears to weigh less than when it is not in the fluid, and as the fluid density increases, the apparent weight decreases. The apparent weight of an object is commonly referred to as the normal force of that object.

Example 2 An object with a mass of 150 kg and a volume of 0.75 m³ is floating in ethyl alcohol, whose density is 800 kg/m³. What fraction of the object's volume is above the surface of the fluid?

Solution. The density of the object is

$$\rho_{object} = \frac{m}{V} = \frac{150\text{ kg}}{0.75\text{ m}^3} = 200\,\frac{\text{kg}}{\text{m}^3}$$

The ratio of the object's density to the fluid's density is

$$\frac{\rho_{\text{object}}}{\rho_{\text{fluid}}} = \frac{200 \text{ kg/m}^3}{800 \text{ kg/m}^3} = \frac{1}{4}$$

This means that 1/4 of the object's volume is *below* the surface of the fluid; therefore, the fraction *above* the surface is 1 – (1/4) = 3/4. Make sure you know exactly what the question is asking. Typically you use this equation to find the volume submerged, but this question (Example 2) asks for the volume *not* submerged.

Example 3 A brick, of density 2,000 kg/m^3 and volume 1.5×10^{-3} m^3, is dropped into a swimming pool full of water, with a density of 1,000 kg/m^3.

(a) Explain briefly why the brick will sink.

(b) When the brick is lying on the bottom of the pool, what is the apparent weight of the brick?

Solution.

(a) The brick has a greater density than the surrounding fluid (water), so it will sink.

(b) When the brick is lying on the bottom surface of the pool, it is totally submerged, so $V_{\text{sub}} = V$; this means the buoyant force on the brick is

$$\begin{aligned} F_{\text{buoy}} = \rho_{\text{fluid}} V_{\text{sub}} g = \rho_{\text{water}} Vg \\ &= (1{,}000 \text{ kg/m}^3)(1.5 \times 10^{-3} \text{ m}^3)(10 \text{ N/kg}) \\ &= 15 \text{ N} \end{aligned}$$

The weight of the brick is

$$\begin{aligned} F_g = mg = \rho_{\text{brick}} Vg \\ &= (2{,}000 \text{ kg/m}^3)(1.5 \times 10^{-3} \text{ m}^3)(10 \text{ N/kg}) \\ &= 30 \text{ N} \end{aligned}$$

When the brick is lying on the bottom of the pool, the net force it feels is zero.

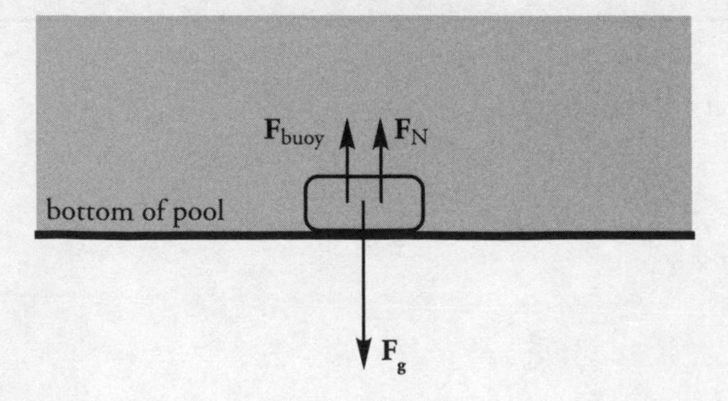

Therefore, we must have $F_{buoy} + F_N = F_g$, so $F_N = F_g - F_{buoy} = 30\text{ N} - 15\text{ N} = 15\text{ N}$. So the apparent weight of the brick, that is, the normal force it feels, is 15 N.

Example 4 A helium balloon has a volume of 0.03 m^3. Ignoring the weight of the plastic of the balloon, calculate the net force on the balloon if it's surrounded by air. (Note: The density of helium is 0.2 kg/m^3, and the density of air is 1.2 kg/m^3.)

Solution. The balloon will feel a buoyant force upward, and the force of gravity—the balloon's weight—downward. Because the balloon is completely surrounded by air, $V_{sub} = V$, and the buoyant force is

$$\begin{aligned} F_{buoy} = \rho_{fluid} V_{sub} g = \rho_{air} V g \\ = (1.2\text{ kg/m}^3)(0.03\text{ m}^3)(10\text{ N/kg}) \\ = 0.36\text{ N} \end{aligned}$$

The weight of the balloon is

$$\begin{aligned} F_g = mg = \rho_{helium} V g = (0.2\text{ kg/m}^3)(0.03\text{ m}^3)(10\text{ N/kg}) \\ = 0.06\text{ N} \end{aligned}$$

Because $F_{buoy} > w$, the net force on the balloon is upward and has magnitude

$$F_{net} = F_{buoy} - F_g = 0.36\text{ N} - 0.06\text{ N} = 0.3\text{ N}$$

This is why a helium balloon floats away if you let go of its string.

THE CONTINUITY EQUATION: FLOW RATE AND CONSERVATION OF MASS

CED Unit 1.7
Conservation of Mass
Flow Rate in Fluids

Consider a pipe through which fluid is flowing. The **volume flow rate**, f, is the volume of fluid that passes a particular point per unit time. In SI units, flow rate is expressed in m^3/s. To find the volume flow rate, all we need to do is multiply the size of the pipe at a particular location by the average speed of the flow at that point. In this context, the size of the pipe means the area of a cross-section of the pipe that is perpendicular to the direction of the flow.

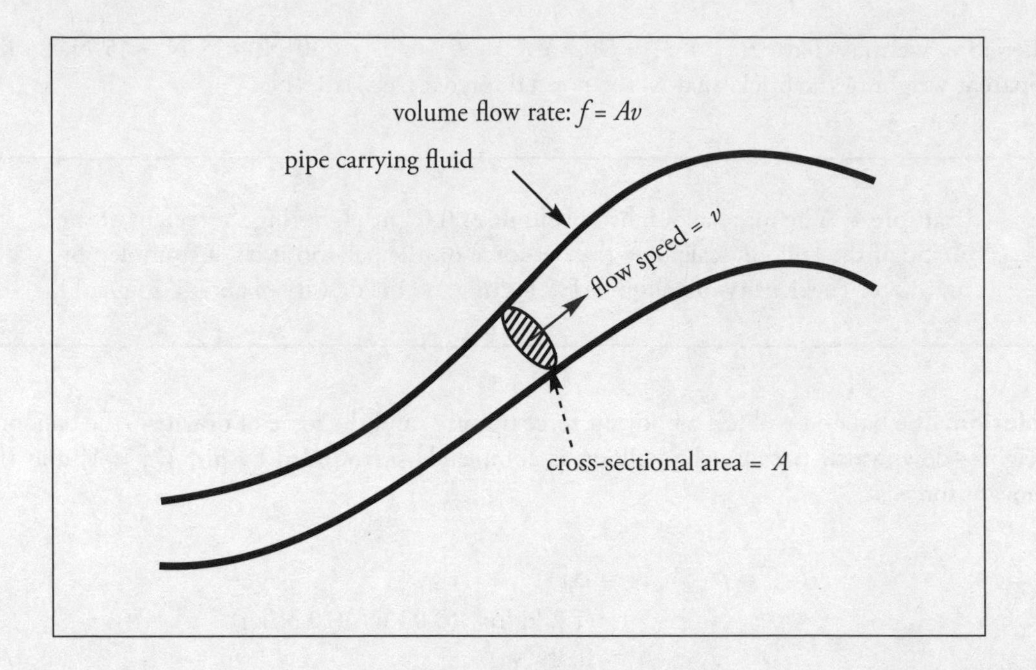

Be careful not to confuse *volume flow rate* with *flow speed*; volume flow rate tells us how *much* volume of fluid flows per unit time; flow speed tells us how *fast* it's moving. The volume flow rate is measured in cubic meters per second, while the flow speed is measured in meters per second.

An incompressible fluid has a density that does not change. For an incompressible fluid, if we imagine some fluid moving past a particular point in the pipe during some small amount of time, then we can conclude the same amount of mass must flow past another point in the pipe in the same amount of time because mass is conserved. The amount of mass m flowing for distance Δl past any point with area A during a time interval Δt is given by $\Delta m/\Delta t = \rho\Delta V/\Delta t = \rho A(\Delta l/\Delta t) = \rho A\Delta v$. The final two terms in this equation are simply the volume flow rate we described above. Because the density of the fluid is constant, we get the **Continuity Equation:**

Equation Sheet

$$A_1v_1 = A_2v_2$$

Because the product Av is a constant, the flow speed will increase where the pipe narrows (decreases in area) and decrease where the pipe widens (increases in area). In fact, we can say that the flow speed is inversely proportional to the cross-sectional area—or to the square of the radius (in the case of a circular cross-section)—of the pipe.

Example 5 A circular pipe of non-uniform diameter carries water. At one point in the pipe, the radius is 2 cm and the flow speed is 6 m/s.

(a) What is the volume flow rate?

(b) Without performing a new flow rate calculation, explain what the volume flow rate and flow speed will be if, at a later point, the radius of the pipe decreases to 1 cm.

Solution.

(a) At any point, the volume flow rate, *f*, is equal to the cross-sectional area of the pipe multiplied by the flow speed

$$f = Av = \pi r^2 v = \pi (2 \times 10^{-2}\ \text{m})^2 (6\ \text{m/s}) \approx 75 \times 10^{-4}\ \text{m}^3/\text{s} = 7.5 \times 10^{-3}\ \text{m}^3/\text{s}$$

(b) The volume flow rate will be unchanged. Then, from the Continuity Equation, we know that v, the flow speed, is inversely proportional to A, the cross-sectional area of the pipe. If the pipe's radius decreases by a factor of 2 (from 2 cm to 1 cm), then A decreases by a factor of 4, because A is proportional to r^2.

If A decreases by a factor of 4, then v will increase by a factor of 4. Therefore, the flow speed at a point where the pipe's radius is 1 cm will be $4 \cdot (6\ \text{m/s}) = 24\ \text{m/s}$.

BERNOULLI'S EQUATION: CONSERVATION OF ENERGY IN FLUIDS

CED Unit 1.6
Conservation of Energy in Fluid Flow

The most important equation in fluid mechanics is **Bernoulli's Equation**, which is the statement of conservation of energy for ideal fluid flow. First, let's describe the conditions that make fluid flow *ideal*.

- *The fluid is incompressible.*

  This works very well for liquids and also applies to gases if the pressure changes are small.

- *The fluid's viscosity is negligible.*

  Viscosity is the force of cohesion between molecules in a fluid; think of viscosity as internal friction for fluids. For example, maple syrup is sticky and has a greater viscosity than water: there's more resistance to a flow of maple syrup than to a flow of water. While Bernoulli's Equation would give good results when applied to a flow of water, it would not give good results if it were applied to a flow of maple syrup.

- *The flow is streamline.*

  In a tube carrying a flowing fluid, a **streamline** is just what it sounds like: it's a *line* in the *stream*. If we were to inject a drop of dye into a clear glass pipe carrying, say, water, we'd see a streak of dye in the pipe, indicating a streamline.

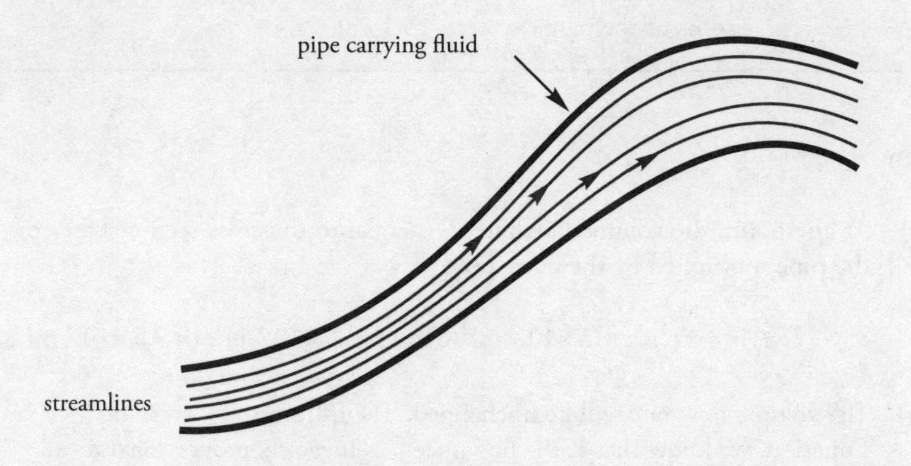

  When the flow is streamline, the fluid moves smoothly through the tube. (The opposite of streamline flow is **turbulent** flow, which is characterized by rapidly swirling whirlpools; such chaotic flow is unpredictable.)

If the three conditions described above hold, and the flow rate, f, is steady (meaning it doesn't change with time), Bernoulli's Equation can be applied to any pair of points along a streamline within the flow. Let ρ be the density of the fluid that's flowing. Label the points we want to compare as Point 1 and Point 2. Choose a horizontal reference level, and let y_1 and y_2 be the heights of these points above this level. If the pressures at Points 1 and 2 are P_1 and P_2, and if the flow speeds at these points are v_1 and v_2, then Bernoulli's Equation says

Equation Sheet

$$P_1 + \rho g y_1 + \tfrac{1}{2}\rho v_1^2 = P_2 + \rho g y_2 + \tfrac{1}{2}\rho v_2^2$$

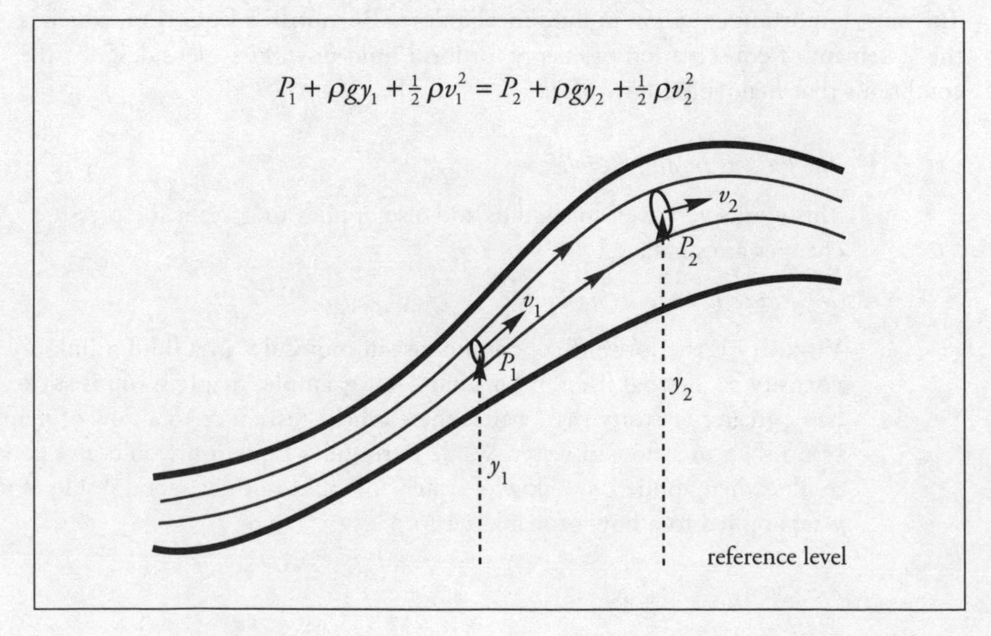

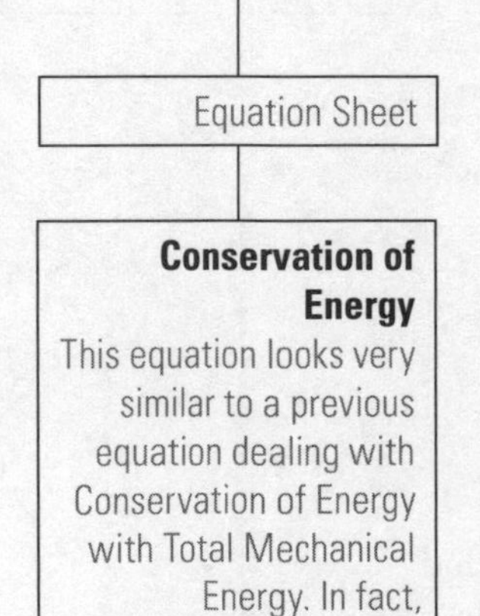

Conservation of Energy
This equation looks very similar to a previous equation dealing with Conservation of Energy with Total Mechanical Energy. In fact, Bernoulli's Equation deals with the Conservation of Energy for fluids.

An alternative, but equivalent, way of stating Bernoulli's Equation is to say that the quantity

$$P + \rho g y + \tfrac{1}{2}\rho v^2$$

is constant along a streamline. We mentioned earlier that Bernoulli's Equation is a statement of conservation of energy. You should notice the similarity between $\rho g y$ and mgh (gravitational potential energy) as well as between $\frac{1}{2}\rho v^2$ and $\frac{1}{2}mv^2$ (kinetic energy). The term P, indicating the pressure due to external fluids, is similar to the work done by an external force in a statement of conservation of energy.

Example 6 In the figure below, a pump forces water at a constant flow rate through a pipe whose cross-sectional area, A, gradually decreases: at the exit point, A has decreased to 1/3 its value at the beginning of the pipe. If y = 60 cm and the flow speed of the water just after it leaves the pump (Point 1 in the figure) is 1 m/s, what is the gauge pressure at Point 1?

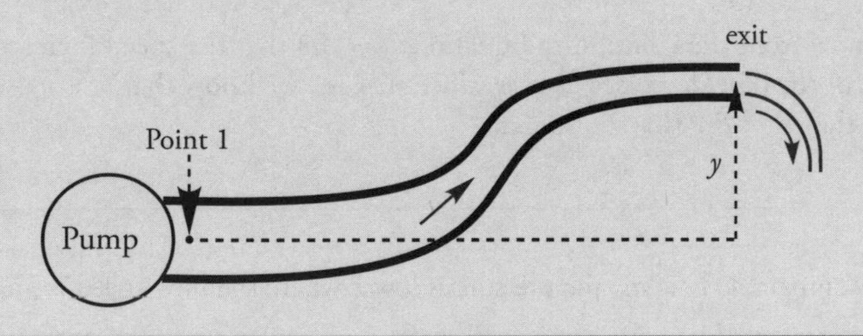

Solution. We'll apply Bernoulli's Equation to Point 1 and the exit point, Point 2. We'll choose the level of Point 1 as the horizontal reference level; this makes $y_1 = 0$. Now, because the cross-sectional area of the pipe decreases by a factor of 3 between Points 1 and 2, the Continuity Equation tells us that the flow speed must increase by a factor of 3; that is, $v_2 = 3v_1$. Since the pressure at Point 2 is P_{atm}, Bernoulli's Equation becomes

$$P_1 + \tfrac{1}{2}\rho v_1^2 = P_{atm} + \rho g y_2 + \tfrac{1}{2}\rho v_2^2$$

Now, P_1 is the total pressure at Point 1. Recall that gauge pressure is $P_{tot} - P_{atm}$, so $P_{gauge} = P_1 - P_{atm}$.

Therefore,

$$\begin{aligned} P_1 - P_{atm} &= \rho g y_2 + \tfrac{1}{2}\rho v_2^2 - \tfrac{1}{2}\rho v_1^2 \\ &= \rho g y_2 + \tfrac{1}{2}\rho (3v_1)^2 - \tfrac{1}{2}\rho v_1^2 \\ &= \rho(g y_2 + 4v_1^2) \\ &= (1{,}000\ \text{kg/m}^3)[(10\ \text{m/s}^2)(0.6\ \text{m}) + 4(1\ \text{m/s})^2] \\ &= 10^4\ \text{Pa} \end{aligned}$$

The Bernoulli Effect

Consider the two points labeled in the pipe shown below:

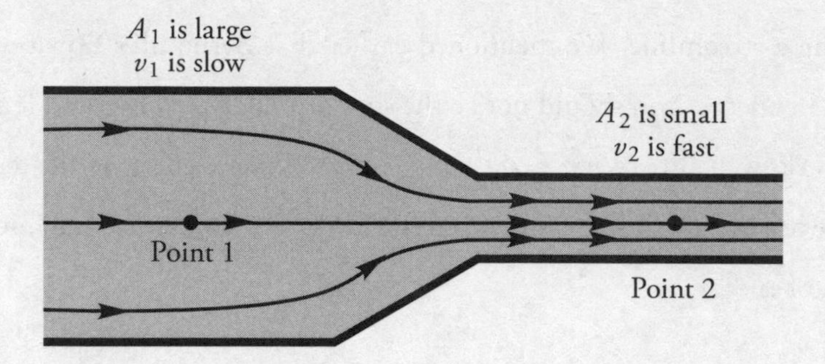

Since the height of the fluid flow is constant in this case, the terms in Bernoulli's Equation that involve height will cancel, leaving us with

$$P_1 + \frac{1}{2}\rho v_1^2 = P_2 + \frac{1}{2}\rho v_2^2$$

We already know from the Continuity Equation ($f = Av$) that the speed increases as the cross-sectional area of the pipe decreases; that is, since $A_2 < A_1$, we know that $v_2 > v_1$, so the equation above tells us that $P_2 < P_1$. This shows that

> At comparable heights, the pressure is lower where the flow speed is greater.

This is known as the **Bernoulli** (or **Venturi**) **Effect**, and is illustrated in the figure below.

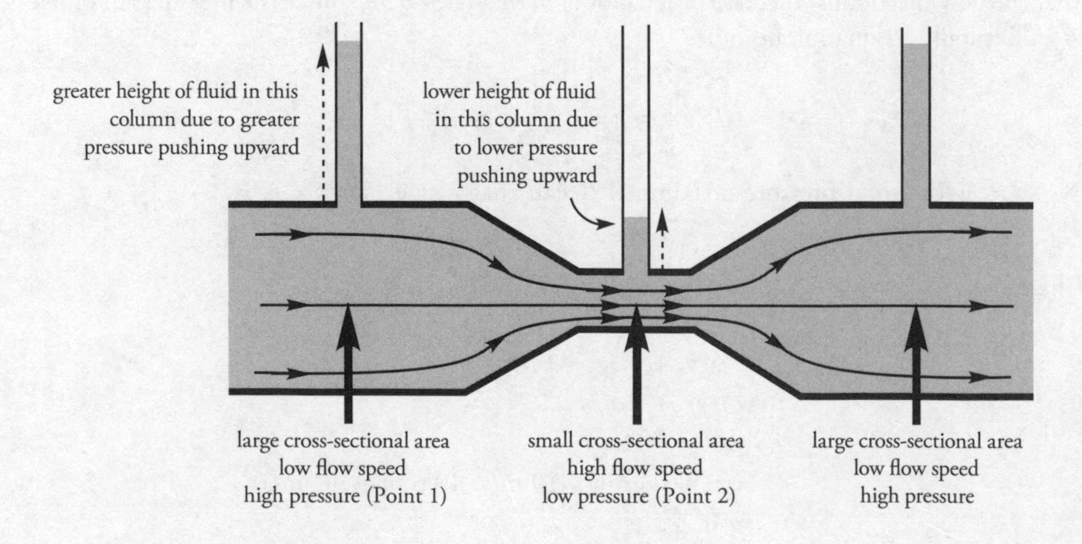

The height of the liquid column above Point 2 is less than the height of the liquid column above Point 1, because the pressure at Point 2 is lower than the pressure at Point 1. This is because the flow speed at Point 2 is greater than the flow speed at Point 1.

The Bernoulli Effect also accounts for many everyday phenomena. It's what allows airplanes to fly, curve balls to curve, and tennis balls hit with topspin to drop quickly. You may have seen skydivers or motorcycle riders wearing a jacket that seems to puff out as they move rapidly through the air. The essentially stagnant air trapped inside the jacket is at a much higher pressure than the air whizzing by outside, and as a result, the jacket expands outward. Yet another example is the drastic drop in air pressure that accompanies the high winds in a hurricane or tornado. In fact, if high winds streak across the roof of a house whose windows are closed, the outside air pressure can be reduced so much that the air pressure inside the house (where the air speed is essentially zero) can be great enough to blow the roof off.

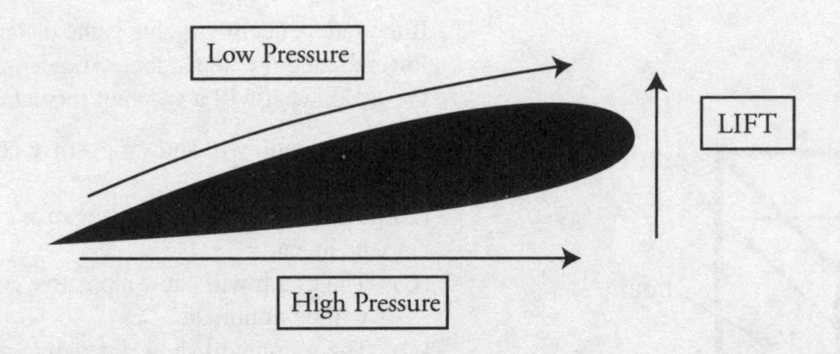

Air flow: The upper side of the wing is longer, so air takes more time to reach the edge. If it takes the same time (but longer distance) [$\uparrow d = \uparrow v\, t$], then the top air is moving faster than the bottom air. The air on the bottom has greater pressure and pushes up on the wing, giving airplanes lift force.

Chapter 3 Review Questions

Solutions can be found in Chapter 11.

Section I: Multiple Choice

1. A large tank is filled with water to a depth of 6 m. If Point X is 1 m from the bottom and Point Y is 2 m from the bottom, how does P_X, the hydrostatic pressure due to the water at Point X, compare to P_Y, the hydrostatic pressure due to the water at Point Y?

(A) $P_X = 2P_Y$
(B) $2P_X = P_Y$
(C) $5P_X = 4P_Y$
(D) $4P_X = 5P_Y$

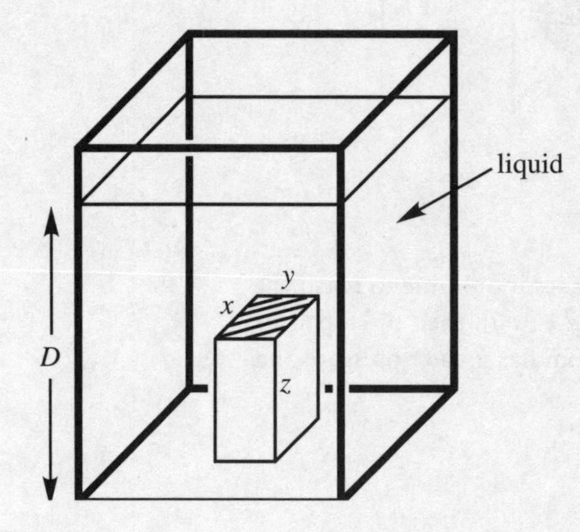

2. In the figure above, a box of dimensions *x*, *y*, and *z* rests on the bottom of a tank filled to depth *D* with a liquid of density *ρ*. If the tank is open to the atmosphere, what is the force on the (shaded) top of the box?

(A) $xy(P_{atm} + \rho gD)$
(B) $xyz[P_{atm} - \rho g(z - D)]$
(C) $xy[P_{atm} + \rho g(D - z)]$
(D) $xyz[P_{atm} + \rho gD]$

Questions 3 and 4 refer to the following material.

An experiment is performed in which a cube is suspended from a spring scale. The cube is lowered into a beaker of water.

3. If the independent variable is the distance the cube is lowered, and the dependent variable is the reading on the scale, what will a graph of the data show?

(A) The graph will show a positive correlation and be linear.
(B) The graph will show a negative correlation and be linear.
(C) The graph will show a positive correlation, but will be nonlinear.
(D) The graph will show a negative correlation, but will be nonlinear.

4. If the cube is replaced with a sphere and the same experiment is performed, which correctly describes the graph of the variables for the new experiment?

(A) The graph will be the same for both experiments because both experiments have the same independent and dependent variables.
(B) The graph will be different for both experiments because both experiments have the same independent and dependent variables.
(C) The graph will be different for both experiments because they have different independent variables but the same dependent variable.
(D) The graph will be different for both experiments because they have different dependent variables but the same independent variable.

5. A block of Styrofoam, with a density of ρ_S and volume V, is pushed completely beneath the surface of a liquid whose density is ρ_L, and released from rest. Given that $\rho_L > \rho_S$, which of the following expressions gives the magnitude of the block's initial upward acceleration?

(A) $(\rho_L - \rho_S)g$

(B) $\left(\dfrac{\rho_L}{\rho_S} - 1\right)g$

(C) $\left(\dfrac{\rho_L}{\rho_S} + 1\right)g$

(D) $\left[\left(\dfrac{\rho_L}{\rho_S}\right)^2 - 1\right]g$

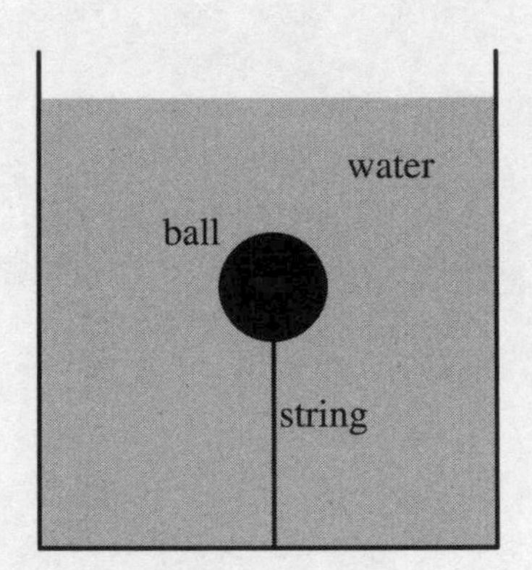

6. A ball is tied to a string and placed in a container of water, as shown. The water is slowly drained from the container until the water level is below the position of the ball shown, but the draining stops while there is still water in the container. The tension in the string is measured while this occurs. Which describes why the tension measurements decrease?

(A) The tension decreases because there is less pressure on the ball when the water level drops because of the change in the water density.
(B) The tension decreases because there is less pressure on the ball when the water level drops because the water is moving at a faster speed.
(C) The tension decreases because the force of gravity on the ball decreases.
(D) The tension decreases because the buoyant force on the ball decreases.

7. An object with a density of 2,000 kg/m^3 weighs 100 N less when it's weighed while completely submerged in water than when it's weighed in air. What is the actual weight of this object?

(A) 200 N
(B) 300 N
(C) 400 N
(D) 600 N

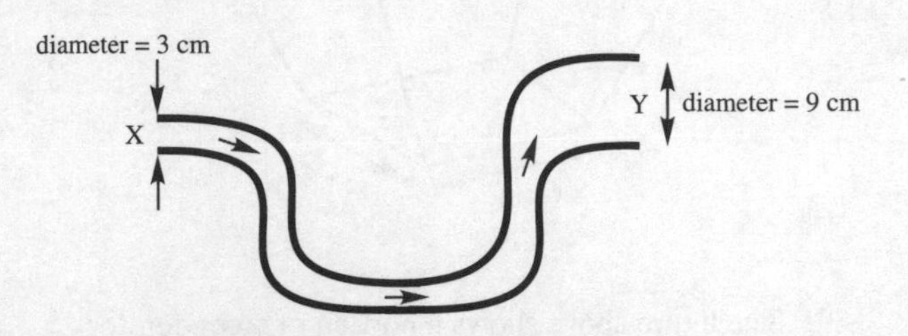

8. In the pipe shown above, which carries water, the flow speed at Point X is 6 m/s. What is the flow speed at Point Y ?

(A) $\dfrac{2}{3}$ m/s
(B) 2 m/s
(C) 18 m/s
(D) 54 m/s

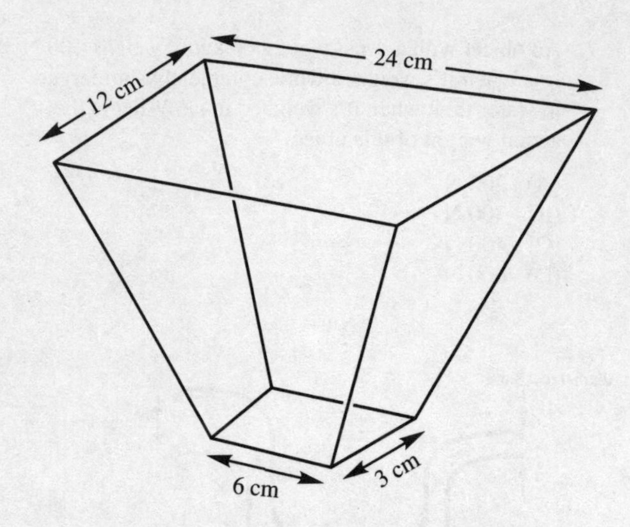

9. The figure above shows a portion of a conduit for water, one with rectangular cross sections. If the flow speed at the top is v, what is the flow speed at the bottom?

(A) $4v$
(B) $8v$
(C) $12v$
(D) $16v$

10. A pump is used to send water through a hose, the diameter of which is 10 times that of the nozzle through which the water exits. If the nozzle is 1 m higher than the pump, and the water flows through the hose at 0.4 m/s, what is the difference in pressure between the pump and the atmosphere?

(A) 108 kPa
(B) 260 kPa
(C) 400 kPa
(D) 810 kPa

Section II: Free Response

1. The figure below shows a tank open to the atmosphere and filled to depth D with a liquid of density ρ_L. Suspended from a string is a block of density ρ_B (which is greater than ρ_L), whose dimensions are x, y, and z (meters). The top of the block is at depth h meters below the surface of the liquid.

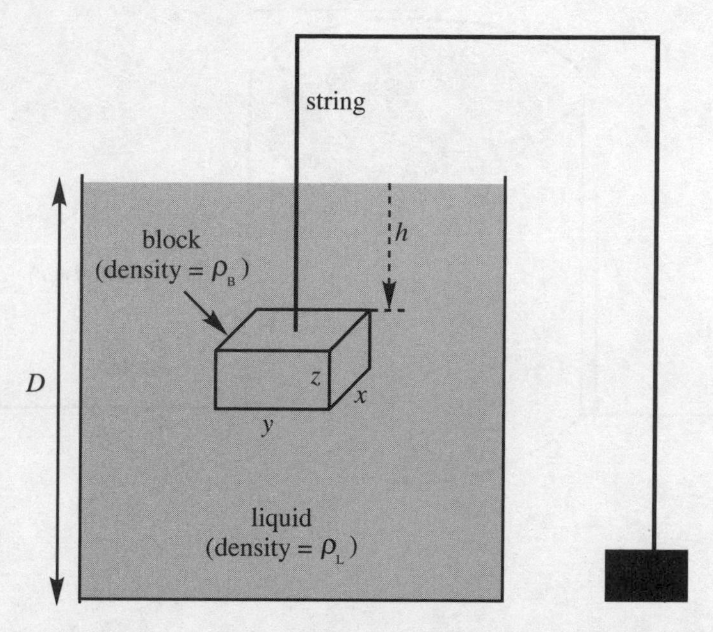

In each of the following, write your answer in simplest form in terms of ρ_L, ρ_B, x, y, z, h, D, and g.

(A) Find the force due to the pressure on the top surface of the block and on the bottom surface. Sketch these forces in the diagram below:

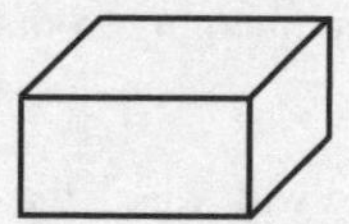

(B) What are the average forces due to the pressure on the other four sides of the block? Sketch these forces in the diagram above.

(C) What is the total force on the block due to the pressure?

(D) Find an expression for the buoyant force on the block. How does your answer here compare to your answer to part (C)?

(E) What is the tension in the string?

2. The figure below shows a large, cylindrical tank of water, open to the atmosphere, filled with water to depth D. The radius of the tank is R. At a depth h below the surface, a small circular hole of radius r is punctured in the side of the tank, and the point where the emerging stream strikes the level ground is labeled X.

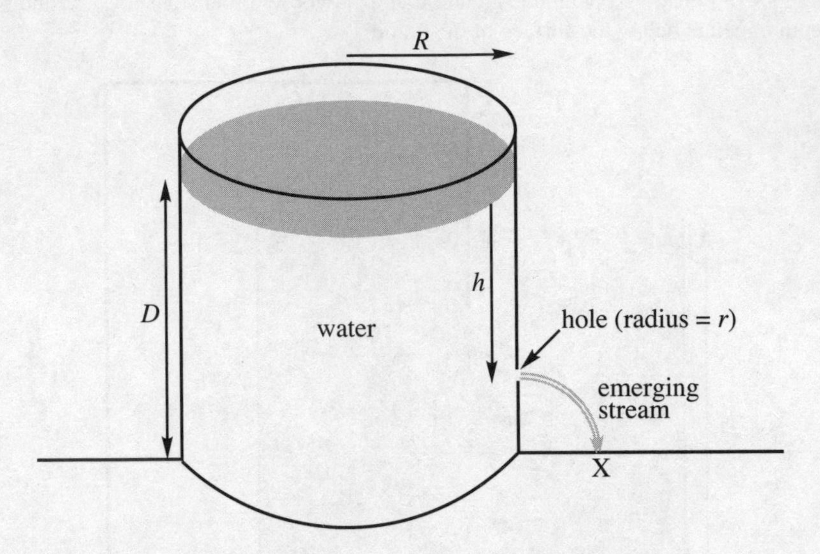

In parts (A) through (C), assume that the speed with which the water level in the tank drops is negligible.

(A) At what speed does the water emerge from the hole?

(B) How far is point X from the edge of the tank?

(C) Assume that a second small hole is punctured in the side of the tank, at a distance of $h/2$ directly above the hole shown in the figure. If the stream of water emerging from this second hole also lands at Point X, find h in terms of D.

(D) For this part, do *not* assume that the speed with which the water level in the tank drops is negligible, and derive an expression for the speed of efflux from the hole punctured at depth h below the surface of the water. Write your answer in terms of r, R, h, and g.

3. The figure below shows a pipe fitted with a Venturi U-tube. Fluid of density ρ_F flows at a constant flow rate and with negligible viscosity through the pipe, which constricts from a cross-sectional area A_1 at Point 1 to a smaller cross-sectional area A_2 at Point 2. The upper portion of both sides of the Venturi U-tube contains the same fluid that's flowing through the pipe, while the lower portion is filled with a fluid of density ρ_V (which is greater than ρ_F). At Point 1 in the pipe, the pressure is P_1 and the flow speed is v_1 ; at Point 2 in the pipe, the pressure is P_2 and the flow speed is v_2. All the fluid within the Venturi U-tube is stationary.

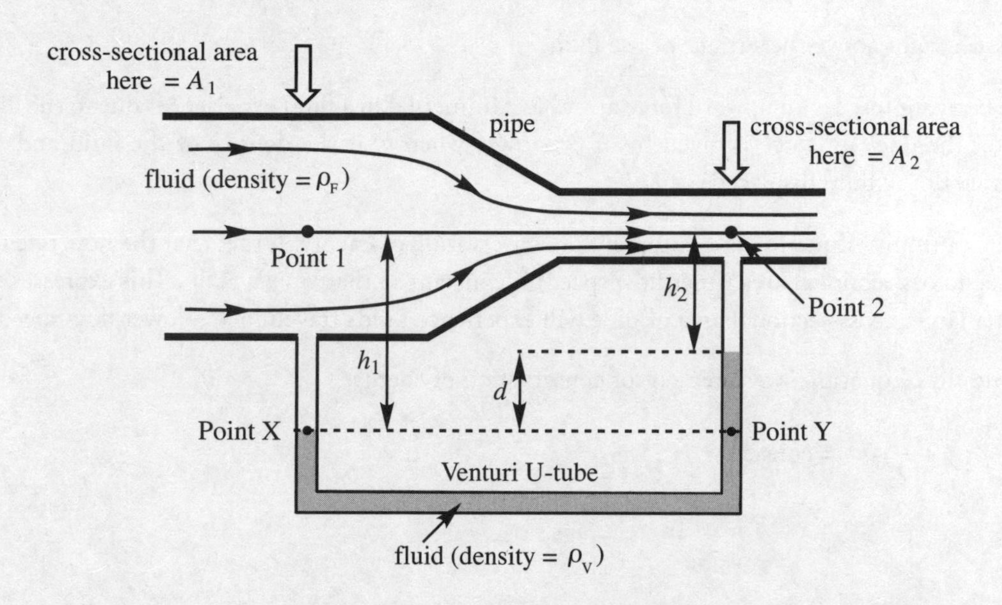

(A) What is P_X, the hydrostatic pressure at Point X? Write your answer in terms of P_1, ρ_F, h_1, and g.

(B) What is P_Y, the hydrostatic pressure at Point Y? Write your answer in terms of P_2, ρ_F, ρ_V , h_2, d, and g.

(C) Write down the result of Bernoulli's Equation applied to Points 1 and 2 in the pipe, and solve for $P_1 - P_2$.

(D) Since $P_X = P_Y$, set the expressions you derived in parts (A) and (B) equal to each other, and use this equation to find $P_1 - P_2$.

(E) Derive an expression for the flow speed, v_2, and the flow rate, f, in terms of A_1, A_2, d, ρ_F, ρ_V, and g. Show that v_2 and f are proportional to $\sqrt{d}$.

Summary

- Density is given by $\rho = \frac{m}{V}$. Pressure is given by $P = \frac{F}{A}$. Hydrostatic pressure can be found using $P = P_0 + \rho gh$, where ρgh is the pressure at a given depth below the surface of the fluid and P_0 is the pressure right above the surface of the fluid.
- The buoyant force is an upward force any object immersed in a fluid experiences due to the displaced fluid. The buoyant force is given by $F_{buoy} = \rho Vg$, where ρ is the density of the fluid and V is the volume of the fluid displaced.
- The Continuity Equation is a statement of conservation of energy. It says that the flow rate through a pipe (cross-sectional area times flow speed) is constant so that $A_1v_1 = A_2v_2$. This expresses the idea that a larger cross-sectional area of pipe will experience fluids traveling at a lower flow speed.
- Bernoulli's Equation is a statement of conservation of energy.

  $$P + \rho gy + \frac{1}{2}\rho v^2 = \text{constant}$$

Chapter 4
Thermodynamics

This chapter deals with CED Unit 2.1: Thermodynamic Systems.

INTRODUCTION

This chapter looks at heat and temperature, concepts that seem familiar from our everyday experience. Technically, **heat** is defined as thermal energy transmitted from one body to another. While an object can contain **internal energy** (due to the random motion of its molecules), an object doesn't *contain* heat; heat is energy *in transit.* **Temperature**, on the other hand, is a measure of an object's internal energy. To understand how these ideas are related, we need to examine physics at a microscopic level.

THE KINETIC THEORY OF GASES

Unlike the condensed phases of matter—solid and liquid—the atoms or molecules that make up a gas do not move around relatively fixed positions. Rather, the molecules of a gas move freely and rapidly, in a chaotic swarm. The molecules of the gas, which are constantly in motion, bump against one another. To study such a gas, it would need to be put into a container, and the random motion of the molecules would cause the molecules to bump into not just one another but also the walls of the container. The size of the container is the volume (V) of the thermodynamic system, and as the molecules bump against the walls, they create a pressure (P) where pressure is defined the same way as in the study of fluids as $P = F/A$. The **Kinetic Theory of Gases** deals with the relationship between the volume, pressure, and temperature of a gas, as well as the **Ideal Gas Law**.

CED Unit 2.8
Thermodynamics and Elastic Collisions

One of the difficulties in describing the behavior of gases is that they are comprised of an extremely large number of atoms or molecules. For example, at room temperature, air has a volume of 24 liters per mole. That means an empty one-liter soda bottle will contain 25 sextillion (2.5×10^{22}) air molecules. That is 25,000,000,000 trillions of molecules. Each of these molecules, at any given instant, will have its own position and be moving with its own momentum. A confined gas exerts a force on the walls of its container, and as the molecules are zipping around inside the container, striking the walls and rebounding, they create pressure within the container. It is impossible to apply Newton's laws to each molecule and determine the amount of force that each molecule would apply to the container of the bottle to determine the total pressure. To avoid dealing with individual molecules, the Kinetic Theory of Gases applies the mathematics of statistics to develop the concept of temperature as a measurement relating to the "average" kinetic energy of molecules in a gas sample and the relationship between temperature, pressure, and volume.

CED Unit 2.9
Thermodynamics and Inelastic Collisions

When considering the microscopic world, as the molecules move around in the sample and collide with one another, every collision between molecules and every collision between a molecule and the walls of the container is assumed to be a perfectly elastic collision. This means that the sum of the kinetic energy before the collision is equal to the sum of the kinetic energy after the collision. For inelastic collisions—for example, when a lump of clay is dropped and hits the floor—a certain amount of kinetic energy from the clay is used to generate "thermal energy," resulting in deforming the clay and increasing its temperature and the temperature of the floor. When considering collisions between individual molecules or between a molecule and the wall, the kinetic energy of the molecules is no longer distinct from the thermal energy. It is the motion of the molecules (kinetic energy) that is the thermal energy, and we can no longer use the convenient label of "thermal energy" to account for a decrease in mechanical energy as we did when looking at macroscopic objects. Energy will be contained within the gas as **internal**

energy, which will determine the temperature. Transfers of energy into or out of the gas will be a result of **work**, a volume change in the container, or **heating**, a flow of energy that naturally occurs between systems at unequal temperatures.

The Ideal Gas Law

CED Unit 2.2
Pressure, Thermal Equilibrium, and the Ideal Gas Law

Three physical properties—pressure (P), volume (V), and temperature (T)—describe a gas. At low densities, all gases approach *ideal* behavior; this means that these three variables are related by the equation

$$PV = nRT$$

Equation Sheet

where n is the number of moles of gas and R is a constant (8.31 J/mol•K) called the **universal gas constant**. This equation is known as the **Ideal Gas Law**. It can also be written as $PV = Nk_BT$, where N is the number of atoms (or molecules) of gas and k_B is Boltzmann's constant ($k_B = 1.38 \times 10^{-23}$ J/K). The Ideal Gas Law tells us how these variables are related to one another for a gas with no intermolecular forces. This is a bad assumption for a charged gas, where the individual charges would interact electrically (a strong interaction), but it is a good assumption for a neutral gas, where the atoms would only interact gravitationally (a very weak interaction).

CED Unit 2.2
Pressure, Thermal Equilibrium, and the Ideal Gas Law

An important consequence of this equation is that, for a fixed volume of gas, an increase in P gives a proportional increase in T. The pressure increases when the gas molecules strike the walls of their container with more force, which occurs if they move more rapidly. The macroscopic variables of pressure, volume, and temperature can be related to the average kinetic energy of any particular molecule within the gas using Newton's Second Law.

We can find that the pressure exerted by N molecules of gas in a container is related to the average kinetic energy by the equation $PV = \frac{2}{3}NK_{\text{avg}}$. Comparing this with the Ideal Gas Law, we see that $\frac{2}{3}NK_{\text{avg}} = nRT$. We can rewrite this equation in the form $\frac{2}{3}N_AK_{\text{avg}} = RT$, since, by definition, $N = nN_A$. The ratio R/N_A is a fundamental constant of nature called **Boltzmann's constant** ($k_B = 1.38 \times 10^{-23}$ J/K). Our equation becomes

$$K_{\text{avg}} = \frac{3}{2}k_BT$$

Equation Sheet

To convert from degrees Celsius to Kelvin, add 273.15.

This tells us that the average translational kinetic energy of the gas molecules is directly proportional to the absolute temperature of the sample. Remember, this means you must use kelvins as your temperature unit.

Since the average kinetic energy of the gas molecules is $K_{avg} = \frac{1}{2}m(v^2)_{avg}$, the equation above becomes $\frac{1}{2}m(v^2)_{avg} = \frac{3}{2}k_B T$, so

$$\sqrt{(v^2)_{avg}} = \sqrt{\frac{3k_B T}{m}}$$

The quantity on the left-hand side of this equation, the square root of the average of the square of v, is called the **root-mean-square velocity**, v_{rms}, so

$$v_{rms} = \sqrt{\frac{3k_B T}{m}}$$

It's important to realize that the molecules in the container have a wide range of speeds; some are much slower and others are much faster than v_{rms}. The root-mean-square speed is important because it gives us a type of average speed that's easy to calculate from the temperature of the gas.

Example 1 In order for the rms velocity of the molecules in a given sample of gas to double, what must happen to the temperature?

Solution. Temperature is a measure of the average kinetic energy. The velocity is determined from the following equation:

$$v_{rms} = \sqrt{\frac{3k_b T}{m}}$$

Since v_{rms} is proportional to the square root of T, the temperature must quadruple, again, assuming the temperature is given in kelvins.

CED Unit 2.5
Thermodynamics and Contact Forces

Example 2 A cylindrical container of radius 15 cm and height 30 cm contains 0.6 mole of gas at 433 K. How much force does the confined gas exert on the lid of the container?

Solution. The volume of the cylinder is $\pi r^2 h$, where r is the radius and h is the height. Since we know V and T, we can use the Ideal Gas Law to find P. Because pressure is force per unit area, we can find the force on the lid by multiplying the gas pressure by the area of the lid.

$$P = \frac{nRT}{V} = \frac{(0.6 \text{ mol})(8.31 \text{ J/mol}\cdot\text{K})(433 \text{ K})}{\pi(0.15 \text{ m})^2(0.30 \text{ m})} = 1.018 \times 10^5 \text{ Pa}$$

So, since the area of the lid is πr^2, the force exerted by the confined gas on the lid is

$$F = PA = (1.018 \times 10^5 \text{ Pa}) \cdot \pi(0.15 \text{ m})^2 = 7{,}200 \text{ N}$$

This is about 1,600 pounds of force, which seems like a lot. Why doesn't this pressure pop the lid off? Because, while the bottom of the lid is feeling a pressure (due to the confined gas) of 1.018×10^5 Pa that exerts a force upward, the top of the lid feels a pressure of 1.013×10^5 Pa (due to the atmosphere) that exerts a force downward. The net force on the lid is 35 N (shown below), which is only about 8 pounds.

$$F_{\text{net}} = (\Delta P)A = (0.005 \times 10^5 \text{ Pa}) \cdot \pi(0.15 \text{ m})^2 = 35 \text{ N}$$

THE MAXWELL-BOLTZMANN DISTRIBUTION

The Kinetic Theory of Gases applies to large numbers of particles, which is why the term "average" continually appears in the previous section when describing the energy or speed of any particular molecule. The gas will have a variety of molecules with different energies and therefore different speeds. Some of the molecules will be moving much faster than average, and some much slower. A graph of the distribution of the speeds of all of the molecules is called the Maxwell-Boltzmann distribution.

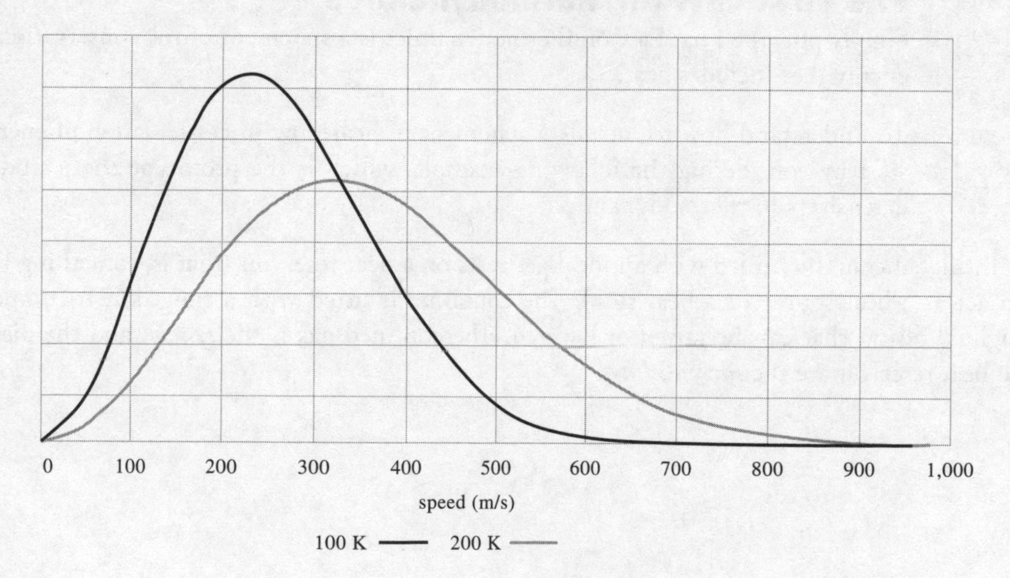

The important features of this type of graph are the position of the peak and the spread of the curve. As a gas sample is heated, the peak (called the most probable speed, which is actually always a bit lower than the v_{rms}) moves to the right, indicating the most probable speed for any particular molecule in the sample is increasing. The graph also spreads out and grows a longer tail on the right-hand side, indicating that there will be more molecules moving at much higher speed than average for hotter gas samples, which intuitively makes sense.

THE LAWS OF THERMODYNAMICS

We've learned about two ways in which energy may be transferred between a system and its environment. One is work, which takes place when a force acts over a distance. The other is heat, which takes place when energy is transferred due to a difference in temperature. The study of the energy transfers involving work and heat, and the resulting changes in internal energy, temperature, volume, and pressure is called **thermodynamics**.

The Zeroth Law of Thermodynamics

When two objects are brought into contact, heat will flow from the warmer object to the cooler one until they reach thermal equilibrium. This property of temperature is expressed by the Zeroth Law of Thermodynamics.

The Zeroth Law of Thermodynamics

If Objects 1 and 2 are each in thermal equilibrium with Object 3, then Objects 1 and 2 are in thermal equilibrium with each other.

CED Unit 2.6
Heat and Energy Transfer

CED Unit 2.7
Internal Energy and Energy Transfer

The First Law of Thermodynamics

Simply put, the First Law of Thermodynamics is a statement of the conservation of energy that includes heat.

To understand how to include a statement of heat flow in conservation of energy, start by considering the following example, which is the prototype that's studied extensively in thermodynamics.

An insulated container filled with an ideal gas rests on a heat reservoir (that is, something that can act as a heat source or a heat sink). The container is fitted with a snug, but frictionless, weighted piston that can be raised or lowered. The confined gas is the *system*, and the piston and heat reservoir are the *surroundings*.

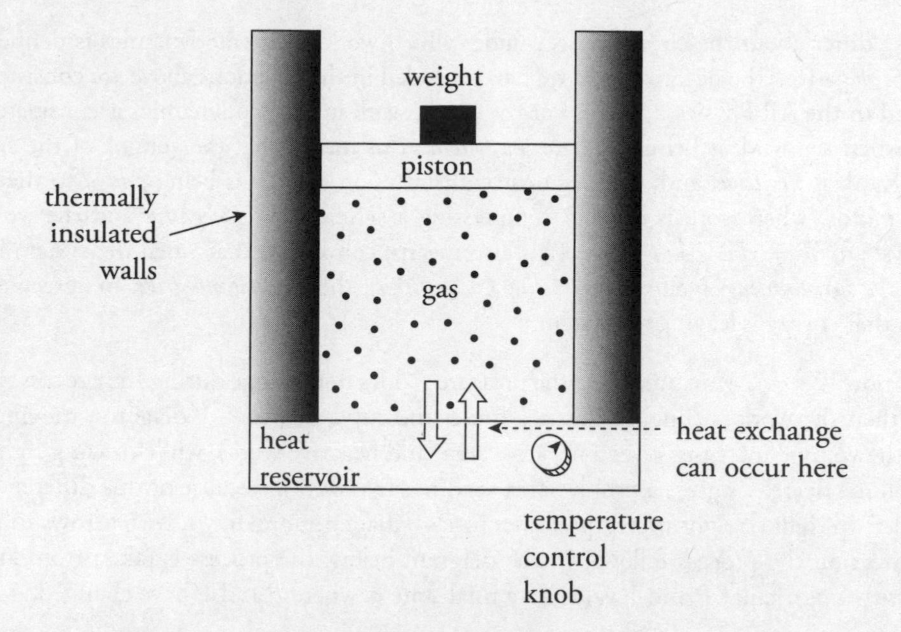

The **state** of the gas is given once its pressure, volume, and temperature are known, and the equation that connects these state variables is the Ideal Gas Law, $PV = nRT$. We'll imagine performing different experiments with the gas, such as heating it or allowing it to cool or increasing or decreasing the weight on the piston, and study the energy transfers (work and heat) and the changes in the state variables. If each process is carried out such that, at each moment, the system and its surroundings are in thermal equilibrium, we can plot the pressure (P) versus the volume (V) on a diagram. By following the path of this ***P–V* diagram**, we can study how the system is affected as it moves from one state to another.

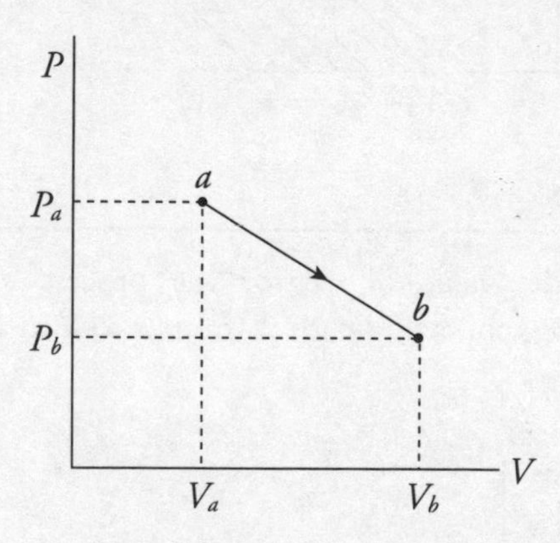

Work is done on or by the system when the piston is moved and the volume of the gas changes. For example, imagine that the weight pushes the piston downward a distance d, causing a decrease in volume. Assume that the pressure stays constant at P. (Heat must be removed via the reservoir to accomplish this.) We can calculate the work done on the gas during this compression as $W = -Fd$, but since $F = PA$, we have $W = -PAd$, and because $Ad = \Delta V$, we have

$$W = -P\Delta V$$

Equation Sheet

Textbooks differ about the circumstances under which work in thermodynamics is defined to be positive or negative. The negative signs we have included in the equations above are consistent with those used in the AP Physics 2 Exam. For the exam, work in thermodynamics is considered to be positive when the work is being done *on the system.* This means that the volume of the system is *decreasing,* ΔV is *negative*, and, in agreement with intuition, energy is being *added* to the system. In other words, when work is done in compressing a system, ΔV is *negative* and the work done on the system, $W = -P\Delta V$, is *positive.* This also means, conversely, that when the system is doing work *on the surroundings* (volume *increasing,* ΔV *positive*), the work is *negative*, in agreement with intuition that energy is leaving the system.

The equation $W = -P\Delta V$ assumes that the pressure P does not change during the process. If P *does* change, then the work is equal to the area under the curve in the P–V diagram; moving left to right as the volume increases gives a negative area (and negative work), while moving right to left as the volume decreases gives a positive area (and positive work). Because of this difference when moving left to right or right to left, processes in P–V diagrams are drawn with arrows to indicate which direction the process follows. In the diagram below, the process is taken from an initial state a, with a particular P and V value, to a final state b, where P and V have changed.

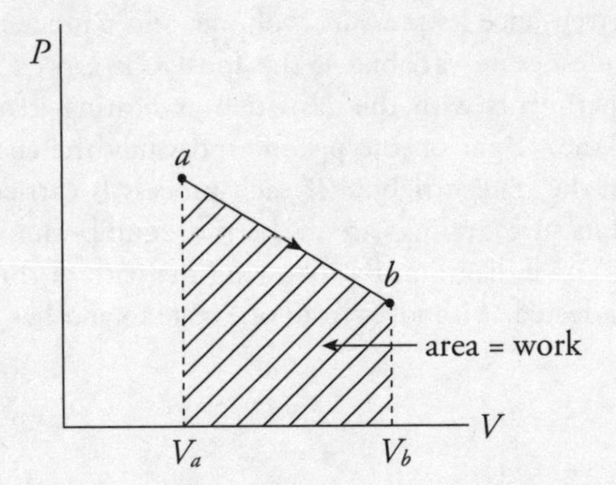

Example 3 What's the value of W for the process ab following path 1 and for the same process following path 2 (from a to d to b), shown in the P–V diagram below?

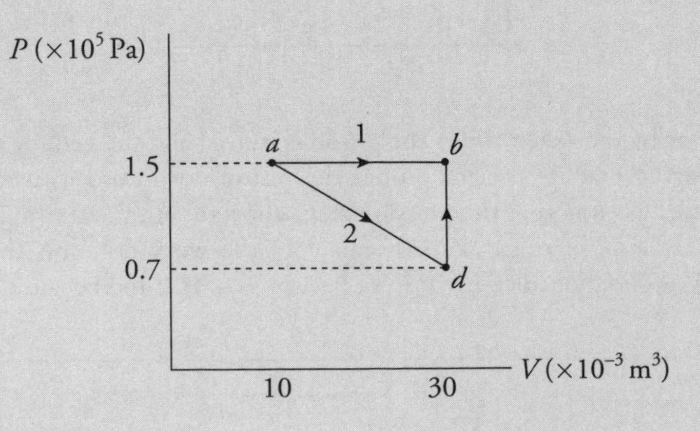

Solution:

Path 1. Since, in path 1, P remains constant, the work done is just $-P\Delta V$:

$$W = -P\Delta V = -(1.5 \times 10^5 \text{ Pa})[(30 \times 10^{-3} \text{ m}^3) - (10 \times 10^{-3} \text{ m}^3)] = -3{,}000 \text{ J}$$

Path 2. If the gas is brought from state a to state b, along path 2, then work is done only along the part from a to d. From d to b, the volume of the gas does not change, so no work can be performed. The area under the graph from a to d is

$$\begin{aligned} W &= -\frac{1}{2}h(b_1 + b_2) = -\frac{1}{2}(\Delta V)(P_a + P_d) \\ &= -\frac{1}{2}(20 \times 10^{-3} \text{ m}^3)[(1.5 \times 10^5 \text{ Pa}) + (0.7 \times 10^5 \text{ Pa})] \\ &= -2{,}200 \text{ J} \end{aligned}$$

As this example shows, the amount of work done by a gas depends not only on the initial and final states of the system, but also on the path between the two. In general, different paths give different values for W.

Experiments have shown that the value of the sum of the total heat added (or removed), denoted by Q, and the total work done on (or by) the system, denoted by W, does not depend on the path taken by a thermodynamic process; it depends only on the initial and the final state of the system, so it describes a change in a fundamental property. This property is called the system's **internal energy**, denoted U, and the change in the system's internal energy, ΔU, is equal to $Q + W$. This is true regardless of the process that brought the system from its initial to final state. This statement is known as

The First Law of Thermodynamics

$$\Delta U = Q + W$$

Equation Sheet

As stated above, the internal energy, U, of a gas depends only on the state of the gas. As we saw in the section on the Ideal Gas Law, the average kinetic energy of a molecule in a gas sample is directly proportional to the temperature. Because of this, the change in internal energy, ΔU, which arises in the First Law of Thermodynamics, is directly proportional to a change in the temperature of the gas.

This statement of the First Law is consistent with the interpretation of work ($W = -P\Delta V$) explained above. The First Law identifies W and Q as separate physical mechanisms for adding to or removing energy from the system. The signs of both W and Q are defined consistently: both are positive when they are adding energy to the system and negative when they are removing energy from the system.

Example 4 A 0.5 mol sample of an ideal gas is brought from state *a* to state *b* when 7,500 J of heat is added along the path shown in the following *P–V* diagram:

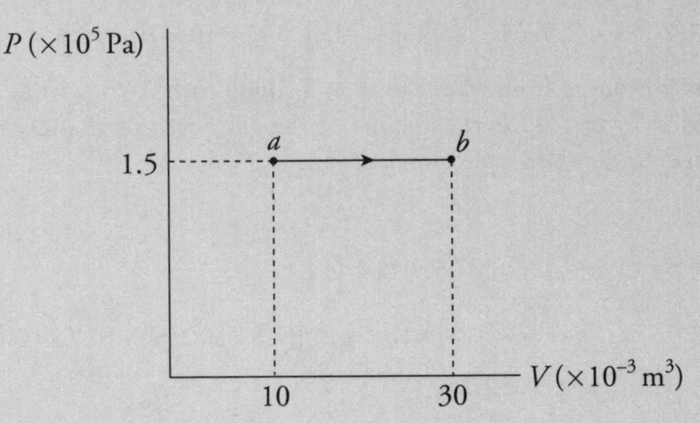

What are the values of each of the following?

(a) the temperature at *a*
(b) the temperature at *b*
(c) the work done by the gas during process *ab*
(d) the change in the internal energy of the gas

Solution.

(a, b) Both of these questions can be answered using the Ideal Gas Law, $T = PV/(nR)$:

$$T_a = \frac{P_a V_a}{nR} = \frac{(1.5\times10^5 \text{ Pa})(10\times10^{-3} \text{ m}^3)}{(0.5 \text{ mol})(8.31 \text{ J/mol}\cdot\text{K})} = 360 \text{ K}$$

$$T_b = \frac{P_b V_b}{nR} = \frac{(1.5\times10^5 \text{ Pa})(30\times10^{-3} \text{ m}^3)}{(0.5 \text{ mol})(8.31 \text{ J/mol}\cdot\text{K})} = 1{,}080 \text{ K}$$

(c) Since the pressure remains constant during the process, we can use the equation $W = -P\Delta V$. Because $\Delta V = (30 - 10) \times 10^{-3}$ m³ $= 20 \times 10^{-3}$ m³, we find that

$W = -P\Delta V = (1.5 \times 10^5 \text{ Pa})(20 \times 10^{-3} \text{ m}^3) = -3{,}000 \text{ J}$

The expanding gas did negative work against its surroundings, pushing the piston upward. Important note: if the pressure remains constant (which is designated by a horizontal line in the *P–V* diagram), the process is called **isobaric**.

(d) By the First Law of Thermodynamics,

$\Delta U = Q + W = 7{,}500 - 3{,}000 \text{ J} = 4{,}500 \text{ J}$

Example 5 A 0.5 mol sample of an ideal monatomic gas is brought from state *a* to state *b* along the path shown in the following *P*–*V* diagram:

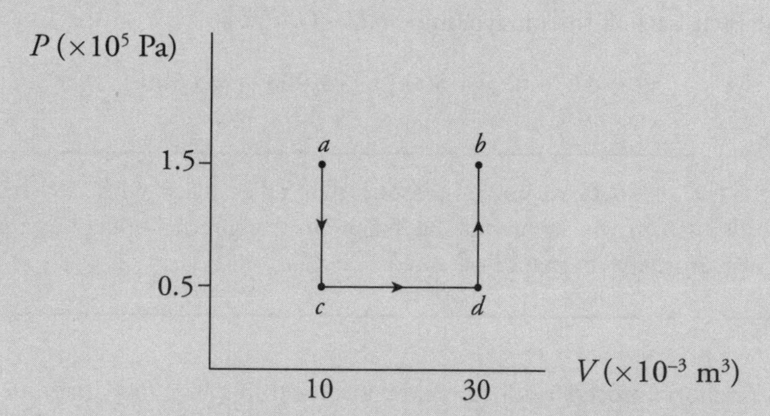

What are the values of each of the following?

(a) the work done by the gas during process *ab*

(b) the change in the internal energy of the gas

(c) the heat added to the gas during process *ab*

Solution. Note that the initial and final states of the gas are the same as in the preceding example, but the path is different.

(a) Let's break the path into 3 pieces:

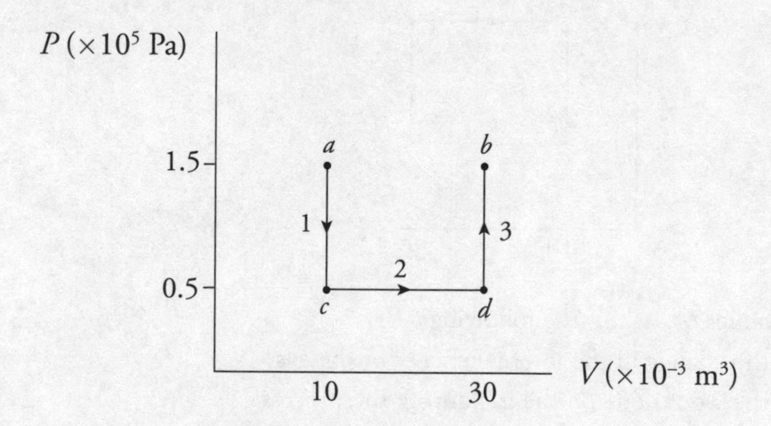

Over paths 1 and 3, the volume does not change, so no work is done. Work is done only over path 2:

$$W = -P\Delta V = -(0.5 \times 10^5 \text{ Pa})(20 \times 10^{-3} \text{ m}^3) = -1{,}000 \text{ J}$$

Once again, the expanding gas does negative work against its surroundings, pushing the piston upward.

(b) Because the initial and final states of the gas are the same here as they were in the preceding example, the change in internal energy, ΔU, *must* be the same. Therefore, $\Delta U = 4{,}500$ J.

(c) By the First Law of Thermodynamics, $\Delta U = Q + W$, so

$$Q = \Delta U - W = 4{,}500 \text{ J} - (-1{,}000 \text{ J}) = 5{,}500 \text{ J}$$

Example 6 An **isochoric** process is one that takes place with no change in volume. What can you say about the change in the internal energy of a gas if it undergoes an isochoric change of state?

Solution. An isochoric process is illustrated by a vertical line in a P–V diagram and, since no change in volume occurs, $W = 0$. By the First Law of Thermodynamics, $\Delta U = Q + W = Q$. Therefore, the change in internal energy is entirely due to (and equal to) the heat transferred. If heat is transferred into the system (positive Q), then ΔU is positive; if heat is transferred out of the system (negative Q), then ΔU is negative.

Example 7 A 0.5 mol sample of an ideal gas is brought from state *a* back to state *a* along the path shown in the following P–V diagram:

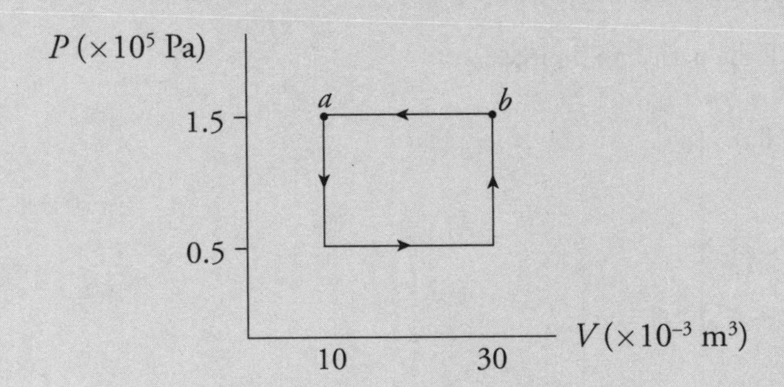

What are the values of each of the following?

(a) the change in the internal energy of the gas

(b) the work done on the gas during the process

(c) the heat added to the gas during the process

Solution. A process such as this, which begins and ends at the same state, is said to be cyclical.

(a) Because the final state is the same as the initial state, the internal energy of the system cannot have changed, so $\Delta U = 0$.

(b) The total work involved in the process is equal to the work done from c to d plus the work done from b to a,

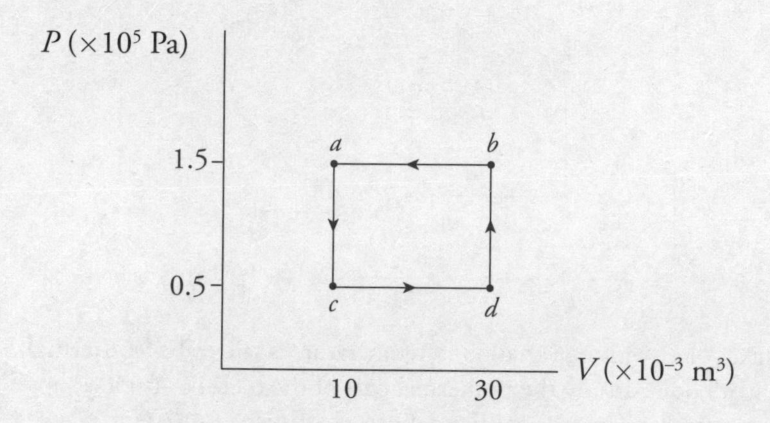

because only along these paths does the volume change. Along these portions, we find that

$$W_{cd} = -P\Delta V_{cd} = (0.5 \times 10^5 \text{ Pa})(+20 \times 10^{-3} \text{ m}^3) = -1{,}000 \text{ J}$$

$$W_{cd} = -P\Delta V_{ba} = (1.5 \times 10^5 \text{ Pa})(-20 \times 10^{-3} \text{ m}^3) = +3{,}000 \text{ J}$$

So the total work done is $W = +2{,}000$ J. The fact that W is positive means that, overall, work was done *on* the gas by the surroundings. Notice that for a cyclical process, the total work done is equal to the area enclosed by the loop, with clockwise travel taken as negative and counterclockwise travel taken as positive.

(c) The First Law of Thermodynamics states that $\Delta U = Q + W$. Since $\Delta U = 0$, it must be true that $Q = -W$ (which will always be the case for a cyclical process), so $Q = -2{,}000$ J.

Example 8 A 0.5 mol sample of an ideal gas is brought from state a to state d along an **isotherm**, and then isobarically to state c and isochorically back to state a, as shown in the following P–V diagram:

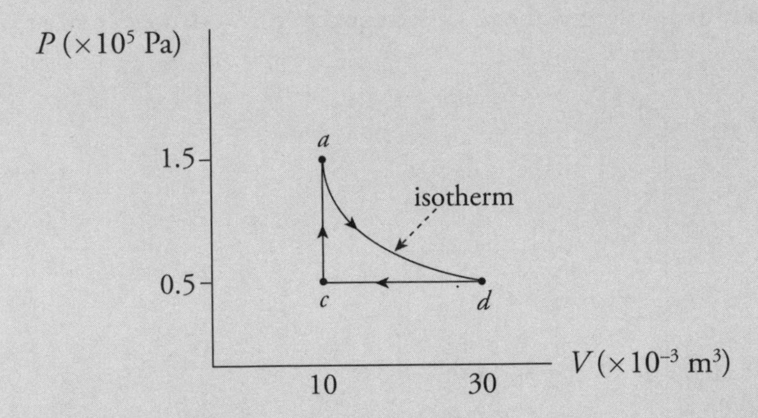

A process that takes place with no variation in temperature is said to be **isothermal**. Given that the work done during the isothermal part of the cycle is –1,650 J, how much heat is transferred during the isothermal process, from a to d?

Solution. Be careful that you don't confuse *isothermal* with *adiabatic*. A process is isothermal if the *temperature* remains constant; a process is **adiabatic** if $Q = 0$. You might ask, *How could a process be isothermal without also being adiabatic at the same time?* Remember that the temperature is determined by the internal energy of the gas, which is affected by changes in Q, W, or both. Therefore, it's possible for U to remain unchanged even if Q is not 0 (because there can be an equal but opposite W to cancel it out). In fact, this is the key to this problem. Since T doesn't change from a to d, neither can the internal energy, which depends entirely on T. Because $\Delta U_{ad} = 0$, it must be true that $Q_{ad} = -W_{ad}$. Since W_{ad} equals –1,650 J, Q_{ad} must be +1,650 J. The gas absorbs heat from the reservoir and uses all this energy to do negative work as it expands, pushing the piston upward.

The Second Law of Thermodynamics

CED Unit 2.11
Probability, Thermal Equilibrium, and Entropy

The Second Law of Thermodynamics describes how systems evolve over time. Conservation of energy states that system A and system B will have the same energy without any external work, but it never describes how a system changes from state A into state B. One description of this evolution is about **entropy**, an equivalent description is about **spontaneous heat flow**, and a third equivalent description is about **efficiency** in engines.

Entropy

Consider a box containing two pure gases separated by a partition. What would happen if the partition were removed? The gases would mix, and the positions of the gas molecules would be random.

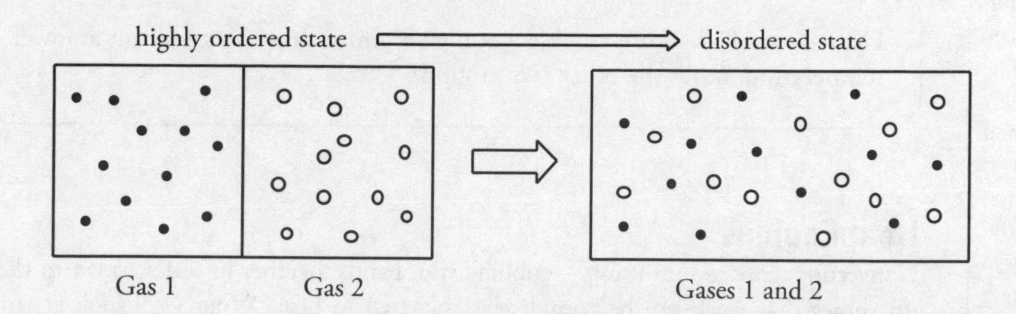

A closed system that shows a high degree of order tends to evolve in such a way that its degree of order decreases. In other words, disorder (or, as it's technically called, entropy) increases. The term "disorder" often makes the concept of entropy sound negative. Rather, entropy is better described as **increasing molecular freedom**. This is the reason why broken glass does not put itself back together, and the reason why chemical reactions take place. Suppose we started with the box on the right, containing the mixture of the gases. It would be virtually impossible that, at any later time, all the molecules of Gas 1 would happen to move to the left side of the box and, at the same time, the molecules of Gas 2 would spontaneously move to the right side of the box. If we were to watch a video of this process, and saw the mixed-up molecules suddenly separate and move to opposite sides of the box, we'd assume that the video was running backward. In a way, the Second Law of Thermodynamics defines the direction of time. Time flows in such a way that ordered systems become disordered. Disordered states do not spontaneously become ordered without any other changes taking place. The following is the essence of one form of the Second Law of Thermodynamics:

> The total amount of disorder—the total entropy—of a system plus its surroundings will never decrease.

It is possible for the entropy of a system to decrease, but it will always be at the expense of a greater increase in entropy in the surroundings. For example, when water freezes, its entropy decreases. The molecules making up an ice crystal have a more structured order than the random collection of water molecules in the liquid phase, so the entropy of the water decreases when it freezes. But when water freezes, it releases heat energy into its environment, which creates disorder in the surroundings. If we were to figure out the total change in entropy of the water plus its surroundings, we would find that although the entropy of the water itself decreased, it was more than compensated by a greater amount of entropy increase in the surroundings. So, the total entropy of the system and its surroundings increased, in agreement with the Second Law of Thermodynamics.

Entropy and Heat
An isolated system never decreases in entropy over time. A non-idealized (real-world) isolated system always increases in entropy over time. This entropy is usually in the form of heat given off. When a system is not isolated, it is possible for the entropy of the system to decrease as long as the surroundings to the system increase in entropy by a greater amount than the decrease in the system. As a result, the entropy of the universe never decreases, and the entropy of the universe will increase over time as real-world systems transfer heat and work.

There are several equivalent statements of the second law. In addition to the entropy form, another form of the Second Law of Thermodynamics says that heat always flows from hot to cold, never cold to hot. Another form, which will be considered more fully in a moment, says it is impossible to convert heat completely into work.

Heat always flows from an object at higher temperature to an object at lower temperature, never the other way around.

Heat Engines

Converting work to heat is easy—rubbing your hands together in order to warm them up shows that work can be completely converted to heat. What we'll look at is the reverse process: how efficiently can heat be converted into work? A device that uses heat to produce useful work is called a **heat engine**. The internal-combustion engine in a car is an example. In particular, we're interested only in engines that take their working substance (a mixture of air and fuel in this case) through a cyclic process, so that the cycle can be repeated. The basic components of any cyclic heat engine are simple: energy in the form of heat comes into the engine from a high-temperature source, some of this energy is converted into useful work, the remainder is ejected as exhaust heat into a low-temperature sink, and the system returns to its original state to run through the cycle again.

All engines function to convert heat into work by exchanging energy from a reservoir at higher temperature to reservoirs of lower temperature. A good example of this is a refrigerator. In a fridge, a liquid moves through tubes to remove heat from the internal chamber (such as the freezer) to the outer air. Thus, cold is not actually a concept, but the sensation of low heat content. This is why the back of a refrigerator feels hot.

Since we're looking at cyclic engines only, the system returns to its original state at the end of each cycle, so ΔU must be 0. Therefore, by the First Law of Thermodynamics, $Q_{net} = -W$. That is, the net heat absorbed by the system is equal to the work performed by the system. The heat absorbed from the high-temperature source is denoted Q_H (H for *hot*), and the heat that is discharged into the low-temperature reservoir is denoted Q_C (C for *cold*). Because heat coming *in* is positive and heat going *out* is negative, Q_H is positive and Q_C is negative, and the net heat absorbed is $Q_H + Q_C$. Instead of writing Q_{net} in this way, it's customary to write it as $Q_H - |Q_C|$, to show explicitly that Q_{net} is less than Q_H.

This is one of the forms of the Second Law of Thermodynamics:

The Second Law of Thermodynamics
For any cyclic heat engine, some exhaust heat is always produced. It's impossible to completely convert heat into useful work.

Example 9 A heat engine draws 800 J of heat from its high-temperature source and discards 450 J of exhaust heat into its cold-temperature reservoir during each cycle. How much work does this engine perform per cycle?

Solution. The absolute value of the work output per cycle is equal to the difference between the heat energy drawn in and the heat energy discarded:

$$|W| = Q_H - |Q_C| = 800 \text{ J} - 450 \text{ J} = 350 \text{ J}$$

HEAT TRANSFER

There are three principal modes by which energy can be transferred: conduction, convection, and radiation.

CED Unit 2.6
Heat and Energy Transfer

Conduction

An iron skillet is sitting on a hot stove, and you accidentally touch the handle. You notice right away that there's been a transfer of thermal energy to your hand. The process by which this happens is known as conduction. The highly agitated atoms in the handle of the hot skillet bump into the atoms of your hand, making them vibrate more rapidly, thus heating up your hand.

CED Unit 2.10
Thermal Conductivity

Heat conducts from one point to another only if there is a temperature difference between the two objects. The rate at which heat is transferred is given by

$$\frac{Q}{\Delta t} = \frac{kA\Delta T}{L}$$

Equation Sheet

The constant k in the conduction equation is the "thermal conductivity" of an object. This is a property of the object and explains why things like metals get hot faster than plastics. It is also responsible for double-paned windows being "more energy efficient," as the layer of air between the two sheets of glass has a much smaller thermal conductivity than a single sheet of glass.

Convection

As the air around a candle flame warms, it expands, becomes less dense than the surrounding cooler air, and thus rises due to buoyancy. As a result, heat is transferred away from the flame by the large-scale (from the atoms' point of view, anyway) motion of a fluid (in this case, air). This is convection.

Radiation

Sunlight on your face warms your skin. Radiant energy from the Sun's fusion reactions is transferred across millions of kilometers of essentially empty space via electromagnetic waves. Absorption of the energy carried by these light waves defines heat transfer by radiation.

Chapter 4 Review Questions

Solutions can be found in Chapter 11.

Section I: Multiple Choice

1. A container holds a mixture of two gases, CO_2 and H_2, in thermal equilibrium. Let K_C and K_H denote the average kinetic energy of a CO_2 molecule and an H_2 molecule, respectively. Given that a molecule of CO_2 has 22 times the mass of a molecule of H_2, the ratio K_C / K_H is equal to

   (A) 1/22
   (B) 1
   (C) $\sqrt{22}$
   (D) 22

2. If the temperature and volume of a sample of an ideal gas are both doubled, then a *P–V* diagram will show what sort of shape?

   (A) A horizontal line
   (B) A vertical line
   (C) A curve indicating a positive correlation
   (D) A curve indicating a negative correlation

3. In three separate experiments, a gas is transformed from state P_i, V_i to state P_f, V_f along the paths (1, 2, and 3) illustrated in the figure below:

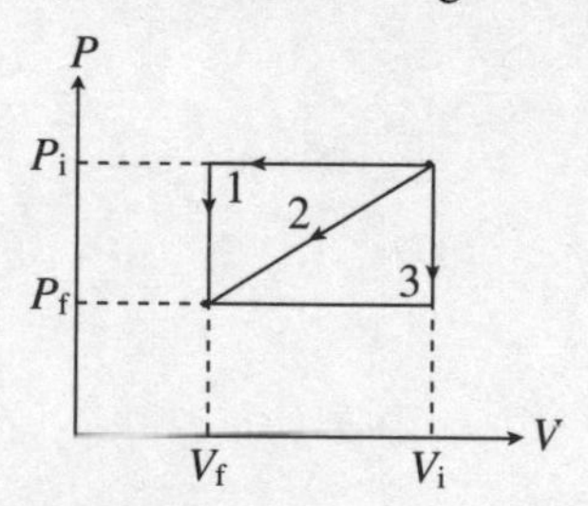

   The work done on the gas is

   (A) greatest for path 1
   (B) least for path 2
   (C) greatest for path 2
   (D) the same for all three paths

4. An ideal gas is compressed isothermally from 20 m^3 to 10 m^3. During this process, 5 J of work is done to compress the gas. What is the change in internal energy for this gas?

   (A) –10 J
   (B) –5 J
   (C) 0 J
   (D) 5 J

Questions 5 and 6 refer to the following material.

An ideal gas is confined in a container with a fixed volume. The amount of gas, *n*, is slowly increased in the container. This experiment is done in such a way that the temperature of the gas remains constant. Pressure data is collected.

5. Which describes a graph with pressure on the vertical axis and amount of gas on the horizontal axis?

   (A) The graph will be linear, and a fit line will go through the origin. (0, 0) will be a data point.
   (B) The graph will be linear, and a fit line will go through the origin. (0, 0) will not be a data point.
   (C) The graph will be nonlinear. (0, 0) will be a data point.
   (D) The graph will be nonlinear. (0, 0) will not be a data point.

6. How could the experiment be done so that as the amount of gas is increased, the temperature remains constant?

   (A) Allow the pressure to change while *n* is changed so that *T* will remain constant.
   (B) Allow work to be added or removed from the gas so that *T* will remain constant.
   (C) Allow heat to be added or removed from the gas so that *T* will remain constant.
   (D) Allow internal energy to be added or removed from the gas so that *T* will remain constant.

7. Through a series of thermodynamic processes, the internal energy of a sample of confined gas is increased by 560 J. If the net amount of work done on the sample by its surroundings is 320 J, how much heat was transferred between the gas and its environment?

   (A) 240 J absorbed
   (B) 240 J dissipated
   (C) 880 J absorbed
   (D) 880 J dissipated

8. What is the total work performed on the gas as it is transformed from state *a* to state *c*, along the path indicated?

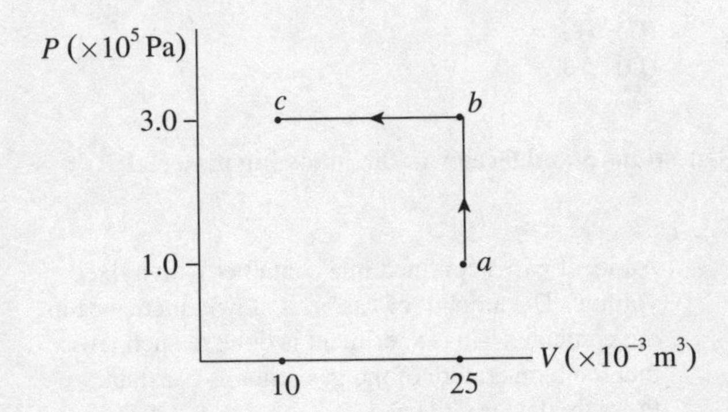

   (A) 1,500 J
   (B) 3,000 J
   (C) 4,500 J
   (D) 9,500 J

9. In one of the steps of the Carnot cycle, the gas undergoes an isothermal expansion. Which of the following statements is true concerning this step?

   (A) No heat is exchanged between the gas and its surroundings, because the process is isothermal.
   (B) The temperature decreases because the gas expands.
   (C) The internal energy of the gas remains constant.
   (D) The internal energy of the gas decreases due to the expansion.

10. A cup of hot coffee is sealed inside a perfectly thermally insulating container. A long time is allowed to pass. Which of the following correctly explains the final thermal configuration within the box?

    (A) The coffee has not changed temperature because the container is perfectly insulating.
    (B) The coffee has gotten warmer, and the air in the container has gotten cooler, because of an exchange of thermal energy between the air and the coffee.
    (C) The coffee has gotten cooler, and the air in the container has gotten warmer, because of an exchange of thermal energy between the air and the coffee.
    (D) The coffee has gotten cooler, but the air in the container has not changed its temperature. The energy from the coffee has caused an increase in entropy within the box.

Section II: Free Response

1. When a system is taken from state *a* to state *b* along the path *acb* shown in the figure below, 70 J of heat flows into the system, and the system does 30 J of work.

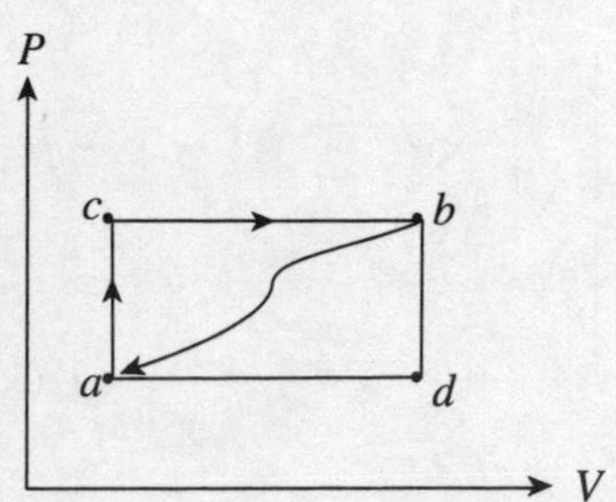

(A) When the system is returned from state *b* to state *a* along the curved path shown, 60 J of heat flows out of the system. Does the system perform work on its surroundings or do the surroundings perform work on the system? How much work is done?

(B) If the system does 10 J of work in transforming from state *a* to state *b* along path *adb*, does the system absorb or does it emit heat? How much heat is transferred?

(C) If $U_a = 0$ J and $U_d = 30$ J, determine the heat absorbed in the processes *db* and *ad*.

(D) For the process *adbca*, identify each of the following quantities as positive, negative, or zero:

W = ________________ Q = ________________ ΔU = ________________

2. A 0.4 mol sample of an ideal diatomic gas undergoes slow changes from state *a* to state *b* to state *c* and back to *a* along the cycle shown in the *P*–*V* diagram below:

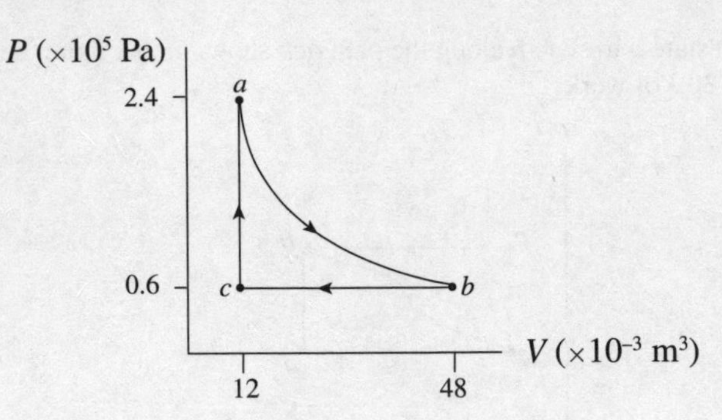

Path *ab* is an isotherm, and it can be shown that the work done by the gas as it changes isothermally from state *a* to state *b* is given by the equation

$$W_{ab} = -nRT \times \ln \frac{V_b}{V_a}$$

(A) What's the temperature of

i. state *a*?

ii. state *b*?

iii. state *c*?

(B) In order for step *ab* to be isothermal, the gas must be kept in thermal equilibrium with its surroundings at all times. Because heat flows spontaneously from hot to cold systems, does this imply that there is no heat flow during step *ab*? Explain.

(C) How much work, W_{ab}, is done by the gas during step *ab*?

(D) What is the total work done over cycle *abca*?

Summary

- For gases, there are a few important ideas to understand:

    - Pressure is the result of the molecules colliding with one another and with the sides of the container. It is defined as the force per unit area ($P = F/A$) on the walls of the container.
    - The Ideal Gas Law is expressed as either $PV = nRT$ or $PV = Nk_BT$.
    - The average kinetic energy of the gas molecules is given by $K_{avg} = \frac{3}{2}k_bT$. This is related to the most likely speed a particle will be moving, but the particles in the gas will be moving with a large distribution of various speeds.

- The work done can be found by finding the area under a P–V graph. If there is a volume change while the pressure remains constant, then the equation $W = -P\Delta V$ can be used to calculate the work.
- The First Law of Thermodynamics is $\Delta U = Q + W$, where ΔU depends only on the temperature change.

    - W positive means energy is being added to the system, so work is done **on** the system **by** the surroundings. A negative W means that energy is being subtracted from the system, so work is done **by** the system **on** the surroundings.
    - Q positive (or negative) means energy is being added to (or subtracted from) the system by means of a flow of heat from the higher temperature surroundings (or system) to the lower temperature system (or surroundings).

Note that some textbooks define work in thermodynamics in a different way: work is considered to be positive when work is done on the surroundings. This is consistent with the idea that the overall objective of a heat engine is to produce external (positive) work. Under this definition, the First Law of Thermodynamics must be written as $U = Q - W$ (or $U + W = Q$), and W must then be interpreted differently from Q. That is, while Q is still positive when heat is being added to the system, W is now positive when work is being done by the system on the surroundings (thus decreasing the internal energy of the system).

- The rate at which heat is transferred is given by $\Delta Q/\Delta t = \dfrac{kA\Delta T}{L}$, where k is the thermal conductivity (a property of the material), A is the cross-sectional area, ΔT is the temperature difference between the two sides, and L is the thickness or distance between the two ends of the material.

Chapter 5
Electric Forces and Fields

INTRODUCTION

The existence of mass in an object results in that object having the ability to experience the gravitational force, and thereby interact with other objects that also have mass. All objects in the universe that have been discovered thus far have mass (although the particle known as the neutrino has such a small mass that it is practically massless). Another fundamental property of objects is charge. A charged object will interact with another charged object through the electromagnetic force, similar to massive objects. The property of charge is as fundamental to physics as the property of mass.

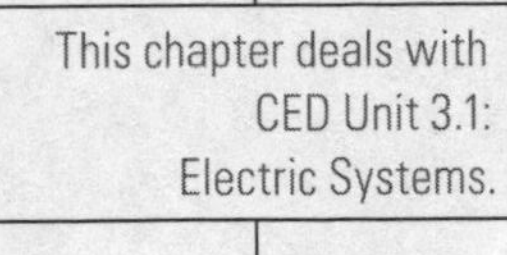

CED Unit 3.2
Electric Charge

CED Unit 3.3
Conservation of Electric Charge

ELECTRIC CHARGE

The basic components of atoms are protons, neutrons, and electrons. Protons and neutrons form the nucleus (and are referred to collectively as *nucleons*), while the electrons keep their distance, swarming around the nucleus. Most of an atom consists of empty space. In fact, if a nucleus were the size of the period at the end of this sentence, then the electrons would be 5 meters away. So what holds such an apparently tenuous structure together? One of the most powerful forces in nature: the *electromagnetic force*. Protons and electrons have a quality called **electric charge** that gives them an attractive force. Electric charge comes in two varieties: positive and negative. A positive particle always attracts a negative particle, and particles of the same charge always repel each other. Protons are positively charged, and electrons are negatively charged.

Protons and electrons are intrinsically charged, but bulk matter is not. This is because the amount of charge on a proton exactly balances the charge on an electron, which is quite remarkable considering that protons and electrons are very different particles. Since most atoms contain an equal number of protons and electrons, their overall electric charge is 0, because the negative charges cancel out the positive charges. Therefore, in order for matter to be **charged**, an imbalance between the numbers of protons and electrons must exist. This can be accomplished by either the removal or addition of electrons (that is, by the **ionization** of some of the object's atoms). If you remove electrons, then the object becomes positively charged, while if you add electrons, then it becomes negatively charged. Furthermore, charge is **conserved**. For example, if you rub a glass rod with a piece of silk, then the silk will acquire a negative charge, and the glass will be left with an *equal* positive charge. *Net charge cannot be created or destroyed.* (*Charge* can be created or destroyed—it happens all the time—but *net* charge cannot.)

The magnitude of charge on an electron (and therefore on a proton) is denoted e. This stands for **elementary charge** because it's the basic unit of electric charge. The charge of an ionized atom must be a whole number times e because charge can be added or subtracted only in lumps of size e. For this reason, we say that charge is **quantized**. To remind us of the quantized nature of electric charge, the charge of a particle (or object) is denoted by the letter q. In the SI system of units, charge is expressed in **coulombs** (abbreviated **C**). One coulomb is a tremendous amount of charge, about 10^{18} electrons. The value of e is about 1.6×10^{-19} C.

COULOMB'S LAW

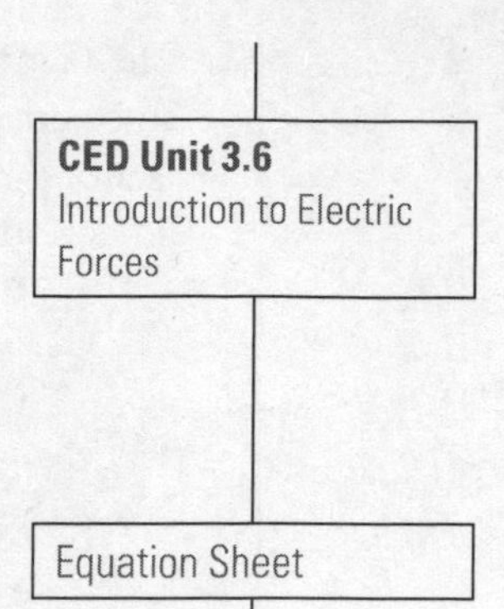

The electric force between two charged particles obeys a law that is very similar to that describing the gravitational force between two masses: they are both inverse-square laws. The **electric force** between two particles with charges of q_1 and q_2, separated by a distance r, has a magnitude given by the equation

$$\left|\vec{F}_E\right| = \frac{1}{4\pi\varepsilon_0}\frac{|q_1 q_2|}{r^2}$$

This is **Coulomb's Law**. If we were to leave off the absolute value bars, we would interpret a negative F_E as an attraction between the charges and a positive F_E as a repulsion. The value of the proportionality constant, k, depends on the material between the charged particles. In empty space (vacuum)—or air, for all practical purposes—it is called **Coulomb's constant** and has the approximate value $k_0 = 9 \times 10^9$ N·m²/C². The expression k_0 is sometimes written in terms of a fundamental constant known as the **permittivity of free space**, denoted ε_0:

$$k_0 = \frac{1}{4\pi\varepsilon_0}$$

You should notice a similarity between the form of Coulomb's Law and Newton's Law of Universal Gravitation you learned in AP Physics 1. Both of these force laws depend on the product of the material property that generates the force (charge or mass), and both are inversely proportional to the square of the separation distance. These two laws are therefore known as **inverse square laws**.

CED Unit 3.9
Gravitational and Electromagnetic Forces

> **Example 1** Consider the proton and electron in hydrogen. The proton has a mass of 1.6×10^{-27} kg and a charge of $+e$. The electron has a mass of 9.1×10^{-31} kg and a charge of $-e$. In hydrogen, they are separated by a Bohr radius, about 0.5×10^{-10} m. What is the electric force between the proton and electron in hydrogen? What about the gravitational force between the proton and electron?

Solution. The electric force between the proton and the electron is given by Coulomb's Law:

$$F_E = \frac{1}{4\pi\varepsilon_0}\frac{q_1 q_2}{r^2} = (9\times10^9\ \text{N}\cdot\text{m}^2/\text{C}^2)\frac{(1.6\times10^{-19}\ \text{C})(-1.6\times10^{-19}\ \text{C})}{(0.5\times10^{-10}\ \text{m})^2} = -9.24\times10^{-8}\ \text{N}$$

The fact that F_E is negative means that the force is one of *attraction*, which we naturally expect, since one charge is positive and the other is negative. The force between the proton and electron is along the line that joins the charges, as we've illustrated below. The two forces shown form an action/reaction pair.

$$q_1 \oplus \xrightarrow{\mathbf{F}_E} \qquad \xleftarrow{\mathbf{F}_E} \ominus q_2$$

Now, the gravitational force between the two charges is given by Newton's Law of Gravitation:

$$\mathbf{F}_G = G \times \frac{(m_1 m_2)}{r^2} = (6.67 \times 10^{-11} \frac{\text{kg} \cdot \text{m}^2}{\text{kg}^2}) \times \frac{(1.67 \times 10^{-27}\ \text{kg})(9.11 \times 10^{-31}\ \text{kg})}{(0.5 \times 10^{-10}\ \text{m})^2} = 4.06 \times 10^{-47}\ \text{N}$$

Now, compare the orders of magnitude of the electric force to the gravitational force. The electric force has an order of magnitude of 10^{-8} N, but the gravitational force has an order of magnitude of 10^{-47} N! This means that the electric force is something like 10^{39} times larger! This is the reason that, in problems in which we calculate electrostatic forces, we often neglect gravitational forces.

CED Unit 3.7
Electric Forces and Free-Body Diagrams

CED Unit 3.8
Describing Electric Forces

Addition of Electric Forces

Consider three point charges: q_1, q_2, and q_3. The total electric force acting on, say, q_2 is simply the sum of $\mathbf{F}_{1\text{-on-}2}$, the electric force on q_2 due to q_1, and $\mathbf{F}_{3\text{-on-}2}$, the electric force on q_2 due to q_3:

$$\mathbf{F}_{\text{on } 2} = \mathbf{F}_{1\text{-on-}2} + \mathbf{F}_{3\text{-on-}2}$$

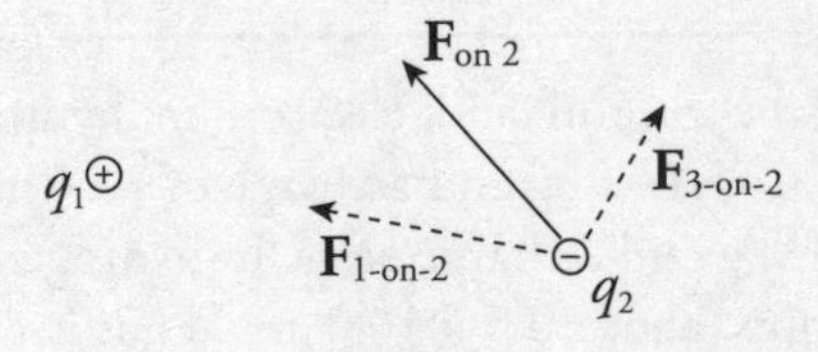

Example 2 Consider four equal, positive point charges that are situated at the vertices of a square. Find the net electric force on a negative point charge placed at the square's center.

Solution. Refer to the diagram below. The attractive forces due to the two charges on each diagonal cancel out: $F_1 + F_3 = 0$, and $F_2 + F_4 = 0$, because the distances between the negative charge and the positive charges are all the same and the positive charges are all equivalent. Therefore, by symmetry, the net force on the center charge is zero.

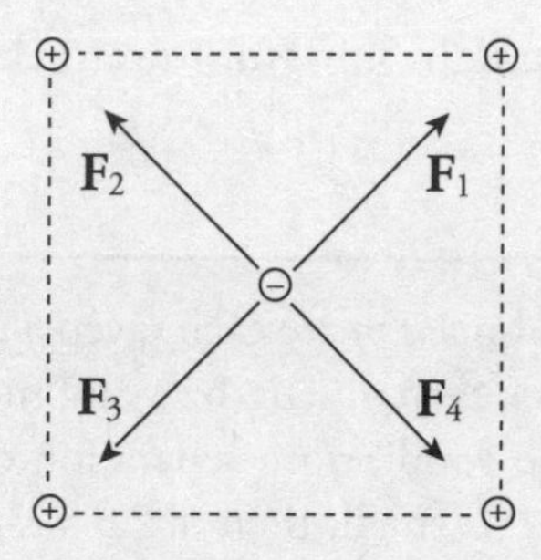

Example 3 If the two positive charges on the bottom side of the square in the previous example were removed, what would be the net electric force on the negative charge? Assume that each side of the square is 4.0 cm, each positive charge is 1.5 ∝C, and the negative charge is –6.2 nC.

Solution. If we break down $\mathbf{F}_1$ and $\mathbf{F}_2$ into horizontal and vertical components, then by symmetry the two horizontal components will cancel each other out, and the two vertical components will add:

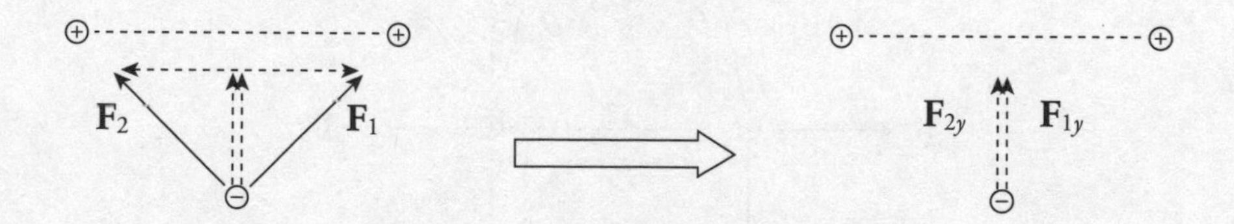

Since the diagram on the left shows the components of $\mathbf{F}_1$ and $\mathbf{F}_2$ making right triangles with legs each of length 2 cm, it must be that $F_{1y} = F_1 \sin 45°$ and $F_{2y} = F_2 \sin 45°$. Also, the magnitude of $\mathbf{F}_1$ equals that of $\mathbf{F}_2$. So the net electric force on the negative charge is $F_{1y} + F_{2y} = 2F \sin 45°$, where F is the strength of the force between the negative charge and each of the positive charges.

If s is the length of each side of the square, then the distance r between each positive charge and the negative charge is $r = \frac{1}{2}s\sqrt{2}$ and

$$\begin{aligned}F_E &= 2F\sin 45^\circ = 2\frac{1}{4\pi\varepsilon_0}\frac{q_1q_2}{r^2}\sin 45^\circ \\ &= 2(9\times10^9\ \text{N}\cdot\text{m}^2/\text{C}^2)\frac{(1.5\times10^{-6}\ \text{C})(6.2\times10^{-9}\ \text{C})}{(\frac{1}{2}\cdot4.0\times10^{-2}\cdot\sqrt{2}\ \text{m})^2}\sin 45^\circ \\ &= 0.15\ \text{N}\end{aligned}$$

The direction of the net force is straight upward, toward the center of the line that joins the two positive charges.

Example 4 Two pith balls of mass m are each given a charge of $+q$. They are hung side-by-side from two threads each of length L, and move apart as a result of their electrical repulsion. Find the equilibrium separation distance x in terms of m, q, and L. (Use the fact that if θ is small, then $\tan\theta \approx \sin\theta$.)

Solution. Three forces act on each ball: weight, tension, and electrical repulsion:

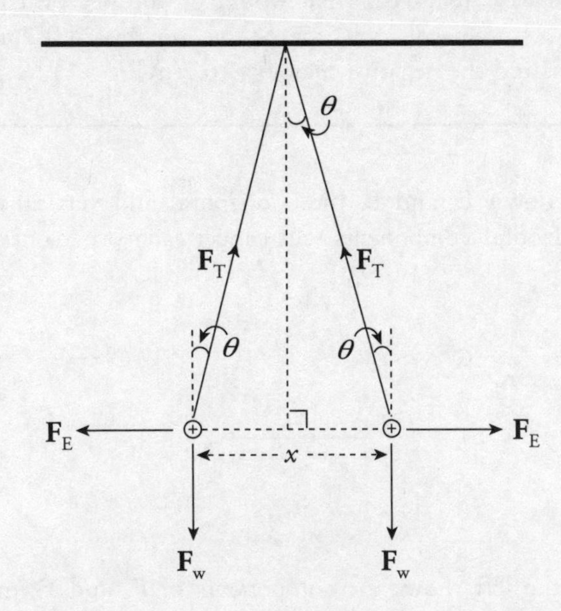

When the balls are in equilibrium, the net force that each feels is zero. Therefore, the vertical component of $\mathbf{F}_T$ must cancel out $\mathbf{F}_w$ and the horizontal component of $\mathbf{F}_T$ must cancel out $\mathbf{F}_E$:

$$F_T\cos\theta = F_w \qquad \text{and} \qquad F_T\sin\theta = F_E$$

Dividing the second equation by the first, we get $\tan\theta = F_E/F_w$. Therefore,

$$\tan\theta = \frac{k\dfrac{q^2}{x^2}}{mg} = \frac{kq^2}{mgx^2}$$

Now, to approximate: if θ is small, $\tan\theta \approx \sin\theta$, and from the diagram, $\sin\theta = \frac{1}{2}x/L$.

Therefore, the equation above becomes

$$\frac{\frac{1}{2}x}{L} = \frac{kq^2}{mgx^2} \Rightarrow \frac{1}{2}mgx^3 = kq^2L \Rightarrow x = \sqrt[3]{\frac{2kq^2L}{mg}}$$

THE ELECTRIC FIELD

CED Unit 3.11
Electric Charges and Fields

The presence of a massive body such as the Earth causes objects to experience a gravitational force directed toward the Earth's center. For objects located outside the Earth, this force varies inversely with the square of the distance and directly with the mass of the gravitational source. A vector diagram of the gravitational field surrounding the Earth looks like this:

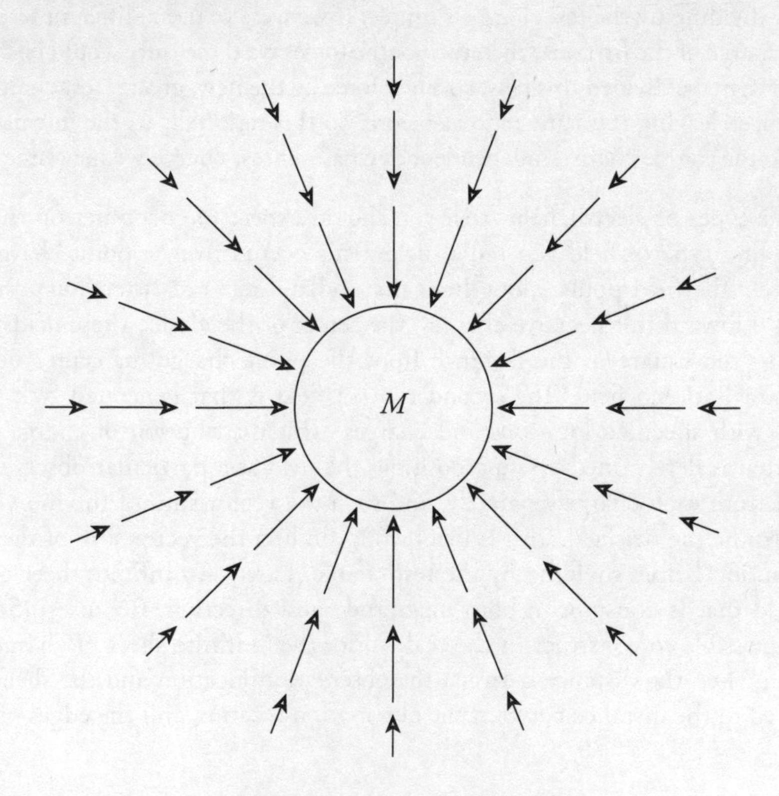

CED Units 3.10 and 5.3
Vector and Scalar Fields

We can think of the space surrounding the Earth as permeated by a **gravitational field** created by the Earth. Any mass placed in this field then experiences a gravitational force due to an interaction with this field.

In general, a field is a physical quantity that has a value at each point in space-time. The values can be scalars (for example, the temperature at each location in a house) or vectors (like the gravitational field in space due to the mass of the Earth). All non-contact forces in physics—such as the gravitational force and the electromagnetic force—are described as the interaction between one object and the field generated by another object.

This is how we describe the electric force. Rather than having two charges reach out across empty space to each other to produce a force, we will instead interpret the interaction in the following way: the presence of a charge creates an **electric field** in the space that surrounds it. Another charge placed in the field created by the first will experience a force due to the field.

Consider a point charge Q in a fixed position and assume that it's positive. Now imagine moving a tiny positive test charge q around to various locations near Q. At each location, measure the force that the test charge experiences, and call it $\mathbf{F}_{\text{on } q}$. Divide this force by the test charge q; the resulting vector is the **electric field vector, E**, at that location:

Equation Sheet

$$\mathbf{E} = \frac{\mathbf{F}_{\text{on } q}}{q}$$

The reason for dividing by the test charge is simple. If we were to use a different test charge with, say, twice the charge of the first one, then each of the forces we'd measure would be twice as much as before. But when we divided this new, stronger force by the new, greater test charge, the factors of 2 would cancel, leaving the same ratio as before. So this ratio tells us the intrinsic strength of the field due to the source charge, independent of whatever test charge we may use to measure it.

There are three types of electric fields that you should expect to encounter on the AP Physics 2 Exam. The first type of field is a radial field. This occurs from a point charge (or from a charged sphere). The field points along lines that radiate outward from (for positive charges) or point inward toward (for negative charges) the center of the circle. These fields are inversely proportional to the square of the distance from the point charge (or center of the sphere), just like the gravitational field. The second type of field is that generated by a collection of point charges with specified locations and charges. This distribution of charge results in an electric field that is determined by superposition; the field at a particular observation location is determined from each charge separately, and then the vector sum of the individual fields is taken to determine the net field. This is identical to finding the vector sum of the forces of the charge distribution before dividing by the test charge. Lastly, an infinite sheet of charge will result in a field that is constant in both magnitude and direction. Because infinite sheets of charge are impossible to construct in the real world, the "infinite sheet of charge" is a useful approximation when the distance between the observation location and the sheet of charge is small compared to the distance between the observation location and the edges of the sheet.

First, let's examine the point charge. Since the test charge used to measure the field is positive, every electric field vector would point radially away from the source charge. *If the source charge is positive, the electric field vectors point away from it; if the source charge is negative, then the field vectors point toward it.* And, since the force decreases as we get farther away from the charge (as $1/r^2$), so does the electric field. This is why the electric field vectors farther from the source charge are shorter than those that are closer.

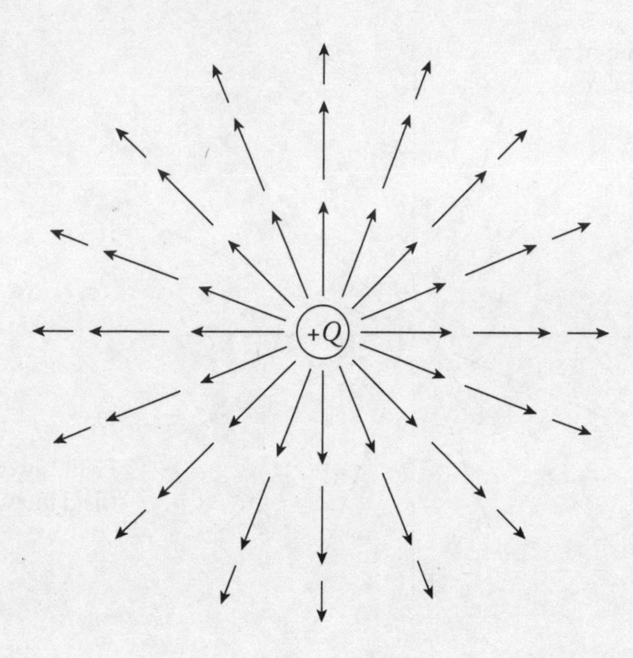

Since the force on any test charge q, due to some source Q, has a strength of $\frac{qQ}{4\pi\varepsilon_0 r^2}$, when we divide this by q, we get the expression for the strength of the electric field created by a point-charge source of magnitude Q:

$$E = \frac{1}{4\pi\varepsilon_0}\frac{Q}{r^2}$$

This is the electric field in the space surrounding a **point charge.**

It is of utmost importance to note that this equation resulted from considering the electric field of a single point charge. Any other distribution of charge, such as placing charge on a sheet to generate a **parallel-plate capacitor** (described later), will have a different electric field and this equation will be invalid. Adding to the confusion, there is one way to distribute a large amount of charge and get the same exact equation for the electric field as having just a single point charge. When the charges were spread out over a sphere (for example, a charged metal ball) instead of a single point charge, at any location outside of the sphere the electric field would be the same as the point charge field.

Sometimes electric fields are drawn, as shown above, with arrows at various locations. The arrows point in the field direction, and their lengths indicate the field strength. However, sometimes they are sketched simply as continuous lines, as shown below, from the source such that the electric field vector is always tangent to the line everywhere it's drawn (a single arrowhead on the line indicates the direction to draw the field line along) and the strength is determined by how close the lines are together.

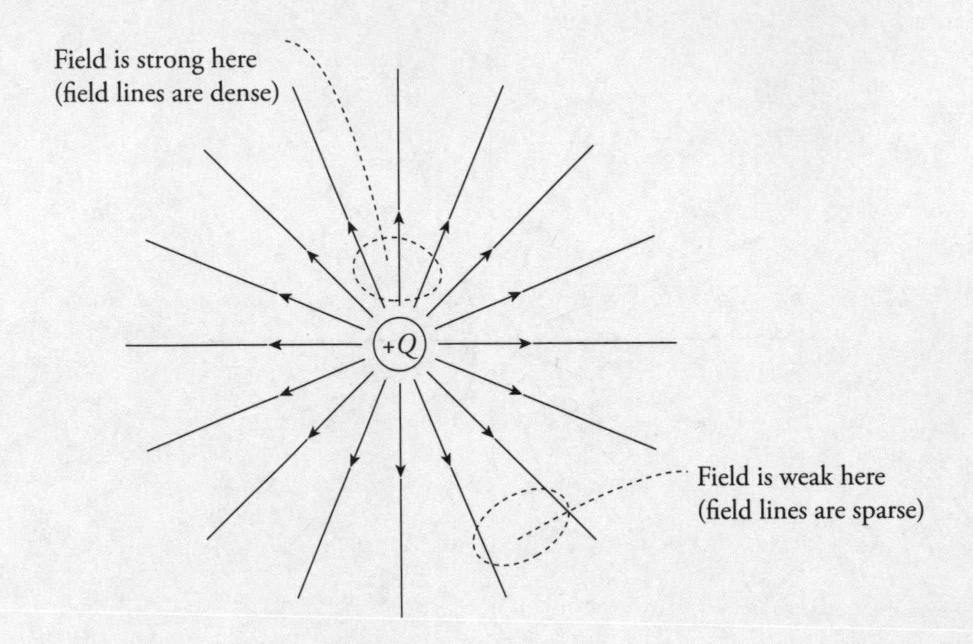

Electric fields obey the same addition properties as the electric force. If we had two source charges, their fields would overlap and effectively add; a third charge would feel the net influence of both charges. At each position in space (referred to as *observation locations*), add the electric field vector due to one of the charges to the electric field vector due to the other charge: $\mathbf{E}_{total} = \mathbf{E}_1 + \mathbf{E}_1$. This extends to any number of source charges. In the diagram below, $\mathbf{E}_1$ is the electric field vector at a particular location due to the charge $+Q$, and $\mathbf{E}_2$ is the electric field vector at that same location due to the other charge, $-Q$. Adding these vectors gives the overall field vector $\mathbf{E}_{total}$ at that location.

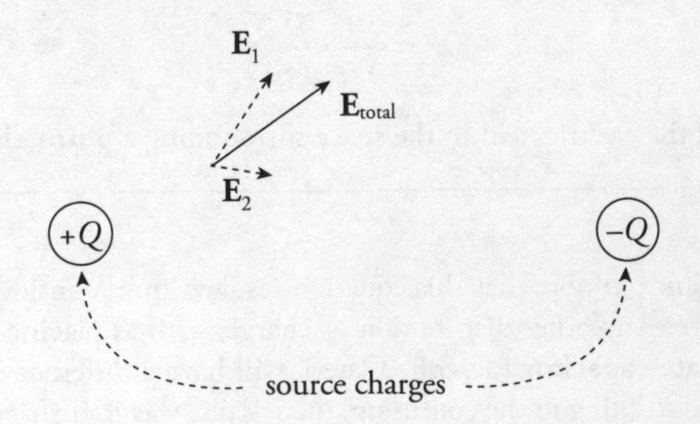

If this is done at enough locations, the electric field lines can be sketched.

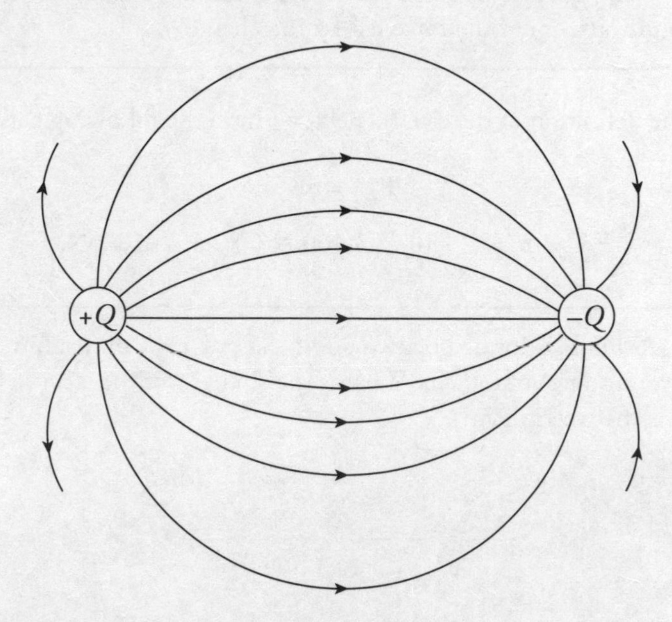

Note that, like electric field vectors, electric field lines always point away from positive source charges and toward negative ones. Two equal but opposite charges, like the ones shown in the diagram above, form a pair called an **electric dipole**.

If a positive charge $+q$ were placed in the electric field above, it would experience a force that is tangent to, and in the same direction as, the field line passing through $+q$'s location. After all, electric field lines indicate the magnitude and direction of the force a positive test charge would experience. On the other hand, if a negative charge $-q$ were placed in the electric field, it would experience a force that is tangent to, but in the direction opposite from, the field line passing through $-q$'s location.

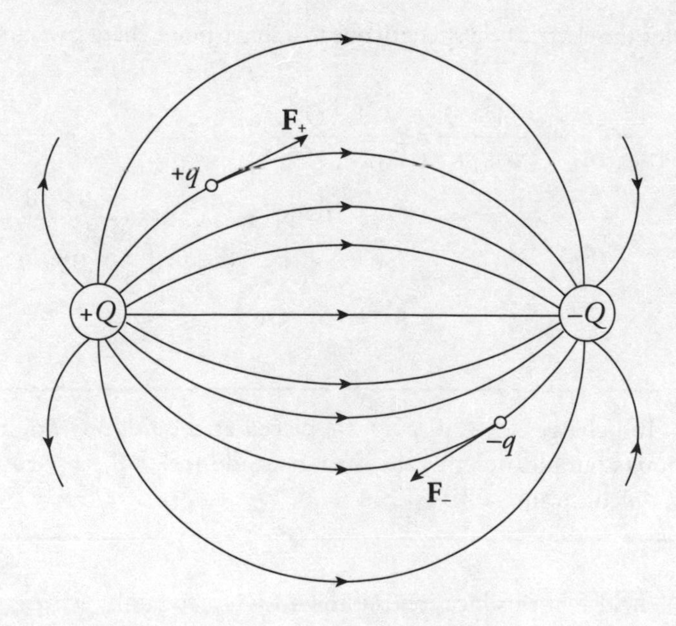

Finally, notice that electric field lines never cross.

Example 5 A charge $q = +3.0$ nC is placed at a location at which the electric field strength is 400 N/C. Find the force felt by the charge q.

Solution. From the definition of the electric field, we have the following equation:

$$\mathbf{F}_{\text{on } q} = q\mathbf{E}$$

Therefore, in this case, $F_{\text{on } q} = qE = (3 \cdot 10^{-9}\text{ C})(400\text{ N/C}) = 1.2 \cdot 10^{-6}\text{ N}$.

Example 6 A dipole is formed by two point charges, each of magnitude 4.0 nC, separated by a distance of 6.0 cm. What is the strength of the electric field at the point midway between them?

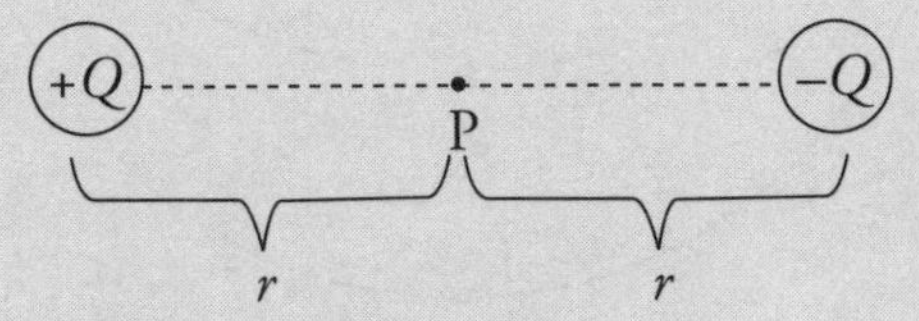

Solution. Let the two source charges be denoted $+Q$ and $-Q$. At Point P, the electric field vector due to $+Q$ would point directly away from $+Q$, and the electric field vector due to $-Q$ would point directly toward $-Q$. Therefore, these two vectors point in the same direction (from $+Q$ to $-Q$), so their magnitudes would add.

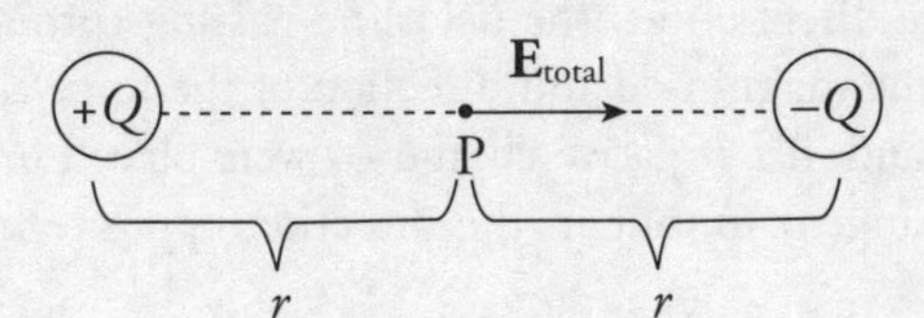

Using the equation for the electric field strength due to a single point charge, we find that

$$\begin{aligned} E_{\text{total}} &= \frac{1}{4\pi\varepsilon_0}\frac{Q}{r^2} + \frac{1}{4\pi\varepsilon_0}\frac{Q}{r^2} = 2\frac{1}{4\pi\varepsilon_0}\frac{Q}{r^2} \\ &= 2(9\times10^{-9}\,\text{N}\cdot\text{m}^2/\text{C}^2)\frac{4.0\times10^{-9}\text{ C}}{\left(\frac{1}{2}(6.0\times10^{-2}\text{ m})\right)^2} \\ &= 8.0\times10^{4}\text{ N/C} \end{aligned}$$

Example 7 If a charge $q = -5.0$ pC were placed at the midway point described in the previous example, describe the force it would feel. ("p" is the abbreviation for "pico-," which means 10^{-12}.)

Solution. Since the field E at this location is known, the force felt by q is easy to calculate:

$$\mathbf{F}_{\text{on } q} = q\mathbf{E} = (-5.0 \cdot 10^{-12}\text{ C})(8.0 \cdot 10^{4}\text{ N/C to the right}) = 4.0 \cdot 10^{-7}\text{ N to the } \textit{left}$$

> **Example 8** What can you say about the electric force that a charge would feel if it were placed at a location at which the electric field were zero?

Solution. Remember that $\mathbf{F}_{\text{on } q} = q\mathbf{E}$. So if $\mathbf{E} = 0$, then $\mathbf{F}_{\text{on } q} = 0$. (Zero field means zero force.)

The Uniform Electric Field

An important subset of problems deals with uniform electric fields. One method of creating this uniform field is to have two large conducting sheets, each storing some charge Q, some distance d apart. Near the edges of each sheet, the field may not be uniform, but near the middle, for all practical purposes, the field is uniform. Having a uniform field means having a constant force (and therefore a constant acceleration), so you can use kinematic equations just as if you had a uniform gravitational field (as you do near the surface of the Earth).

> **Example 9** Positive charge is distributed uniformly over a large, horizontal plate, which then acts as the source of a vertical electric field. An object of mass 5 g is placed at a distance of 2 cm above the plate. If the strength of the electric field at this location is 10^6 N/C, how much charge would the object need to have in order for the electrical repulsion to balance the gravitational pull?

Solution. Clearly, since the plate is positively charged, the object would also have to carry a positive charge so that the electric force would be repulsive.

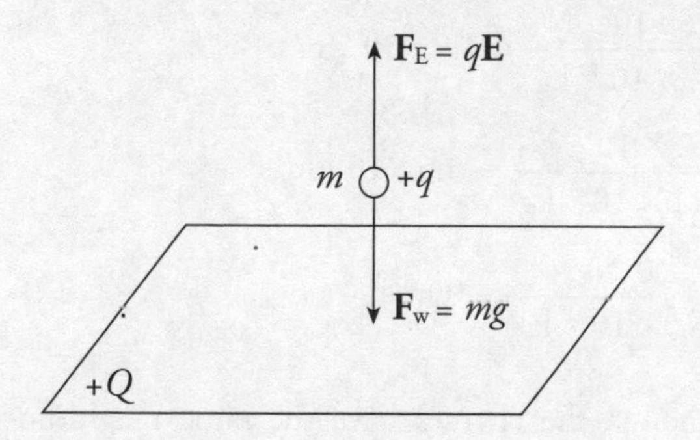

Let q be the charge on the object. Then, in order for F_E to balance mg, we must have

$$qE = mg \quad \Rightarrow \quad q = \frac{mg}{E} = \frac{(5 \times 10^{-3}\ \text{kg})(10\ \text{N/kg})}{10^6\ \text{N/C}} = 5 \times 10^{-8}\ \text{C} = 50\ \text{nC}$$

Example 10 A proton, neutron, and electron are in a uniform electric field of 20 N/C that is caused by two large charged plates that are 30 cm apart. The particles are far enough apart that they don't interact with each other. They are released from rest equidistant from each plate.

(a) What is the magnitude of the net force acting on each particle?
(b) What is the magnitude of the acceleration of each particle?
(c) How much work will be done on the particle as it moves to coincide with one of the charged plates?
(d) What is the speed of each particle when it strikes the plate?
(e) How long does it take to reach the plate?

Solution.

(a) Since $F = qE$, plugging in the values, we get

proton: $F = (1.6 \cdot 10^{-19}\,\text{C})(20\,\text{N/C}) = 3.2 \cdot 10^{-18}\,\text{N}$

electron: $F = (1.6 \cdot 10^{-19}\,\text{C})(20\,\text{N/C}) = 3.2 \cdot 10^{-18}\,\text{N}$

neutron: $F = (0\,\text{C})(20\,\text{N/C}) = 0\,\text{N}$

Note: Because the proton and electron have the same magnitude, they will experience the same force. If you're asked for the direction, the proton travels in the same direction as the electric field and the electron travels in the opposite direction as the electric field.

(b) Since $F = ma$, $a = \dfrac{F}{m}$. Plugging in the values, we get

proton: $a = \dfrac{3.2 \times 10^{-18}\,\text{N}}{1.67 \times 10^{-27}\,\text{kg}} = 1.9 \times 10^{9}\,\text{m/s}^2$

electron: $a = \dfrac{3.2 \times 10^{-18}\,\text{N}}{9.11 \times 10^{-31}\,\text{kg}} = 3.5 \times 10^{12}\,\text{m/s}^2$

neutron: $a = \dfrac{0\,\text{N}}{1.67 \times 10^{-27}\,\text{kg}} = 0\,\text{m/s}^2$

Notice that although the charges have the same magnitude of force, the electron experiences an acceleration almost 2,000 times greater due to its mass being almost 2,000 times smaller than the proton's mass.

(c) Since $W = Fd$, we get $W = qEd$. Plugging in the values, recalling that the charges start midway between the plates, we get

proton: $W = (1.6 \cdot 10^{-19}\,\text{C})(20\,\text{N/C})\,(0.15\,\text{m}) = 4.8 \cdot 10^{-19}\,\text{J}$

electron: $W = (1.6 \cdot 10^{-19}\,\text{C})(20\,\text{N/C})\,(0.15\,\text{m}) = 4.8 \cdot 10^{-19}\,\text{J}$

neutron: $W = (0\,\text{C})\,(20\,\text{N/C})(0.15\,\text{m}) = 0\,\text{J}$

(d) From the Work-Energy Theorem,

$W = \Delta K = 1/2mv^2$, so $v = \sqrt{2W/m}$

Using the answer from part (c) for the work, we get

proton: $v_f = \sqrt{2(4.8 \times 10^{-19}\ \text{J})/(1.67 \times 10^{-27}\ \text{kg})} \rightarrow v_f = 24{,}000$ m/s

electron: $v_f = \sqrt{2(4.8 \times 10^{-19}\ \text{J})/(9.11 \times 10^{-31}\ \text{kg})} \rightarrow v_f = 1.0 \times 10^6$ m/s

neutron: The neutron never hits the plate, so the question about what speed it hits the plate with is not well posed.

Notice that, even though the force is the same and the same work is done on both charges, there is a significant difference in final velocities due to the large mass difference. An alternative solution to this would be using kinematics. You would have obtained the same answers.

(e) Recall the definition of the final velocity of a uniformly accelerated object:

$$v_f = v_i + at \rightarrow t = \frac{v_{fi} - v}{a} \rightarrow t = \frac{v_f}{a}$$

proton: $t = \dfrac{24{,}000\ \text{m/s}}{1.9 \times 10^9\ \text{m/s}^2} \rightarrow 1.3 \cdot 10^{-5}$ s

electron: $t = \dfrac{1 \times 10^6\ \text{m/s}}{3.5 \times 10^{12}\ \text{m/s}^2} \rightarrow 2.9 \cdot 10^{-7}$ s

neutron: The neutron never accelerates, so it will never hit the plate.

CED Unit 3.4
Charge Distribution: Friction, Conduction, and Induction

CONDUCTORS AND INSULATORS

Materials can be classified into broad categories based on their ability to permit the flow of charge. If electrons were placed on a metal sphere, they would quickly spread out and cover the outside of the sphere uniformly. These electrons would be free to flow through the metal and redistribute themselves, moving to get as far away from each other as they could. Materials that permit the flow of excess charge are called **conductors**; they conduct electricity. Metals are the best examples of conductors. Aqueous solutions that contain dissolved electrolytes (such as salt water) are also conductors. Metals conduct electricity because they bind all but their outermost electrons very tightly. That outermost electron is free to move about the metal. This creates a sort of sea of mobile (or conduction) electrons.

Insulators, on the other hand, closely guard their electrons—and even extra ones that might be added. Electrons are not free to roam throughout the atomic lattice. Examples of insulators are glass, wood, rubber, and plastic. If excess charge is placed on an insulator, it stays put. This process, called **charging by friction**, involves simply rubbing the insulator against another material, thereby stripping electrons off one material and depositing them on the other material.

Example 11 A solid sphere of copper is given a negative charge. Discuss the electric field inside and outside the sphere.

Solution. In order for the electrons to eventually come to rest, the net force on any electron *that is free* to move must be 0 N. Copper is a conducting material. Therefore, since any excess charges that reside within the body of the sphere must have a net force of 0 N, we conclude that the electric field inside a conductor is zero. Otherwise, charges within the sphere would experience a force and would move, violating the "static" part of electrostatic equilibrium.

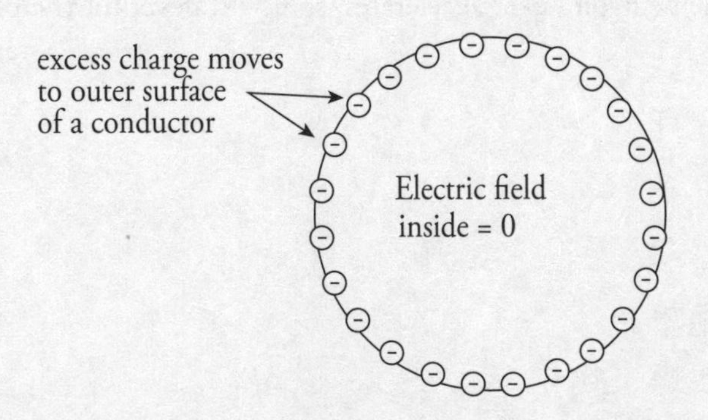

All the excess electrons that are deposited on the sphere arrange themselves on the outer surface, where they are constrained in their motion by the insulating air surrounding the sphere. The electric field within the sphere is 0 N and outside the sphere the field is found to be $E = \frac{1}{4\pi\varepsilon_0}\frac{Q}{r^2}$, which is exactly the same as the field from a single point charge Q located at the center of the conducting sphere.

In fact, you can shield yourself from electric fields simply by surrounding yourself with metal. Charges may move around on the outer surface of your cage, but within the cage, the electric field will be zero. For points outside the sphere, the sphere behaves as if all its excess charge were concentrated at its center. (Remember that this is just like the gravitational field due to a uniform spherical mass.) Also, *the electric field is always perpendicular to the surface, no matter what shape the surface may be.* See the diagram below.

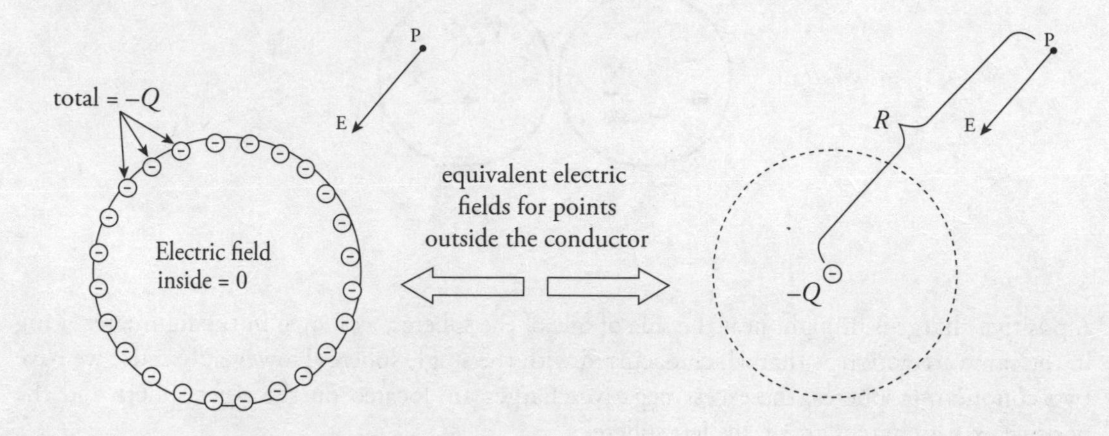

Now let's take our previous sphere and put it in a different situation. Start with a neutral metal sphere and bring a positive charge Q nearby without touching the original metal sphere. What will happen? The positive charge will attract free electrons in the metal, leaving the far side of the sphere positively charged. Since the negative charge is closer to Q than the positive charge, there will be a net attraction between Q and the sphere. So, even though the sphere as a whole is electrically neutral, the separation of charge induced by the presence of Q will create a force of electrical attraction between them. This process for rearranging charges within a conductor is called **charge conduction**.

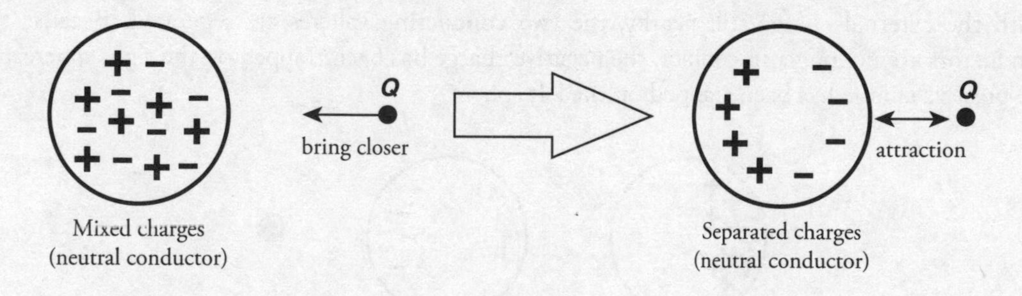

Charging by Induction

The process of charging by induction may be used to redistribute charges among a pair of neutrally charged spheres, so that in the end both spheres are charged. Imagine two neutrally charged spheres that are each set on an insulating stand. The spheres are arranged so that they are in contact with one another.

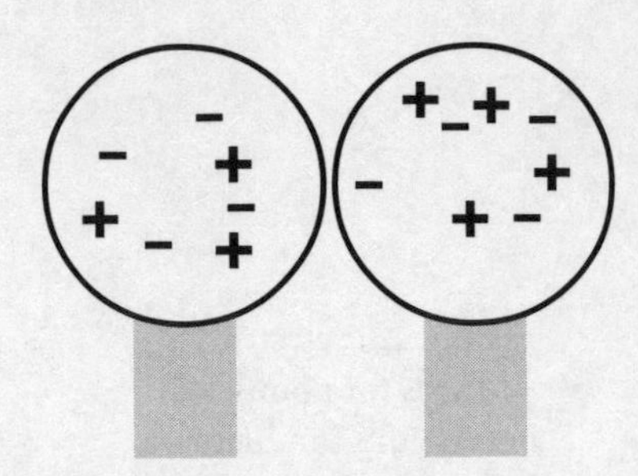

A positive charge is brought near the side of one of the spheres, as shown in the figure, resulting in the same attraction as that which occurred with the single sphere. However, because we have two conducting spheres, the excess negative charges are located on the right sphere and the positive excess charges are on the left sphere.

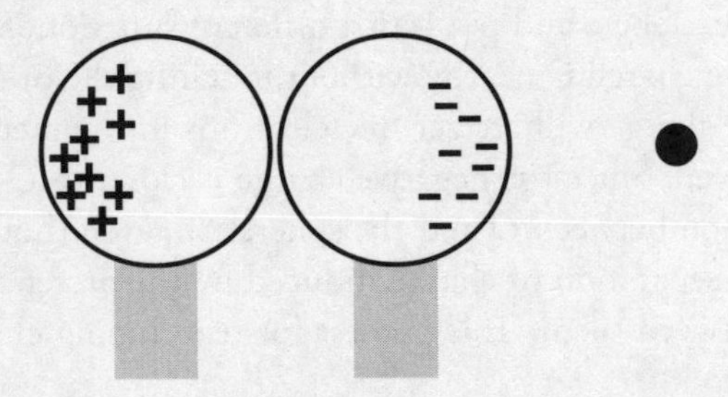

With the external charge still nearby, the two conducting spheres are separated. Because the conductors are no longer in contact, the negative charge has been trapped on the right sphere and the positive charge has been trapped on the left sphere.

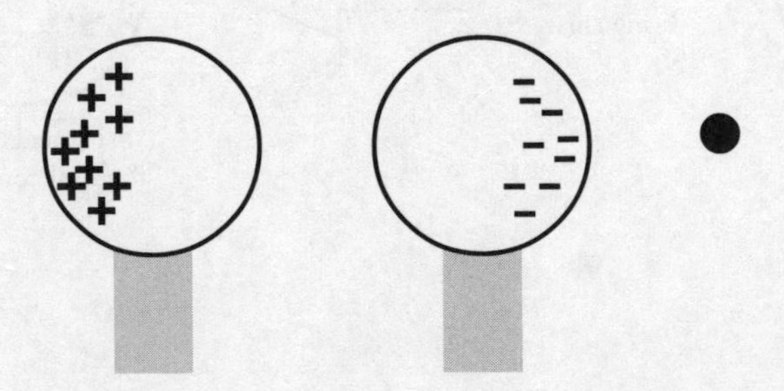

Note that the net charge is unchanged between the two spheres—the distribution of the charges has simply been changed.

Now, what if the sphere was made of glass (an insulator)? Although there aren't free electrons that can move to the near side of the sphere, the atoms that make up the sphere will become polarized. That is, their electrons will feel a tug toward Q, and so will spend more time on the side of the atom closer to Q than on the side opposite Q. This causes the atoms to develop a distribution of charge that is more negative on the side nearby Q (and a partial positive charge on the other side from Q). The effect isn't as dramatic as the mass movement of free electrons in the case of a metal sphere, but the polarization is still enough to cause an electrical attraction between the sphere and Q. For example, if you comb your hair, the comb will pick up extra electrons, making it negatively charged. If you place this electric field source near little bits of paper, the paper will become polarized and will then be attracted to the comb.

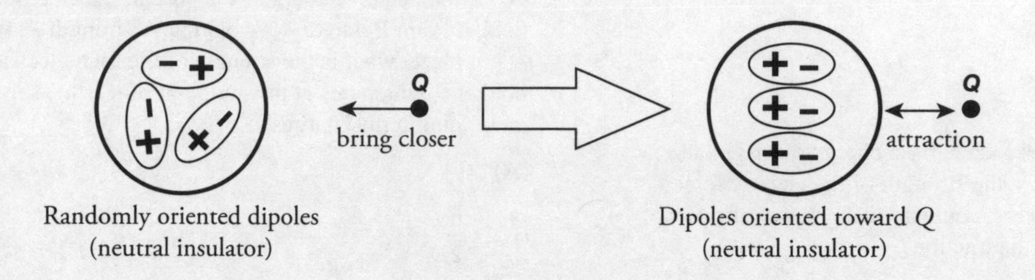

The same phenomenon, in which the presence of a charge tends to cause polarization in a nearby collection of charges, is responsible for a kind of intermolecular force. Dipole-induced forces are caused by a shifting of the electron cloud of a neutral molecule toward positively charged ions or away from negatively charged ions; in either case, the resulting force between the ion and the atom is attractive.

Chapter 5 Review Questions

Solutions can be found in Chapter 11.

Section I: Multiple Choice

1. An experiment is conducted measuring the electrostatic force, F, on a test object at various distances, r. In order to create a plot with a straight line, what should be graphed?

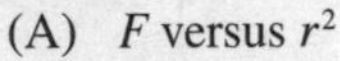

(A) F versus r^2
(B) F versus r
(C) F versus r^{-1}
(D) F versus r^{-2}

2. Two 1 kg spheres each carry a charge of magnitude 1 C. How does F_E, the strength of the electric force between the spheres, compare to F_G, the strength of their gravitational attraction?

(A) $F_E < F_G$
(B) $F_E = F_G$
(C) $F_E > F_G$
(D) If the charges on the spheres are of the same sign, then $F_E > F_G$; but if the charges on the spheres are of the opposite sign, then $F_E < F_G$.

3. The figure below shows three point charges, all positive. If the net electric force on the center charge is zero, what is the value of y/x ?

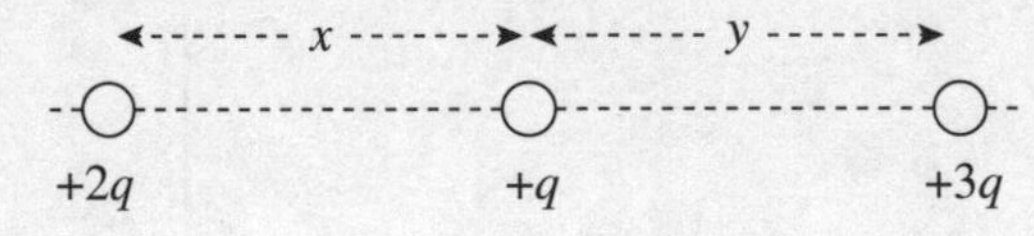

(A) $\frac{4}{9}$
(B) $\sqrt{\frac{2}{3}}$
(C) $\sqrt{\frac{3}{2}}$
(D) $\frac{3}{2}$

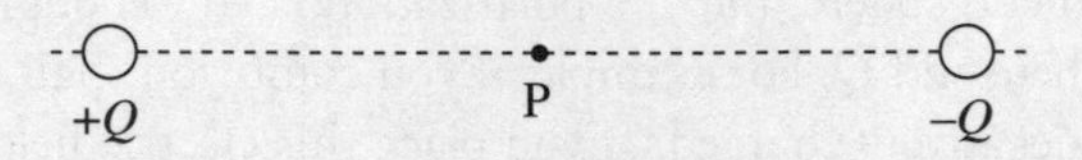

4. The figure above shows two point charges, $+Q$ and $-Q$. If the negative charge were absent, the electric field at Point P due to $+Q$ would have strength E. With $-Q$ in place, what is the strength of the total electric field at P, which lies at the midpoint of the line segment joining the charges?

(A) 0
(B) $\frac{E}{2}$
(C) E
(D) $2E$

5. A sphere of charge $+Q$ is fixed in position. A smaller sphere of charge $+q$ is placed near the larger sphere and released from rest. The small sphere will move away from the large sphere with

(A) decreasing velocity and decreasing acceleration
(B) decreasing velocity and increasing acceleration
(C) increasing velocity and decreasing acceleration
(D) increasing velocity and increasing acceleration

6. An object of charge $+q$ feels an electric force $\mathbf{F}_E$ when placed at a particular location in an electric field, **E**. Therefore, if an object of charge $-2q$ were placed at the same location where the first charge was, it would feel an electric force of

   (A) $\frac{-\mathbf{F}_E}{2}$

   (B) $-2\mathbf{F}_E$

   (C) $-2q\mathbf{F}_E$

   (D) $\frac{-2\mathbf{F}_E}{q}$

7. A charge of $-3Q$ is transferred to a solid metal sphere of radius r. How will this excess charge be distributed?

   (A) $-Q$ at the center, and $-2Q$ on the outer surface

   (B) $-3Q$ at the center

   (C) $-3Q$ on the outer surface

   (D) $-Q$ at the center, $-Q$ in a ring of radius $\frac{1}{2}r$, and $-Q$ on the outer surface

Section II: Free Response

1. In the figure shown, four charges are situated at the corners of a square of side length s. The charges on opposite corners are equal to one another and are labeled Q for corners A and C and labeled q for corners B and D. The charge on Q is positive for all experiments.

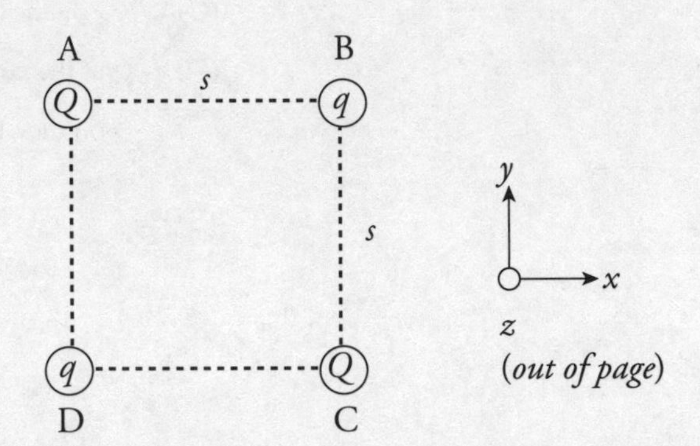

(A) In a first experiment, it is found that the force on the charge at position C is 0 N.

i. Justify the assertion that the charges q cannot have different magnitudes in this experiment.

ii. Derive an expression for the magnitude and sign of charge on q in this experiment.

(B) In a second experiment, the charge at point C is removed. If the charges q are positive, is there anywhere within the boundary square where the electric field could be 0 N/C?

(C) The charges are reassembled into their original positions. In a clear, coherent paragraph-length response, explain why the electric field at the center of the square must be 0 N/C regardless of the magnitudes of the charges q or Q and regardless of their signs.

2. Two charges, $+Q$ and $+2Q$, are fixed in place along the y-axis of an xy-coordinate system, as shown in the figure below. Charge 1 is at the point $(0, a)$, and Charge 2 is at the point $(0, -2a)$.

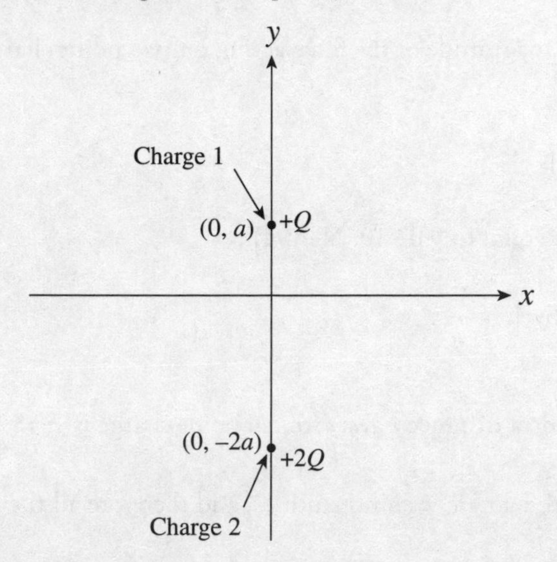

(A) Find the electric force (magnitude and direction) felt by Charge 1 due to Charge 2.

(B) Find the electric field (magnitude and direction) at the origin created by both Charges 1 and 2.

(C) Is there a point on the x-axis where the total electric field is zero? If so, where? If not, explain briefly.

(D) Is there a point on the y-axis where the total electric field is zero? If so, where? If not, explain briefly.

Summary

- Coulomb's Law describes magnitude of the force acting on two point charges and is given by

  $$|\vec{F}_E| = \frac{1}{4\pi\varepsilon_0}\frac{|q_1 q_2|}{r^2}$$

  where $\frac{1}{4\pi\varepsilon_0}$ is a constant equal to $9.0 \cdot 10^9$ N·m²/C².

- The electric field is given by $\mathbf{E} = \frac{\mathbf{F}_E}{q}$.

- The electric field magnitude a distance r away from a point charge is $E = \frac{1}{4\pi\varepsilon_0}\frac{q}{r^2}$.

- Both the electric force and field are vector quantities, and therefore all the rules for vector addition apply.

Chapter 6
Electric Potential and Capacitance

This chapter continues to deal with CED Unit 3.1: Electric Systems.

INTRODUCTION

When an object moves in a gravitational field, it usually experiences a change in kinetic energy and in gravitational potential energy due to the work done on the object by gravity. Similarly, when a charge moves in an electric field, it generally experiences a change in kinetic energy and in electrical potential energy due to the work done on it by the electric field. By exploring the idea of electric potential, we can simplify our calculations of work and energy changes within electric fields.

CED Unit 3.13
Conservation of Electric Energy

ELECTRICAL POTENTIAL ENERGY

When a charge moves in an electric field, unless its displacement is always perpendicular to the field, the electric force does work on the charge. If W_E is the work done by the electric force, then the change in the charge's **electrical potential energy** is defined by

$$\Delta U_E = -W_E$$

Notice that this is the same equation that defined the change in the gravitational potential energy of an object of mass m undergoing a displacement in a gravitational field ($\Delta U_G = -W_G$).

Example 1 A positive charge $+q$ moves from position A to position B in a uniform electric field **E**:

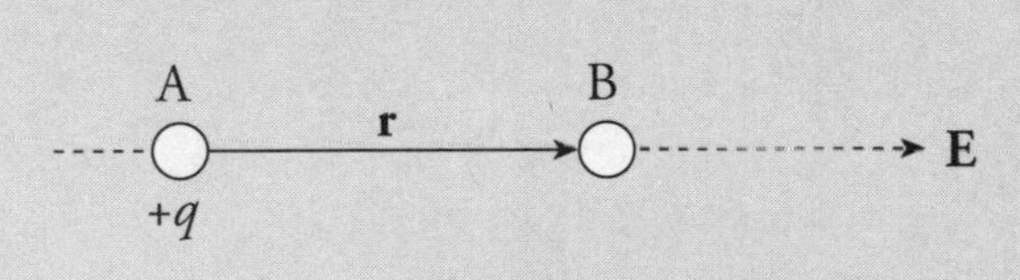

What is its change in electrical potential energy?

Solution. Since the field is uniform, the electric force that the charge feels, $\mathbf{F}_E = q\mathbf{E}$, is constant. Since q is positive, $\mathbf{F}_E$ points in the same direction as **E**, and, as the figure shows, they point in the same direction as the displacement, r. This makes the work done by the electric field equal to $W_E = F_E r = qEr$, so the change in the electrical potential energy is

$$\Delta U_E = -qEr$$

Note that the change in potential energy is negative, which means that potential energy has decreased; this always happens when the field does positive work. It's just like dropping a rock to the ground: gravity does positive work, and the rock loses gravitational potential energy.

Example 2 Do the previous problem, but consider the case of a negative charge, $-q$.

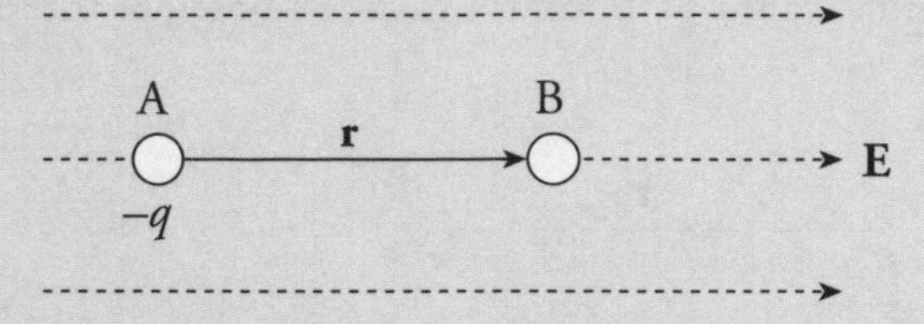

Solution. In this case, an outside agent must be pushing the charge to make it move, because the electric force *naturally* pushes negative charges against field lines. Therefore, we expect that the work done by the electric field is negative. The electric force, $\mathbf{F}_E = (-q)\mathbf{E}$, points in the direction opposite to the displacement, so the work it does is $W_E = -F_E r = -qEr = -qEr$. Thus, the change in electrical potential energy is positive: $\Delta U_E = -W_E = -(-qEr) = qEr$. Because the change in potential energy is positive, the potential energy increases; this always happens when the field does negative work. It's like lifting a rock off the ground: gravity does negative work, and the rock gains gravitational potential energy.

Example 3 A positive charge $+q$ moves from position A to position B in a uniform electric field **E**:

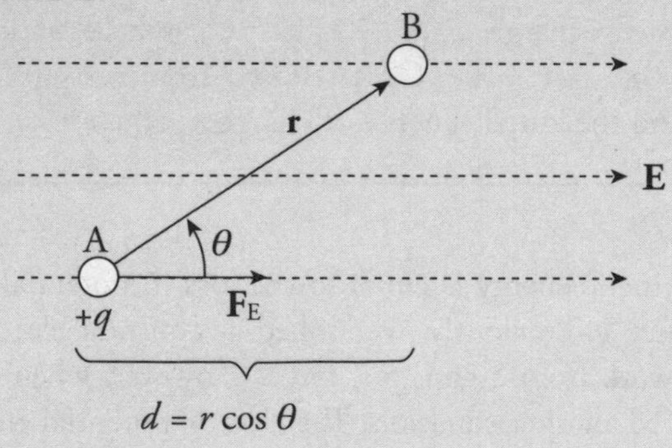

What is its change in electrical potential energy?

Solution. The electric force felt by the charge q is $\mathbf{F}_E = q\mathbf{E}$, and this force is parallel to **E** because q is positive. In this case, because $\mathbf{F}_E$ is not parallel to r (as it was in Example 1), we will use the more general definition of work for a constant force:

$$W_E = \mathbf{F}_E \cdot \mathbf{r} = F_E\, r \cos\theta = qEr\cos\theta$$

But $r\cos\theta = d$, so

$$W_E = qEd \text{ and } \Delta U_E = -W_E = -qEd$$

Because the electric force is a conservative force (which means that the work done does not depend on the path that connects the positions A and B), the work calculated above could have been figured out by considering the path from A to B composed of the segments $\mathbf{r}_1$ and $\mathbf{r}_2$:

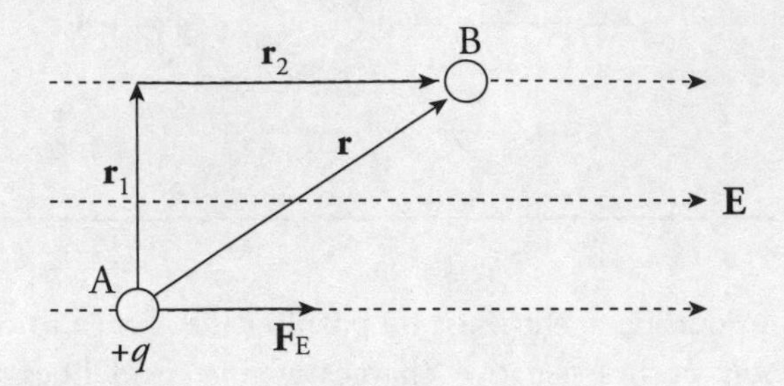

Along $\mathbf{r}_1$, the electric force does no work since this displacement is perpendicular to the force. Thus, the work done by the electric field as q moves from A to B is simply equal to the work it does along $\mathbf{r}_2$. And since the length of $\mathbf{r}_2$ is $d = r \cos \theta$, we have $W_E = F_E d = qEd$, just as before.

Electric Potential Energy of a System of Two Point Charges

Example 4 A positive charge, $q_1 = +2 \cdot 10^{-6}$ C, is held stationary, while a negative charge, $q_2 = -1 \cdot 10^{-8}$ C, is released from rest at a distance of 10 cm from q_1. Find the kinetic energy of charge q_2 when it's 1 cm from q_1.

Solution. The gain in kinetic energy is equal to the loss in potential energy; you know this from Conservation of Energy. Previously, we looked at constant electric fields and were able to use the equation for work from a constant force. However, when the field (and therefore the force) changes, we need another equation. The electric potential energy of a system of two point charges separated by a distance r is given by

$$U_E = \frac{1}{4\pi\varepsilon_0}\left(\frac{q_1 q_2}{r}\right)$$

Therefore, if q_1 is fixed and q_2 moves from r_A to r_B, the change in potential energy is

$$\begin{aligned}\Delta U_E &= U_B - U_A \\ &= \frac{q_1}{4\pi\varepsilon_0}\left(\frac{q_2}{r_B} - \frac{q_2}{r_A}\right) \\ &= \frac{q_1 q_2}{4\pi\varepsilon_0}\left(\frac{1}{r_B} - \frac{1}{r_A}\right) \\ &= (9\times10^9\ \text{N}\cdot\text{m}^2/\text{C}^2)(+2\times10^{-6}\ \text{C})(-1\times10^{-8}\ \text{C})\left(\frac{1}{0.01\ \text{m}} - \frac{1}{0.10\ \text{m}}\right) \\ &= -0.016\ \text{J}\end{aligned}$$

Since q_2 lost 0.016 J of potential energy, the gain in kinetic energy is 0.016 J. Since q_2 started from rest (with no kinetic energy), this is the kinetic energy of q_2 when it's 1 cm from q_1.

Example 5 Two positive charges, q_1 and q_2, are held in the positions shown below. How much work would be required to bring (from infinity) a third positive charge, q_3, and place it so that the three charges form the corners of an equilateral triangle of side length *s*?

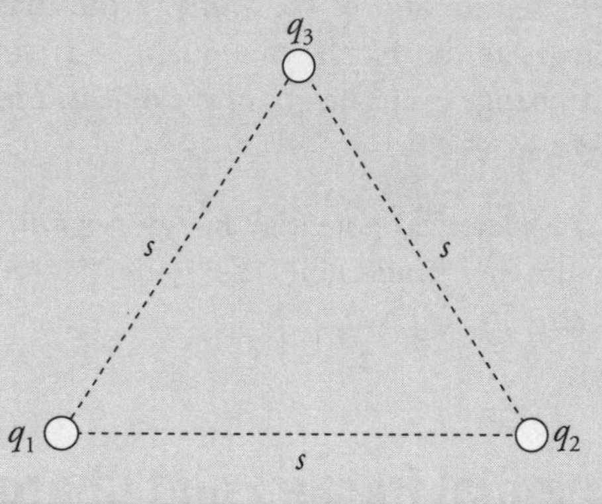

Solution. An external agent would need to do positive work, equal in magnitude to the negative work done by the electric force on q_3 as it is brought into place, so let's first compute this quantity. Let's first compute the work done *by the electric force* as q_3 is brought in. Since q_3 is fighting against both q_1's and q_2's electric field, the total work done on q_3 by the electric force, W_E, is equal to the work done on q_3 by q_1 ($W_{1\text{-}3}$) plus the work done on q_3 by q_2 ($W_{2\text{-}3}$). Using the equation $W_E = -\Delta U_E$ and the one we gave above for ΔU_E, we have

$$\begin{aligned}W_{1\text{-}3} + W_{2\text{-}3} &= -\Delta U_{1\text{-}3} + -\Delta U_{2\text{-}3} \\ &= \left(-\frac{q_1 q_3}{4\pi\varepsilon_0}\frac{1}{s} - 0\right) + \left(-\frac{q_2 q_3}{4\pi\varepsilon_0}\frac{1}{s} - 0\right) \\ &= -\frac{1}{4\pi\varepsilon_0}\frac{q_1 q_3}{s} + \frac{1}{4\pi\varepsilon_0}\frac{q_2 q_3}{s}\end{aligned}$$

Therefore, the work that an external agent must do to bring q_3 into position is

$$-W_E = \frac{1}{4\pi\varepsilon_0}\frac{q_1 q_3}{s} + \frac{1}{4\pi\varepsilon_0}\frac{q_2 q_3}{s}$$

NOT the Same
Electric potential and electric potential energy, although very closely related and having similar names, are not the same thing. Even the units are measured differently—electric potential is measured in joules per coulomb and electric potential energy in joules.

Equation Sheet

ELECTRIC POTENTIAL

Let W_E be the work done by the electric field on a charge q as it undergoes a displacement. If another charge, say $2q$, were to undergo the same displacement, the electric force would be twice as great on this second charge, and the work done by the electric field would be twice as much, $2W_E$. Since the work would be twice as much in the second case, the change in electrical potential energy would be twice as great as well, but the ratio of the change in potential energy to the charge would be the same: $W_E/q = (2W_E)/2q$. This ratio says something about the work done by the field and the *displacement* but not the charge that made the move. The change in **electric potential**, ΔV, is defined as this ratio:

$$\Delta V = \frac{\Delta U_E}{q}$$

There is a similar concept with gravitational fields. The work done by gravity (near the Earth) on an object as it changes its position is $W = mgh$. If twice the mass were moved, the work would have to be $W = (2m)gh$. The ratio of the work to the mass moved is technically the "gravitational potential." But near the Earth, this quantity is simply the gravitational field strength, g, multiplied by the change in the height of the object. *Electric potential is the electrical equivalent of a change in height.*

Potential Difference
Finding the potential at a certain point in space is meaningless. With potential, what is more important is an initial position and final position. Just like in gravity, finding the potential energy at a certain point is meaningless if there is no relation to another point. Once we establish this potential difference, when we move a charge across it, we can generate energy. This will be key in the next chapter, when potential difference is needed in order to have an electrical circuit.

Electric potential is electrical potential energy *per unit charge*; the units of electric potential are joules per coulomb. One joule per coulomb is called one **volt** (abbreviated V), so 1 J/C = 1 V.

Electric Potential from a Point Charge

Consider the electric field created by a point source charge Q. If a charge q moves from a distance r_A to a distance r_B from Q, then the change in the potential energy is

$$U_B - U_A = \frac{Qq}{4\pi\varepsilon_0}\left(\frac{1}{r_B} - \frac{1}{r_A}\right)$$

The difference in electric potential between positions A and B in the field created by Q is

$$V_B - V_A = \frac{U_B - U_A}{q} = \frac{Q}{4\pi\varepsilon_0}\left(\frac{1}{r_B} - \frac{1}{r_A}\right)$$

If we designate $V_A \to 0$ as $r_A \to \infty$ (an assumption that's stated on the AP Physics 2 Exam), then for a point charge Q, the electric potential at a distance r from Q is

$$V = \frac{1}{4\pi\varepsilon_0}\frac{Q}{r}$$

Equation Sheet

Note that the potential depends on the strength of the source charge making the field and the distance from the source charge.

Example 6 Let $Q = 2 \cdot 10^{-9}$ C. What is the potential at a Point P that is 2 cm from Q?

Solution. Relative to $V = 0$ at infinity, we have

$$V = \frac{1}{4\pi\varepsilon_0}\frac{Q}{r} = \left(9\times10^9\ \text{N}\cdot\text{m}^2/\text{C}^2\right)\frac{2\times10^{-9}\ \text{C}}{0.02\ \text{m}} = 900\ \text{V}$$

This means that the work done by the electric field on a charge of q coulombs brought to a point 2 cm from Q would be $-900q$ joules.

Note that, like potential energy, electric potential is a *scalar*. In the preceding example, we didn't have to specify the direction of the vector from the position of Q to the Point P, because it didn't matter. Imagine a sphere with a surface of 2 cm from Q; at any point on that sphere, the potential will be 900 V. These spheres around Q are called **equipotential surfaces**, and they're surfaces of constant potential. The equipotentials are always perpendicular to the electric field lines.

Remember
Equipotential surfaces are often imaginary. Whether we put a metal sphere in that location, or whether we visualize a sphere that isn't really there, doesn't change the electric potential in that region.

Example 7 How much work is done by the electric field as a charge moves along an equipotential surface?

Solution. If the charge always remains on a single equipotential, then, by definition, the potential, V, never changes. Therefore, $\Delta V = 0$, so $\Delta U_E = 0$. Since $W_E = -\Delta U_E$, the work done by the electric field is zero.

Addition of Electric Potential

The formula $V = kQ/r$ tells us how to find the potential due to a single point source charge, Q. Potential is scalar (we will not be concerned with direction, just the sign of charge). When we add up individual potentials, we're simply adding numbers; we're not adding vectors.

Just like Gravitational Potential Energy
The electric force is conservative. All we care about is the change in position when calculating the change in potential energy.

Example 8 How much work would it take to move a charge $q = +1 \cdot 10^{-2}$ C from Point A to Point B (the point midway between $q_1 = 4$ nC and $q_2 = -6$ nC)?

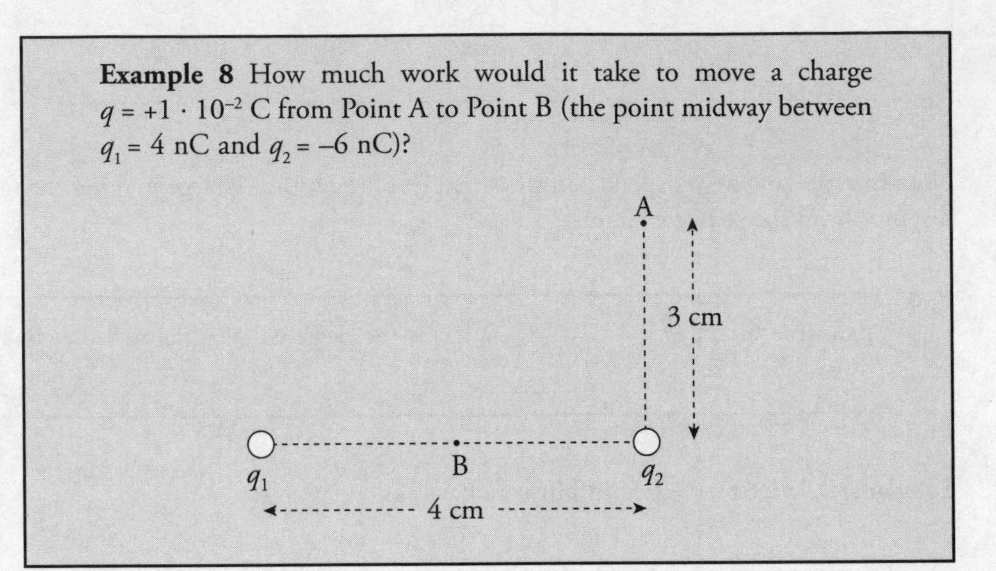

Solution. $\Delta U_E = q\Delta V$, so if we calculate the potential difference between Points A and B and multiply by q, we will have found the change in the electrical potential energy: $\Delta U_{A \to B} = q\Delta V_{A \to B}$. Then, since the work by the electric field is $-\Delta U$, the work required by an external agent is ΔU.

First, we need the potential at point A, V_A. Since there are two charges, q_1 and q_2, contributing to the potential at point A, we calculate the contribution of each using $V = kQ/r$ and sum the results. Remember that the potential is a scalar quantity, so no vector addition is required here. It is very important to keep track of the positive and negative signs, however.

$$V_A = \Sigma k\left(\frac{q_i}{r_r}\right) = k\left(\frac{q_1}{r_1} + \frac{q_2}{r_2}\right) = \left(9 \times 10^9 \frac{\text{N} \cdot \text{m}^2}{\text{C}^2}\right)\left(\frac{4 \times 10^{-9}\text{ C}}{0.05\text{ m}} + \frac{-6 \times 10^{-9}\text{ C}}{0.03\text{ m}}\right)$$

$$V_A = -1{,}080\text{ V}$$

A similar calculation can be carried out for point B, although the distance from the charges will be different for point B than they were for point A.

$$V_B = \Sigma k\left(\frac{q_i}{r_r}\right) = k\left(\frac{q_1}{r_1} + \frac{q_2}{r_2}\right) = \left(9 \times 10^9 \frac{\text{N} \cdot \text{m}^2}{\text{C}^2}\right)\left(\frac{4 \times 10^{-9}\text{C}}{0.02\text{ m}} + \frac{-6 \times 10^{-9}\text{C}}{0.02\text{ m}}\right)$$

$$V_B = -900\text{ V}$$

$\Delta V_{A \to B} = V_B - V_A = (-900\ V) - (-1080\ V) = +180\ V$. This means that the change in electrical potential energy as q moves from A to B is

$$\Delta U_{A \to B} = q\Delta V_{A \to B} = (+1 \cdot 10^{-2}\ C)(+180\ V) = 1.8\ J$$

This is the work required by an external agent to move q from A to B.

EQUIPOTENTIAL CURVES AND EQUIPOTENTIAL MAPS

CED Unit 3.12
Isolines and Electric Fields

Electric field diagrams are a simple way to represent a sometimes very complex result of the influence of a charge distribution created by several charges. A similar image may be formed using isolines of electric potential instead of electric field vectors. In a downward gravitational field, if an object were to move from one position on a line at a constant height to another position on the line, no work would be done because the displacement would be perpendicular to the gravitational field. Electric equipotential curves are similar. A charge may be moved from one position on an equipotential curve to another without any work being required. A drawing of several equipotential curves at various values of the potential for a charge distribution (which may or may not be specified) is called an **equipotential map**. These images may look somewhat familiar as they are very similar to topographical maps, which plot lines of constant height in a uniformly downward directed gravitational field.

Here are the electric fields (which should look very familiar from Chapter 5) and some equipotential lines for isolated positive or negative point charges. The field lines are shown in solid and the equipotentials are shown as dashed lines.

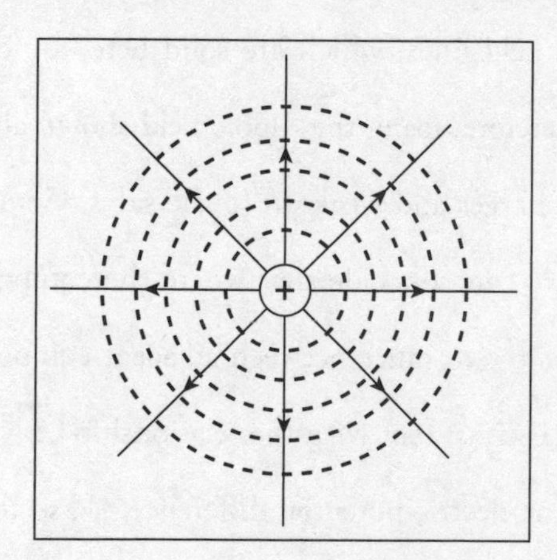

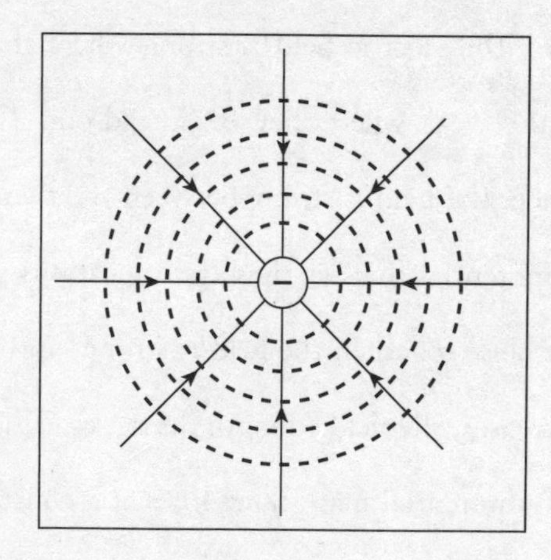

The equipotential map for an isolated positive charge is exactly the same shape as that for an isolated negative charge. Both equipotential maps are made of concentric circles. Notice that everywhere the equipotential curves meet the field lines, the two types of curves meet at a right angle. This must be the case for an equipotential curve to require no work to move along. Imagine if a field component were directed opposite the motion along an equipotential curve. Then an external agent would have to push harder to maintain the same speed along the curve, and that would mean some outside work was being done. Despite the fields for positive and negative charges being identical, it is possible to determine whether the charge at the center is positive or negative whenever two or more of the equipotential lines are labeled with values. By definition, the potential will be higher (or more positive) close to a positive charge, and lower (or more negative) close to a negative charge.

Below is the equipotential map and some electric field lines for an electric dipole.

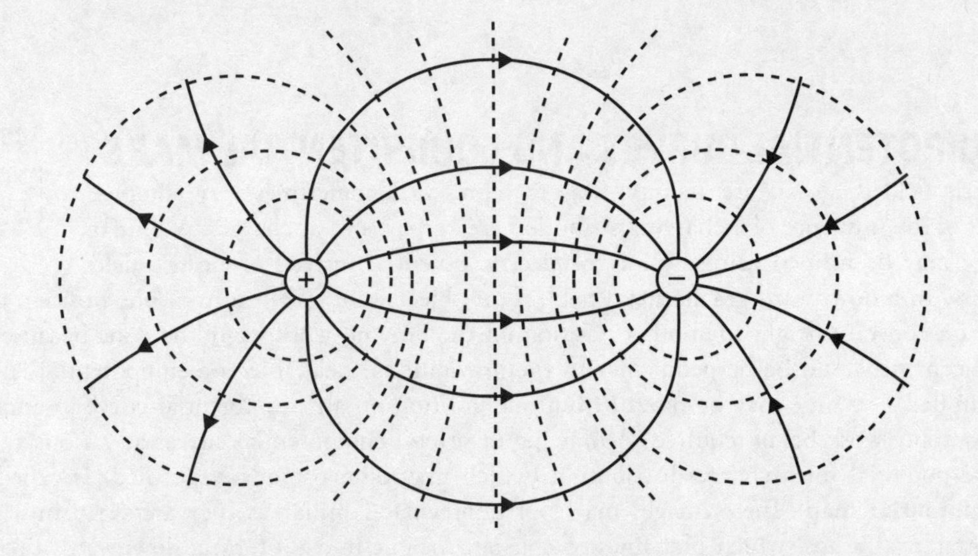

Example 9 Describe the equipotential lines in areas where the electric field is relatively strong compared to the areas where the electric field is relatively weak.

Solution. The electric field is strong where the field lines, which are solid here, are densely packed and weak where they are spread out. Therefore, using the dipole field shown above as a reference, the field is strong between the two charges and weak off to the sides. Comparing the equipotential lines at those two locations, we can conclude that where the equipotential lines are close together, the field is strong, and as the distance between adjacent equipotential lines increases, the field strength decreases. This is consistent with the relationship $|\vec{E}| = \left|\frac{\Delta V}{r}\right|$, as the equipotential map shows lines at a constant electric potential difference ΔV, so the field strength, $|\vec{E}|$, is strong when the distance, r, between the lines is small.

THE ELECTRIC POTENTIAL OF A UNIFORM FIELD

Example 10 Consider a very large, flat plate that contains a uniform surface charge density σ. At points that are not too far from the plate, the electric field is uniform and given by the equation

$$E = \frac{\sigma}{2\varepsilon_0}$$

What is the potential at a point which is a distance d from the sheet (close to the plates), relative to the potential of the sheet itself?

Solution. Let A be a point on the plate and let B be a point a distance d from the sheet. Then

$$V_B - V_A = \frac{-W_{E, A \to B} \text{ on } q}{q}$$

Since the field is constant, the force that a charge q would feel is also constant, and is equal to

$$F_E = qE = q\frac{\sigma}{2\varepsilon_0}$$

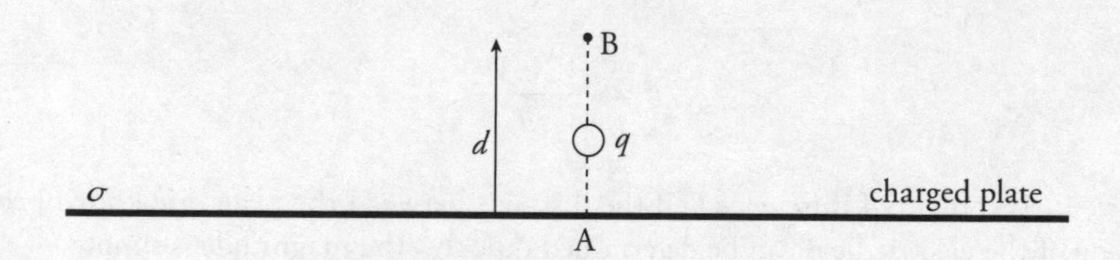

Therefore,

$$W_{E, A \to B} = F_E d = \frac{q\sigma}{2\varepsilon_0}d$$

so applying the definition gives us

$$V_B - V_A = \frac{-W_{E, A \to B}}{q} = -\frac{\sigma}{2\varepsilon_0}d$$

This says that the potential decreases linearly as we move away from the plate.

Example 11 Two large flat plates—one carrying a charge of $+Q$, the other $-Q$—are separated by a distance d. The electric field between the plates, **E**, is uniform. Determine the potential difference between the plates.

Solution. Imagine a positive charge q moving from the positive plate to the negative plate:

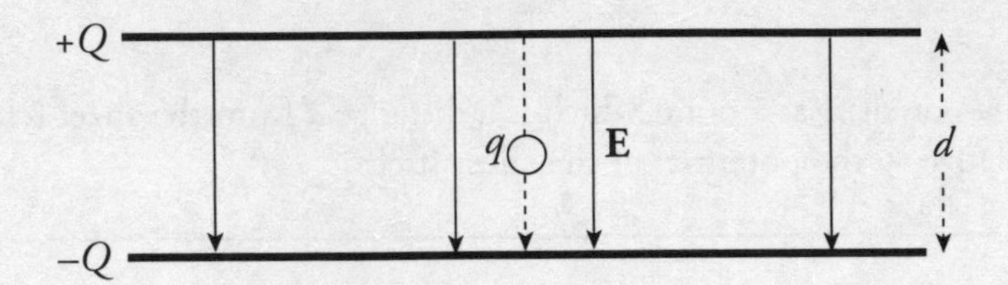

Since the work done by the electric field is

$$W_{E,+\rightarrow-} = F_E d = qEd$$

the potential difference between the plates is

$$V_- - V_+ = \frac{-W_{E,+\rightarrow-}}{q} = \frac{-qEd}{q} = -Ed$$

This tells us that the potential of the positive plate is greater than the potential of the negative plate, by the amount Ed. This equation can also be written as

$$E = -\frac{V_- - V_+}{d}$$

Therefore, if the potential difference and the distance between the plates are known, then the magnitude of the electric field can be determined quickly. The magnitude is simply

Equation Sheet

$$E = \left|\frac{\Delta V}{\Delta r}\right| = \left|\frac{\Delta V}{d}\right|$$

CAPACITORS AND CAPACITANCE

Consider two conductors, separated by some distance, that carry equal but opposite charges, $+Q$ and $-Q$. Such a pair of conductors comprises a system called a **capacitor**. Work must be done to create this separation of charge; as a result, potential energy is stored. Capacitors are basically storage devices for electrical potential energy.

CED Unit 4.3
Resistance and Capacitance

The conductors may have any shape, but the most common conductors are parallel metal plates or sheets. These types of capacitors are called **parallel-plate capacitors**. We'll assume that the distance d between the plates is small compared to the dimensions of the plates since, in this case, the electric field between the plates is uniform. When this is not the case, you must account for the **fringing fields**, which are discussed later in this chapter. The electric field due to *one* such plate, if its surface charge density is $\sigma = Q/A$, is given by the equation $E = \sigma/(2\varepsilon_0)$, with **E** pointing away from the sheet if σ is positive and toward the plate if σ is negative.

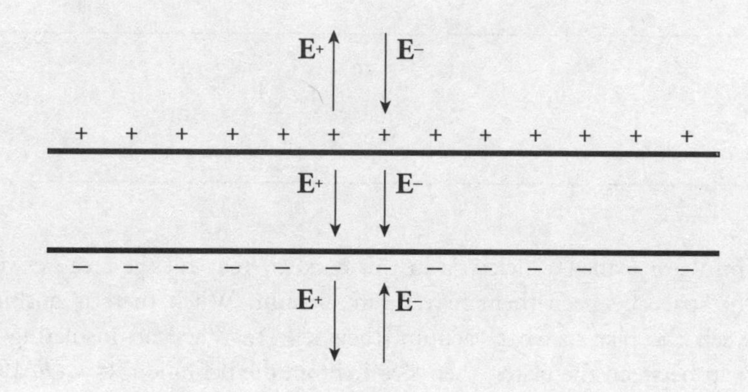

Therefore, with two plates, one with surface charge density $+\sigma$ and the other $-\sigma$, the electric fields combine to give a field that's zero outside the plates and that has the magnitude

$$E_{\text{total}} = \frac{\sigma}{2\varepsilon_0} + \frac{\sigma}{2\varepsilon_0} = \frac{\sigma}{\varepsilon_0}$$

in between.

$$E_{\text{total}} = \frac{Q}{\varepsilon_0 A}$$

Equation Sheet

In Example 11, we learned that the magnitude of the potential difference, ΔV, between the plates satisfies the relationship $\Delta V = Ed$, so combining this with the previous equation, we get

$$E = \frac{\sigma}{\varepsilon_0} \quad \Rightarrow \quad \frac{\Delta V}{d} = \frac{\sigma}{\varepsilon_0} \quad \Rightarrow \quad \frac{\Delta V}{d} = \frac{Q/A}{\varepsilon_0} \quad \Rightarrow \quad \frac{Q}{\Delta V} = \frac{\varepsilon_0 A}{d}$$

Q is the total charge stored on either plate of a capacitor, and ΔV is the potential difference between the plates. The ratio of Q to ΔV, for *any* capacitor, is defined as its **capacitance** (C).

Equation Sheet

$$C = Q/\Delta V$$

The capacitance measures the capacity for holding charge. The greater the capacitance, the more charge can be stored on the plates at a given potential difference. The capacitance of any capacitor depends only on the size, shape, and separation of the conductors, and the "dielectric constant," κ (the Greek letter kappa), that depends on what material is between the plates of the capacitor. For a parallel-plate capacitor, we get

Equation Sheet

$$C = \frac{\kappa \varepsilon_0 A}{d}$$

What Determines Capacitance?
Capacitance does NOT determine the charge on the capacitor, Q, or the potential difference, V, across it. It only shows a relationship between Q and V. The greater the potential difference applied to a capacitor, the greater the amount of charge the capacitor can hold. Think of capacitance as a property of a physical object. Physically, capacitance is determined by three things: the area of the plates, A; the plate separation, d; and the dielectric constant, κ.

An insulator (called a dielectric in this context) may fill the area between the plates, or the space between them may be in vacuum. When there is nothing in the gap between the plates except vacuum, then $\kappa = 1$. When an insulating material is in the gap between the plates, then $\kappa > 1$. From the definition, $C = Q/\Delta V$, the units of C are coulombs per volt. One coulomb per volt is renamed one **farad** (abbreviated F): 1 C/V = 1 F.

Example 12 A 10-nanofarad parallel-plate capacitor holds a charge of magnitude 50 μC on each plate.

(a) What is the potential difference between the plates?

(b) If the plates are separated by vacuum with a distance of 0.2 mm, what is the area of each plate?

Solution.

(a) From the definition, $C = Q/\Delta V$, we find that

$$\Delta V = \frac{Q}{C} = \frac{50 \times 10^{-6}\ \text{C}}{10 \times 10^{-9}\ \text{F}} = 5{,}000\ \text{V}$$

(b) Because $\kappa = 1$, we have the equation $C = \varepsilon_0 \frac{A}{d}$, and can calculate the area, A, of each plate:

$$A = \frac{Cd}{\varepsilon_0} = \frac{(10 \times 10^{-9}\ \text{F})(0.2 \times 10^{-3}\ \text{m})}{8.85 \times 10^{-12}\ \text{C}^2/\text{N} \cdot \text{m}^2} = 0.23\ \text{m}^2$$

ELECTRIC FIELD AND CAPACITORS

For point charges, the electric field created by one or more point source charges varies, depending on the location. For example, as we move farther away from the source charge, the electric field gets weaker. Even if we stay at the same distance from, say, a single source charge, the direction of the field changes as we move around. Therefore, we could never obtain an electric field that was constant in both magnitude and direction throughout some region of space from point-source charges. However, the electric field that is created between the plates of a charged parallel-plate capacitor is constant in both magnitude and direction throughout the region between the plates; in other words, a charged parallel-plate capacitor can create a uniform electric field. The electric field, **E**, always points from the positive plate toward the negative plate, and its magnitude remains the same at every point between the plates, whether we choose a point close to the positive plate, closer to the negative plate, or between them.

"Uniform electric field" is only approximately true. The electric field is only constant everywhere in space for an infinitely large sheet of charge, and such a thing is impossible to construct. However, whenever the observation location is sufficiently close to the plates, or the edges of the plates are sufficiently far away, the constant electric field approximation holds true.

Example 13 The charge on a parallel-plate capacitor is $4 \cdot 10^{-6}$ C. If the distance between the plates is 2 mm and the capacitance is 1 μF, what's the strength of the electric field between the plates?

Solution. Since $C = Q/\Delta V$, we have $\Delta V = Q/C = (4 \cdot 10^{-6}\text{ C})/(10^{-6}\text{ F}) = 4$ V. Now, using the equation $\Delta V = Ed$,

$$E = \Delta V/d = (4\text{ V})/(2 \cdot 10^{-3}\text{ m}) = 2{,}000\text{ V/m}$$

Example 14 The plates of a parallel-plate capacitor are separated by a distance of 2 mm. The device's capacitance is 1 μF. How much charge needs to be transferred from one plate to the other in order to create a uniform electric field whose strength is 10^4 V/m?

Solution. Because $Q = C\Delta V$ and $\Delta V = Ed$, we find that

$$Q = CEd = (1 \cdot 10^{-6}\text{ F})(1 \cdot 10^4\text{ V/m})(2 \cdot 10^{-3}\text{ m}) = 2 \cdot 10^{-5}\text{ C} = 20\ \mu\text{C}$$

Fringing Fields

The mathematical analysis of capacitors up until now assumes that the plates of the capacitor are infinitely large. Because of this approximation, the analysis is very good in the region near the centers of the plates, but is not as precise near the edges of the plates. In the areas beyond the edges of the plates, we can approximate the field as the sum of the capacitor field and a dipole formed by the charges at the edges of the plates.

THE ENERGY STORED IN A CAPACITOR

To figure out the electrical potential energy stored in a capacitor, imagine taking a small amount of negative charge off the positive plate and transferring it to the negative plate. This requires that positive work be done by an external agent, and this is the reason that the capacitor stores energy. If the final charge on the capacitor is Q, then we transferred an amount of charge equal to Q, doing work to move the charge through a potential difference at each stage. If the final potential difference is ΔV, then the average potential difference during the charging process is $\frac{1}{2}\Delta V$; so, using the definition of potential difference, $\Delta V = \Delta U_E / Q$, we can write that for a capacitor $\Delta U_C = Q \cdot \frac{1}{2}\Delta V = \frac{1}{2}Q\Delta V$. At the beginning of the charging process, when there was no charge on the capacitor, we had $U_i = 0$, so $\Delta U_C = U_f - U_i = U_f - 0 = U_f$; therefore, we have

Equation Sheet

$$U_C = \frac{1}{2}Q\Delta V$$

This is the electrical potential energy stored in a capacitor. Because of the definition $C = Q/\Delta V$, the equation for the stored potential energy can be written as

$$U_C = \frac{1}{2}(C\Delta V)\cdot\Delta V = \frac{1}{2}C(\Delta V)^2$$

or

$$U_C = \frac{1}{2}Q\cdot\frac{Q}{C} = \frac{Q^2}{2C}$$

Interestingly, this work is typically done by a battery as the external agent. The battery supplies a total work of $Q\Delta V$, but the capacitor ends up storing only half of this energy. The other 50% of the energy supplied by the battery is dissipated as heat during this charging process.

CAPACITORS AND DIELECTRICS

One method of keeping the plates of a capacitor apart, which is necessary to maintain charge separation and store potential energy, is to insert an insulator (called a **dielectric**) between the plates.

CED Unit 3.5
Electric Permittivity

Insulator?
Why not a conductor? A capacitor stores energy by holding charges apart from one another. If a conductor is placed between the plates, the charges are no longer separated, and the device is no longer a capacitor.

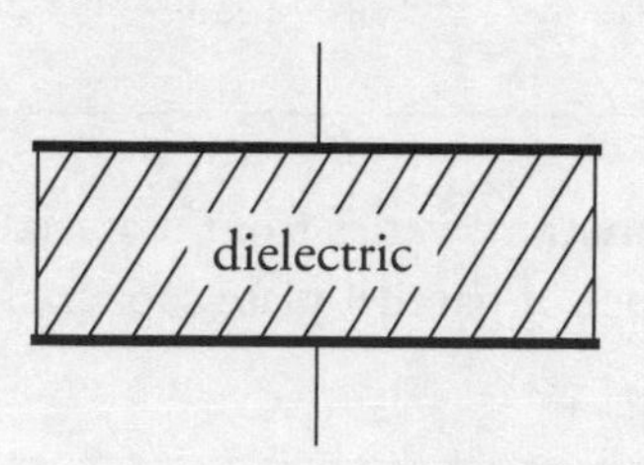

A dielectric always increases the capacitance of a capacitor.

Let's see why this is true. Imagine charging a capacitor to a potential difference of ΔV with charge $+Q$ on one plate and $-Q$ on the other. Now disconnect the capacitor from the charging source and insert a dielectric. What happens? Although the dielectric is not a conductor, the electric field that exists between the plates causes the molecules within the dielectric material to polarize; there is more electron density on the side of the molecule near the positive plate.

The effect of this is to form a layer of negative charge along the top surface of the dielectric and a layer of positive charge along the bottom surface; this separation of charge induces its own electric field ($\mathbf{E}_i$), within the dielectric, which opposes the original electric field, **E**, within the capacitor.

So the overall electric field has been reduced from its previous value: $\mathbf{E}_{total} = \mathbf{E} + \mathbf{E}_i$, and $E_{total} = E - E_i$. Let's say that the electric field has been reduced by a factor of κ from its original value as follows:

$$E_{\text{with dielectric}} = E_{\text{without dielectric}} - E_i = \frac{E}{\kappa}$$

Since $\Delta V = Ed$ for a parallel-plate capacitor, we see that ΔV must have decreased by a factor of κ. But $C = \frac{Q}{\Delta V}$, so if ΔV decreases by a factor of κ, then C increases by a factor of κ:

$$C_{\text{with dielectric}} = \kappa C_{\text{without dielectric}}$$

The value of κ, called the **dielectric constant**, varies from material to material, but it's always greater than 1. In general, the capacitance of parallel-plate capacitors is

Equation Sheet

$$C = \kappa \varepsilon_0 \frac{A}{d}$$

Chapter 6 Review Questions

Solutions can be found in Chapter 11.

Section I: Multiple Choice

1. An experiment is conducted and data is gathered for the electric potential V at various positions r away from a uniformly charged sphere. All measurements are taken outside of the sphere. Which of the following graphs yields a straight line?

(A) V as a function of $\frac{1}{r^2}$

(B) V as a function of $\frac{1}{r}$

(C) V as a function of r

(D) V as a function of r^2

2. Below is shown a section near the center of a parallel-plate capacitor. There are 4 labeled positions between the plates shown as A, B, C, and D. Relative to A, which point has the largest potential difference and why?

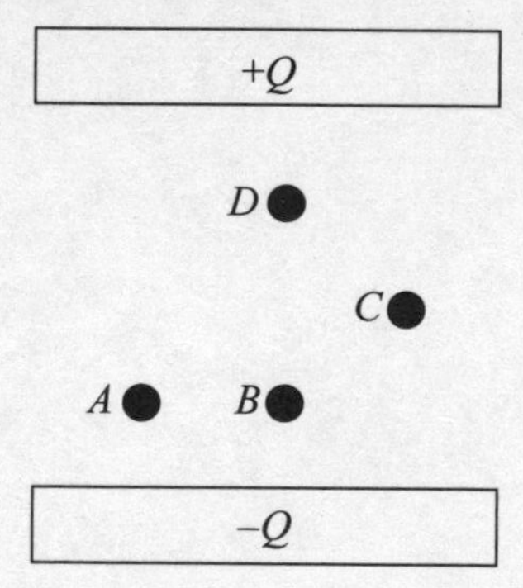

(A) Point A because the potential difference is infinite when the position between points is 0 m
(B) Point B because it is the same distance from the $-Q$ plate as A
(C) Point C because it is farther in distance from Point A
(D) Point D because it is closest to the $+Q$ Plate

3. Negative charges are accelerated by electric fields toward points

(A) of lower electric potential
(B) of higher electric potential
(C) where the electric field is weaker
(D) where the electric field is stronger

4. A charge q experiences a displacement within an electric field from Position A to Position B. The change in the electrical potential energy is ΔU_E, and the work done by the electric field during this displacement is W_E. Then

(A) $V_A - V_B = qW_E$
(B) $V_B - V_A = qW_E$
(C) $V_A - V_B = \Delta U_E/q$
(D) $V_B - V_A = \Delta U_E/q$

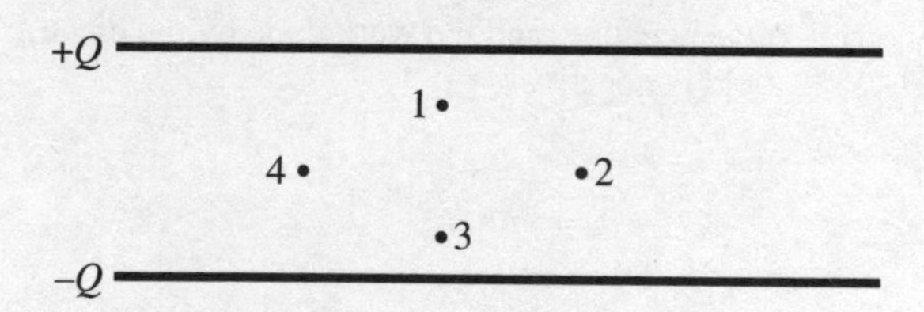

5. Which points in this uniform electric field (between the plates of the capacitor) shown above lie on the same equipotential?

(A) 1 and 3 only
(B) 2 and 4 only
(C) None lie on the same equipotential.
(D) 1, 2, 3, and 4 all lie on the same equipotential since the electric field is uniform.

6. A charge Q creates an electric field through which a second charge q moves, as shown below. q is initially at point B and is moved to point A. The potential from Q at position A is $V_A = 100\ V$ and at B is $V_B = 200\ V$. The charge on q is negative. What is the sign of Q and the sign of the work done by the electric field of Q as q is moved from B to A?

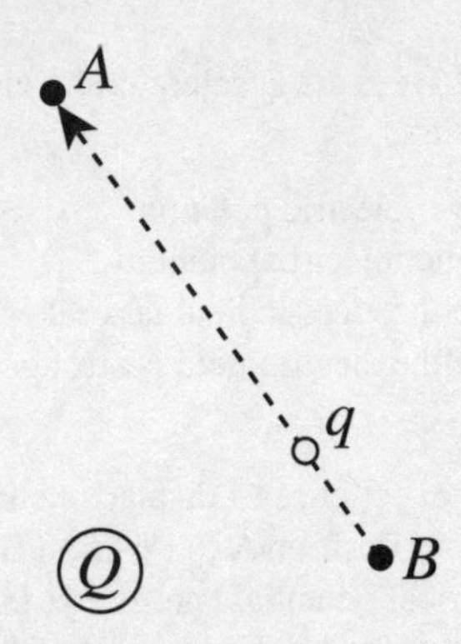

(A) Q is positive, and the work done by the electric field is positive.
(B) Q is positive, and the work done by the electric field is negative.
(C) Q is negative, and the work done by the electric field is positive.
(D) Q is negative, and the work done by the electric field is negative.

Section II: Free Response

1. In the figure shown below, four charges, each of magnitude Q, are situated at the corners of a square with side lengths s. The two charges on the top of the square are positively charged, while the two on the bottom of the square are negatively charged.

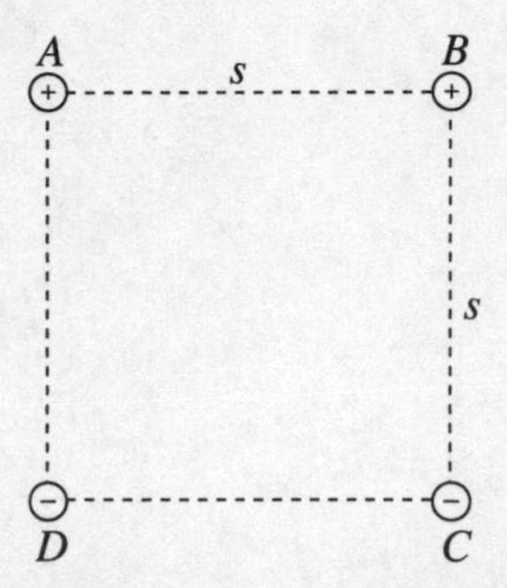

(A) These charges were assembled in order by first bringing in charge A, then bringing in charge B, then bringing in C, and finally bringing in charge D. Rank the amount of energy in the charge distribution in the presence of only charge A, only charges A and B, only charges A, B, and C, and in the presence of all four charges. Negative numbers should be taken as smaller than positive numbers. Justify your answer.

Greatest (most positive) ____ ____ ____ ____ Least (most negative)

(B) Show that the potential at the exact center of the square is 0 V by calculating the potential from each charge at that location.

(C) Sketch (on the diagram) the portion of the equipotential surface that lies in the plane of the figure and passes through the center of the square.

(D) As shown below, solid conducting bars are placed to connect points A to B and also to connect points C to D. The charge is allowed to distribute over these conductors. Sketch the electric field at each of the dots. Explain why the field is constant at the dots in the center but not at the dots on the edges.

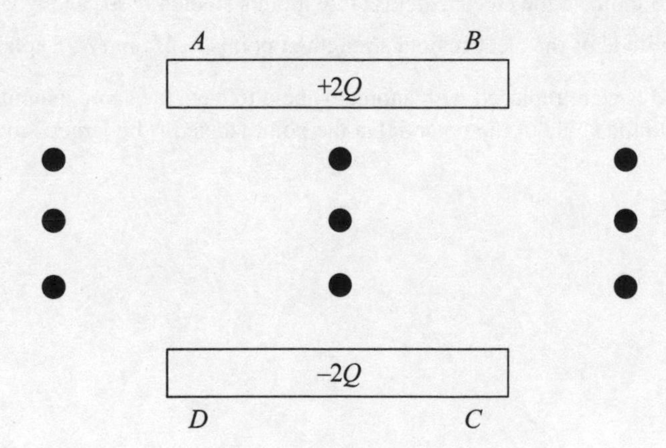

2. The image below shows the isolines of electric potential surrounding two charged spheres, labeled L and R. The spheres carry opposite charges and the potential difference between adjacent pairs of lines is $\Delta V = 15$ V. The isolines with electric potentials of –25 V and –10 V are indicated.

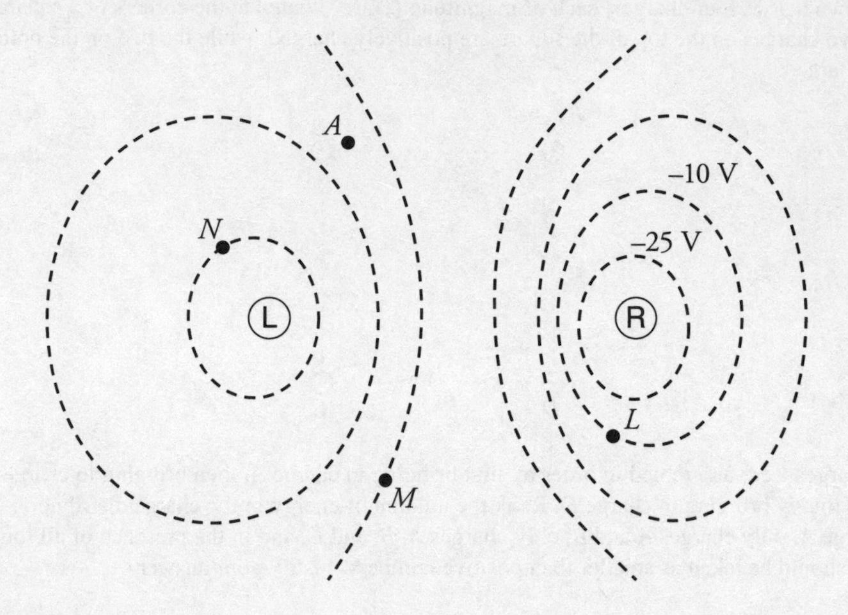

(A) Sketch in the line of potential 0 V on the drawing.

(B) Approximately what is the value of the potential at the point labeled *A*?

(C) Answer the following:

i. Which sphere, L or R, carries a negative charge? Explain your answer.

ii. Which sphere, L or R, carries a greater magnitude of charge? Explain how you know.

(D) Answer the following:

i. Draw arrows to indicate the electric fields at the points labeled *L*, *M*, and *N*.

ii. Rank the magnitude of the electric field strength at points *L*, *M*, and *N*. Explain your answer.

(E) If the sphere labeled L were replaced with another sphere to have the same magnitude of charge but of the opposite sign, would the value of the potential at the point labeled *N* be larger, smaller, or stay the same? Justify your answer.

3. A solid conducting sphere of radius a carries an excess charge of Q.

   (A) Determine the electric field magnitude, $E(r)$, as a function of r, the distance from the sphere's center.

   (B) Determine the potential, $V(r)$, as a function of r. Take the zero of potential at $r = \infty$.

   (C) On the diagrams below, sketch $E(r)$ and $V(r)$. (Cover at least the range $0 < r < 2a$.)

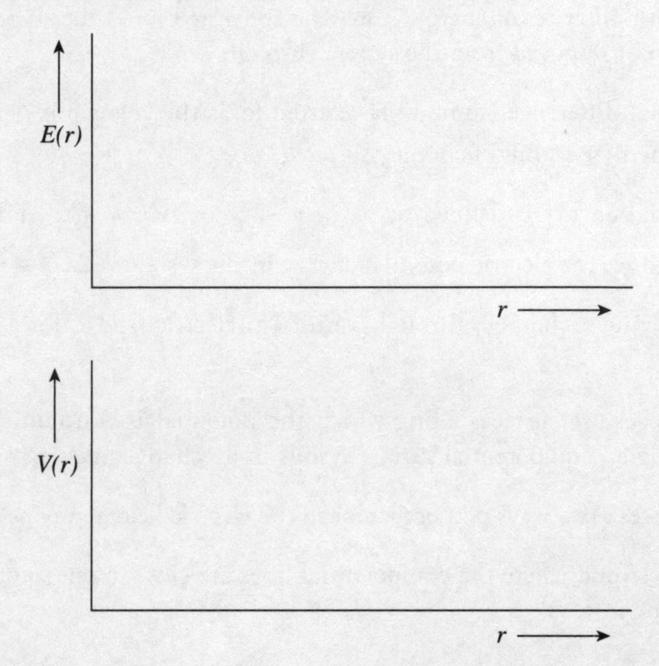

Summary

- Electric potential energy is a type of energy arising from the interaction of a charge and an external electric field. As with all types of energy, it may be converted into other types of energy or may be added to the system or removed from the system through work.
- The electric potential difference (commonly referred to as the voltage) is defined as the change in electric potential energy per unit of charge: $\Delta V = \Delta U_E / q$.
- The electric potential energy is defined by $\Delta U_E = -W_E$ or $\Delta U_E = q\Delta V$. For a pair of interacting point charges q_1 and q_2, the electric potential energy in the system is $\Delta U_E = \frac{1}{4\pi\varepsilon_0}\frac{q_1 q_2}{r}$.
- The work done moving a charge q through a uniform electric field E for a distance d is given by $W = qEd$.
- Equipotential surfaces are surfaces along which the potential is constant. Moving a charge at a constant speed along an equipotential surface results in no change in energy of the charge.
- Equipotential surfaces are always perpendicular to the electric field at any point on the surface.
- The electric field is strong where the equipotential lines are close to one another and weaker where the lines are farther apart.
- Capacitors are devices that store electric potential energy in electric fields. Capacitance is given by $C = \frac{Q}{\Delta V}$. For parallel-plate capacitors, the capacitance is $C = \kappa\varepsilon_0 \frac{A}{d}$.
- A parallel-plate capacitor has a uniform electric field between the plates for regions close to the plates and far from the outer edges of the capacitor.
- The electrical energy stored in a capacitor is given by $U_C = \frac{1}{2}Q\Delta V = \frac{1}{2}C(\Delta V)^2$.
- When a capacitor is filled with a dielectric, its capacitance increases from the capacitance it had when it had a vacuum between its plates.

Chapter 7
Electric Circuits

This chapter deals with CED Unit 4: Electric Circuits.

INTRODUCTION

Electric circuits are comprised of an energy source (typically a battery or a wall outlet), one or more conducting materials (such as wires), and circuit components such as resistors and capacitors. The broad topic of electric circuits is vast, but for the AP Physics 2 Exam, we need to focus on only a small set of components in relatively simple arrangements. We will focus on direct current (DC) voltage supplies in resistor and capacitor (RC) circuits in steady state.

The most basic electric circuit is a collection of conductors, such as wire, and a source of electrical energy, such as a battery. Connecting the components together so that there is no insulating material (such as an air gap) between the components completes the circuit and allows for a continuous flow of charge, known as **current**, in the circuit. In this chapter, we will analyze circuits constructed of one or more batteries, wires, switches, resistors, and capacitors. Other circuit components, such as diodes or inductors, are beyond the scope of AP Physics 2.

CED Unit 4.1
Definition and Conservation of Electric Charge

BATTERIES AND VOLTAGE

A battery is a device that maintains an electric potential difference between the two terminals. When a wire is connected between the two terminals of the battery, there is a potential difference between the ends of the wire. As we saw previously, a potential difference in a conducting material causes an electric field. Previously, when we were looking at electrostatics, the potential difference was always 0 V because we require, as a condition of electrostatics, that the electric field within a wire be 0 N/C. In circuits, when there is a battery present to *maintain* a potential difference, there will be a nonzero electric field within the conductor. As a result, there will be a flow of charge. We call this flow of charge electric current.

A Note on Convention for this Chapter

In the previous chapter, we referred to a change in potential as a voltage, ΔV. In this chapter, we will also use the standard convention, V. Make note of this as you move forward to avoid confusion.

A common misconception is that the battery is the *supply* of the charges that create the current. As we have already seen, conducting materials such as metals contain vast numbers of electrons. The battery supplies the energy needed to arrange the electrons in the wire into a configuration where the current continuously flows. However, the electrons present within the wire constitute the current. The electrons present in the wire also create the electric field, which is the source of the current.

Voltage Is like Electric Pressure
Voltage is a power source that allows for the movement of charges. It is like a region in the atmosphere of high pressure compared to a region of low pressure, which will result in a wind.

Any particular conduction electron will move through the circuit with a **drift velocity** of some millimeters per second, but the influence of the "electricity" moves through the circuit at the speed of light. As we will see, if the battery were the source of the electric charge, not only would it take several hours for light bulbs to turn on, because of the slow drift velocity of electrons, but circuits with capacitors would not function at all.

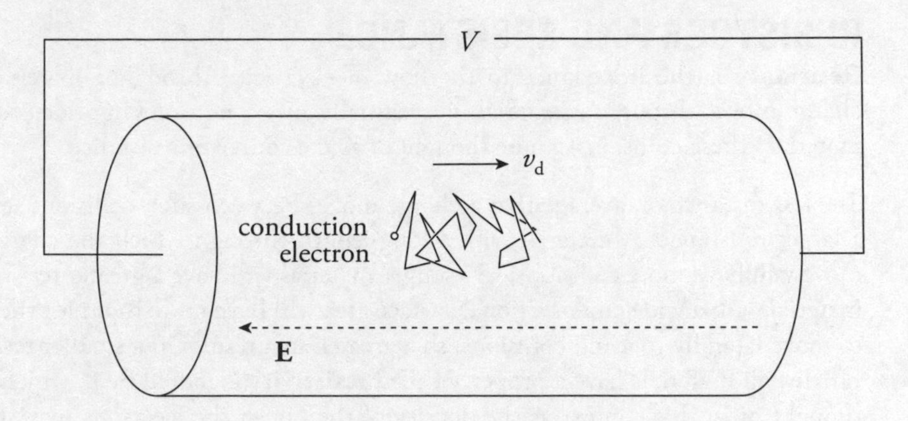

In a circuit diagram, a battery will be represented by one of these two diagrams:

The diagram to the left shows a *single-cell* battery and the one on the right shows a battery with multiple cells. The longer line in the battery diagram represents the positive (higher potential) terminal, and the shorter line is the negative (lower potential) terminal.

The voltage source is a power source. Whether it is a wall plug, a battery, a capacitor, or an electric generator, it supplies energy to the circuit. When two points in a circuit have different voltages, charges will flow from one point to the other. As long as the voltage difference is maintained, the flow of charge will continue.

> A voltage difference between two points in a conducting material causes charge to flow.

Voltage Has Many Names but One Meaning
When scientists were discovering the properties of electricity, it was not clear that all the concepts they were analyzing were the same phenomenon. In different contexts, voltage is referred to as potential difference, voltage drop, voltage, or electromotive force.

The voltage supplied by an ideal battery is often, for historical reasons, referred to as EMF (electromotive force, ε). Whether referred to as the potential difference, voltage drop, voltage, or EMF, the voltage is measured in units of **volts** (abbreviated V).

A battery never changes which pole is at high potential and which pole is at low potential. As a result, the battery always causes electricity to flow in the same direction through the circuit. This is called a **direct current** circuit.

CED Unit 4.2
Resistivity and Resistance

Resistance vs. Resistivity
Resistance and resistivity are NOT the same. Resistance is the extent to which a resistor impedes the passage of electric current. Resistivity is a property of what molecules a material consists of and how the molecules that constitute a material are bound together. The units of resistivity are different as well.

Equation Sheet

Resistors Are like Drinking Straws
A very narrow straw will be harder to drink through than a wide straw. A very long straw will be harder to drink through than a short one.

RESISTORS AND RESISTANCE

Resistance is the impedance to the flow of electricity through a material. As a charge moves through a material, it eventually hits a non-moving nucleus in the material. These collisions can be thought of as the source of resistance.

The less distance on average that a charge moves between such collisions results in a larger resistance. A material with a long length through which the charge must move will have more collisions, so a longer material will have a greater resistance. A material with a wide cross-sectional surface area will have more room for the charge to move laterally to avoid collisions, so a greater area results in a smaller resistance. Finally, all materials have a property called **resistivity**, denoted by ρ, which can be thought of in this context as the density of the nuclei the electrons may strike. A greater density will cause more collisions. The resistance, R, of a material depends on its resistivity and its geometrical configuration, in particular, its length, l, and its cross-sectional area, A.

$$R = \frac{\rho \ell}{A}$$

This equation applies only to shapes with constant cross-sectional area, such as cylinders or rectangular prisms, and will not apply to those with a varying cross-sectional area, such as spheres or cones.

Resistance is measured in **ohms** (Ω, *omega*) and resistivity is measured in **ohm-meters** (Ω·m).

In a circuit diagram, a resistor will be represented by

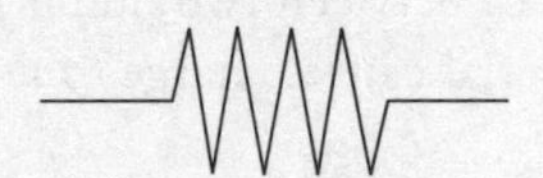

A resistor is awkwardly named because all wires are technically "resistors." Materials with a very low resistivity are referred to as conductors. These include aluminum, silver, and copper, which all have resistivity on the order of 1×10^{-8} Ω·m. Electrical components called "resistors" are made of materials like carbon graphite, which have resistivity on the order of 10×10^{-5} Ω·m, approximately 10,000 times higher than the materials that comprise wire. Materials with a high resistivity, such as glass with resistivity of 10×10^{10} Ω·m, are referred to as insulators.

Example 1 The resistivity of copper is about 1.7×10^{-8} Ω·m. What is the resistance of a 5 m long 16 gauge light-duty extension cord whose diameter is 1.3 mm?

Solution.

$$R = \frac{\rho \ell}{A}$$

To find the cross-sectional area, we use $A = \pi r^2 = (3.14)(1.3 \times 10^{-3}\ \text{m}/2)^2$.

$$R = \frac{(1.7 \times 10^{-8}\ \Omega\cdot\text{m})(5\text{m})}{((3.14)(1.3 \times 10^{-3}\ \text{m}/2)^2)} = 0.064\ \Omega$$

Example 2 A wire with cross-sectional area 1.33 mm^2 has an unknown resistivity.

(a) To find the resistivity, you have a spool of this wire, a meter stick, wire cutters, and a device to measure resistance (such a device is called an Ohmmeter). Briefly describe what data you will collect and how you will determine the resistivity of the wire.

(b) Another student gathers data and makes the following plot. What is the resistivity of the wire in her experiment?

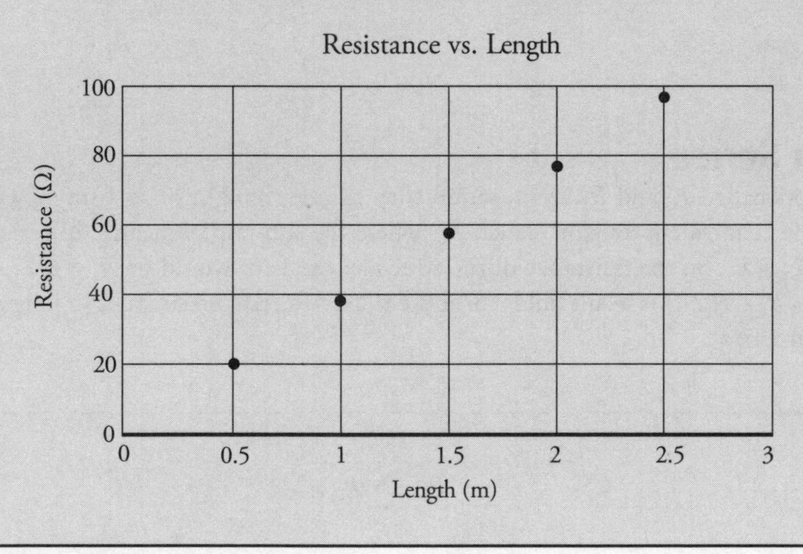

Solution.

(a) Cut a length of the wire with wire cutters. Measure the length of the wire, L, with the meterstick. Measure the resistance of the wire with the Ohmmeter. The resistivity can be found using $\rho = AR/L$.

(b) From the graph, the slope is $(98\ \Omega - 20\ \Omega)/(2.5\ \text{m} - 0.5\ \text{m}) = 39\ \Omega\cdot\text{m}$, which is R/L. Since we determined in (a) that $\rho = AR/L$, we can find the resistivity as $\rho = (1.33 \times 10^{-6}\ \text{m}^2)(39\ \Omega\cdot\text{m}) = 5.19 \times 10^{-5}\ \Omega\cdot\text{m}$.

Multiple Resistors Are like Multiple Straws
If one very narrow straw will be hard to drink through, two very narrow straws side by side will act like a wider straw, decreasing the overall resistance.

COMBINING RESISTORS AND EQUIVALENT RESISTANCE

To deal with arrangements that have multiple resistors, combinations of resistors can be considered. Resistors may be arranged in two ways: one after the other or side by side. In an arrangement in which the resistors are one after the other, we say the resistors are "in series." In an arrangement in which the resistors are side by side, we say the resistors are "in parallel." When two or more resistors are combined mathematically, the resulting resistance is called *equivalent* resistance.

Two resistors R_1 and R_2 arranged in series

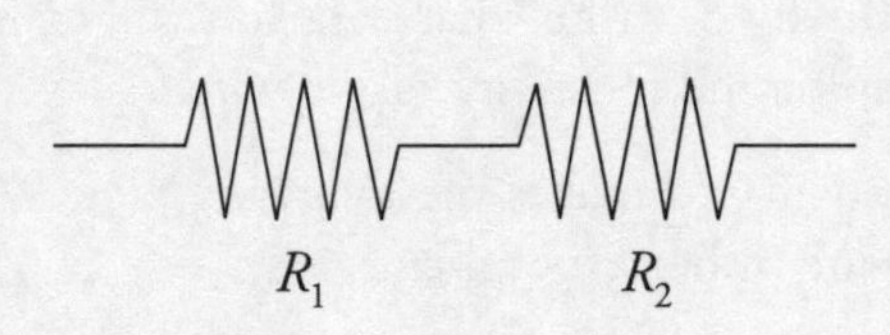

Two resistors R_1 and R_2 arranged in parallel

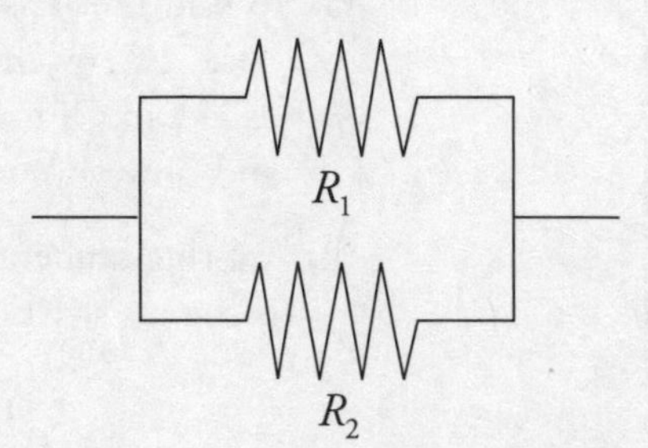

Resistors in Series

When two resistors called R_1 and R_2 are in series, they can be thought of as a single wire of a greater length. The "equivalent resistor" called R_S (where the subscript s means "in series") has a length of $L_{eq} = L_1 + L_2$. So the resistance of the equivalent resistor would be $R_S = \rho(L_1 + L_2)/A = \rho L_1/A + \rho L_2/A = R_1 + R_2$. This result holds for more than two resistors as well, so long as they are all arranged in series.

Equation Sheet

$$R_s = \sum_i R_i$$

Resistors arranged in series result in an overall resistance that is greater than the resistance of any of the individual resistors in the arrangement.

One Over R_P
The equation for equivalent resistance in parallel has R_P in the denominator. Don't forget to take the reciprocal as the final step in calculating the equivalent resistance.

Resistors in Parallel

When the resistors are in parallel, they behave like a wire with a greater cross-sectional area. It would be equivalent to replace those two resistors with a single "equivalent resistor" called R_{eq} of area $A_p = A_1 + A_2$. Then the resistance of the equivalent resistor would be

$$R_p = \rho L/(A_1 + A_2) \rightarrow \frac{1}{R_p} = \frac{(A_1 + A_2)}{\rho L} = \frac{A_1}{\rho L} + \frac{A_2}{\rho L} = \frac{1}{R_1} + \frac{1}{R_2}$$

This result holds for more than two resistors as well, so long as they are all arranged in parallel.

$$\frac{1}{R_p} = \sum_i \frac{1}{R_i}$$

Equation Sheet

Resistors arranged in parallel result in an overall resistance that is less than the resistance of any of the resistors.

When dealing with arrangements of three or more resistors, it is possible for the arrangement to be a mixture of series and parallel.

(a) A pair of resistors R_A and R_B is in series. That combination is in parallel with a third resistor, R_C.

(b) A resistor R_D is in series with a pair of resistors R_E and R_F, which are in parallel.

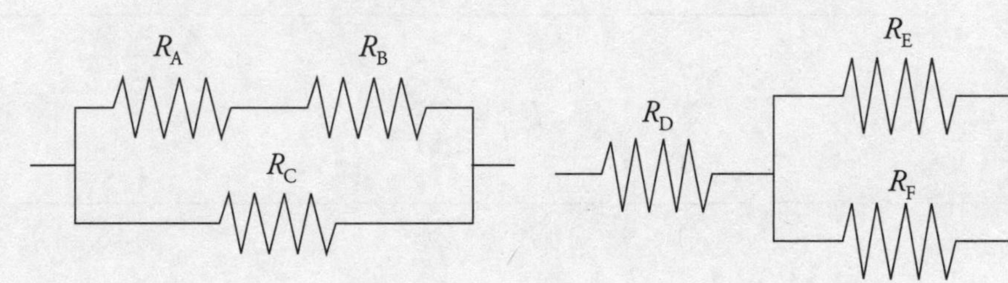

When resistors appear in a mixture of parallel and series, the equivalent resistance is calculated in multiple steps. First, the equivalent resistance of the part of the circuit that is purely series or parallel must be found. Then that resistance is used for further calculations. This process may need to be repeated several times in circuits with many resistors.

Example 3 Calculate the total equivalent resistance of the two arrangements in the figure above.

A Shortcut for Pairs of Parallel Resistors
The equivalent resistance of a pair of resistors in parallel is the product of the resistances over their sum. This shortcut does **not** work for combinations of more than two resistors.

Solution.

(a) The figure labeled (a) shows two resistors in series, R_A and R_B, which are in parallel with the third resistor R_C. The equivalent resistance of the arrangement is found by first combining the series resistors into $R_{eq,AB} = R_A + R_B$ and finally combining that with R_C as a parallel arrangement to get

$$\frac{1}{R_{eq,total}} = \frac{1}{R_A + R_B} + \frac{1}{R_C} \text{ so that } R_{eq,total} = \frac{R_C(R_A + R_B)}{R_A + R_B + R_C}.$$

(b) The figure labeled (b) shows a resistor, R_D, in series with a pair of parallel resistors, R_E and R_F. The equivalent resistance of the arrangement is found by first combining the parallel resistors into $\frac{1}{R_{eq,EF}} = \frac{1}{R_E} + \frac{1}{R_F}$ so that $R_{eq,EF} = \frac{R_E R_F}{R_E + R_F}$. Then that is combined with R_D as a series arrangement to get $R_{eq,total} = R_D + \frac{R_E R_F}{R_E + R_F}$.

Wind as an Analogy to Current

Think of current as the wind. Blowing wind is comprised of many air molecules moving randomly and colliding with one another and with other things the air molecules encounter, but there is an overall flow of air in a particular direction. When talking about wind, the rate of motion of air molecules is considered. When talking about current, the rate of motion of electric charge is under consideration.

CURRENT

Establishing a potential difference across a wire creates an electric field in the wire, and as a result of that electric field, charges move. We can measure how much charge passes any particular point in the wire in a specified unit of time. When an amount of charge of magnitude ΔQ crosses an imaginary plane in a time interval Δt, then the **average current** is

Equation Sheet

$$I_{avg} = \frac{\Delta Q}{\Delta t}$$

Benjamin Franklin

Conventional direction of current is "wrong" because Benjamin Franklin did not have enough information to determine whether charge carriers were positive or negative charges (or both moving in opposite directions) when he was investigating electrostatics. He simply picked a direction based on the flow of positive charges.

Because current is charge per unit time, it's expressed in coulombs per second. One coulomb per second is an **ampere** (abbreviated **A**). Instead of calling the unit an ampere, it is often simply called an amp.

We know that the charge carriers that constitute the current within a metal are electrons. However, the direction of the current is taken to be the direction that positive charge carriers would move. This is referred to as *conventional current*, in contrast with *electron current*, which is the flow of electrons. On the AP Physics 2 Exam, assume all current flow is conventional current.

MEASURING CURRENT AND VOLTAGE IN A CIRCUIT

When a circuit is established, each electrical component in the circuit will have some electricity flowing through it and will have some energy being used up by it.

The current flowing past any point in a circuit is measured using an ammeter. An ammeter is a device with a very low resistance (essentially $R = 0\ \Omega$). To measure the current flowing through a resistor of resistance R, the ammeter is placed in series and the equivalent resistance of the resistor and ammeter is $R_S = R + 0 = R$. Because of the negligible resistance of the ammeter, the resistance of the pair is just R. The fact that this arrangement does not change the equivalent resistance means the ammeter reading is the same as the reading through the resistor when the ammeter is not present. Because current flows through the resistor, the reading will be the same no matter on which side of the resistor the meter is placed.

In a circuit diagram, an ammeter will be represented by

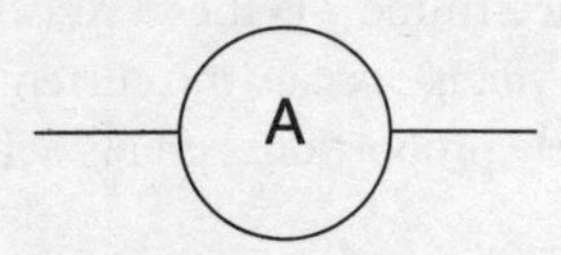

Current Is Not "Used Up"
Current flows through the devices in a circuit. The same amount of current flows into, and then back out of, each resistor.

The electrical energy is not measured directly, but instead the electric potential difference—commonly referred to as voltage or voltage drop—is measured using a voltmeter. A voltmeter has an extremely high resistance (essentially $R = \infty$). One lead is attached to each of the two points where the voltage difference is desired. To measure the voltage across a resistor of resistance R, the voltmeter must be in parallel with the resistor. Because they are in parallel, the equivalent resistance of the pair is found from $\frac{1}{R_P} = \frac{1}{R} + \frac{1}{\infty}$, so $R_P = R$. Because of the divergent resistance of the voltmeter, the resistance of the pair is just R. The fact that this arrangement does not change the equivalent resistance means the voltmeter reading is the same as the reading across the resistor when the meter is not present. In a circuit diagram, a voltmeter will be represented by

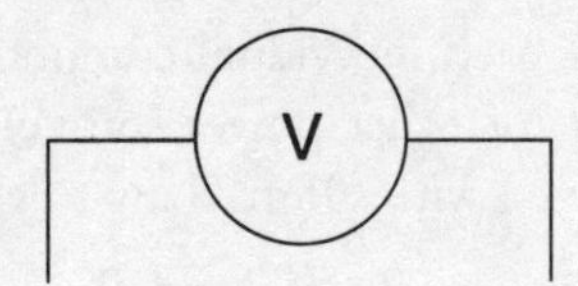

OHM'S LAW

For many resistors, a plot of the current flowing through the resistor for various voltage drops results in a graph showing the relationship between ΔV and I is directly proportional. Moreover, the slope of such a graph is equal to the resistance R of the resistor. Whenever this is the case, the device is said to be an "Ohmic resistor." Ohm's Law states that the voltage drop across an Ohmic resistor is directly proportional to the current flowing through the resistor and that the proportionality constant is the resistance. Because the resistance of a component is determined by the geometric properties of its construction and the voltage supplied is determined by the power source, Ohm's Law tells the amount of current that will flow through the device and is written as

$$I = \frac{\Delta V}{R}$$

Equation Sheet

Not every device obeys Ohm's Law. However, unless a question explicitly states a device is non-Ohmic or asks to explain whether a device is Ohmic or not, every resistor you encounter in AP Physics 2 will be Ohmic. You can determine whether a resistive element in a circuit obeys Ohm's Law or not by making a plot of the voltage versus the current. A nonlinear plot indicates that the resistor is non-Ohmic, while a directly proportional graph will indicate that the resistor is Ohmic.

Ohm's Law can be applied to each individual resistor in a circuit as well as to the overall circuit. Applying Ohm's Law to an individual resistor yields the voltage drop across that resistor and the current flowing through the resistor. Applied to the entire circuit, Ohm's Law tells the voltage drop across the battery and the current that flows through the battery.

POWER DISSIPATION

The power dissipated by a circuit component is given by the product of the current through the component and the voltage drop across it.

Equation Sheet

$$P = I\Delta V$$

Heat and Power
Power in a circuit is very closely associated with the heat it gives off. If we were to touch an incandescent light bulb, we would notice it becomes hot. The brighter the lightbulb is, the greater the heat it gives off.

The power dissipated by each individual component must sum to equal the power supplied by the battery (or other power source). For Ohmic resistors, the power equation may be combined with Ohm's Law to yield

$$P = I^2R \text{ and } P = V^2/R$$

CED Unit 4.4-5
Kirchhoff's Loop Rule, Kirchhoff's Junction Rule, and the Conservation of Electric Charge

KIRCHHOFF'S RULES

In addition to Ohm's Law, Kirchhoff's rules are used when analyzing circuits. These two laws are statements of conservation laws applied to circuit analysis. The **Loop Rule** states that the voltage drop across *any complete loop* in a circuit is 0 V. This statement follows from conservation of energy when applied to circuits. A corollary to the Loop Rule is that any pair of resistors in parallel will have identical voltage drops across them. The **Junction Rule** states that the sum of all current flowing into any junction is equal to the current flowing out of the junction. This statement follows from conservation of charge. A corollary of the Junction Rule is that for two resistors in series, the current must be the same through each because there is no junction.

A Very Simple Circuit

A very simple circuit contains one battery with voltage V and one resistor of resistance R connected by wires.

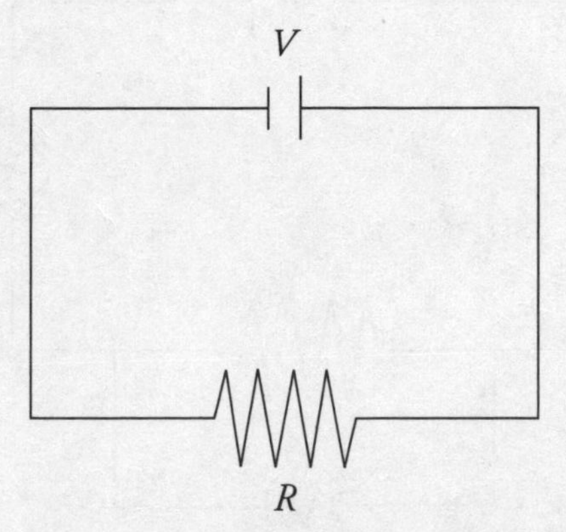

Let's consider how to apply Kirchhoff's Loop Rule to this circuit. First, assign the direction of current flow for all segments of the circuit. There are no junctions in this circuit, so one current flows throughout, a current I flows from the positive terminal of the battery to the negative terminal, clockwise around the circuit. Then, choose a loop. In this simple circuit, there is only a single loop. Next, choose a starting point and a direction to traverse the loop. It is generally advisable to choose the direction of current, so start at point S and go clockwise.

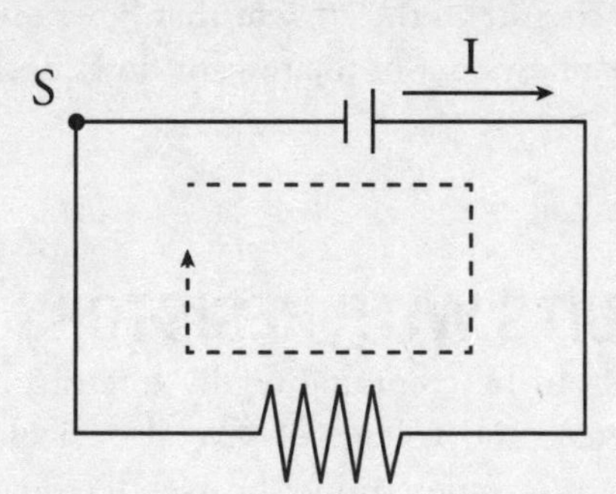

As we traverse the loop clockwise from point S, the first change in electric potential we experience is the increase in electric potential $+V$ from the battery. The next change in potential in the loop is a voltage drop across the resistor, $-IR$, after which we're back at starting point S. The loop must have an overall change in voltage of zero, so $V + (-IR) = 0$, so $V = IR$. Note that if the loop went opposite the direction of current, the battery would correspond to a voltage drop and the resistor a voltage gain—this is why it's important to assign the direction of current first. Add a pair of voltmeters and a pair of ammeters to the circuit in their proper positions to measure the current and voltage of the battery and the resistor.

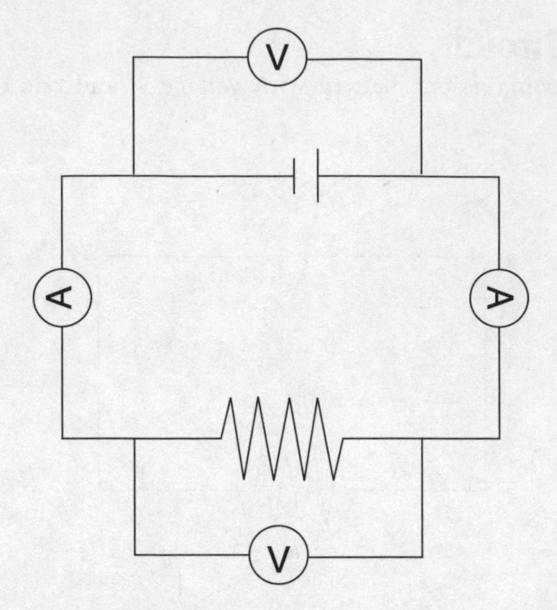

The reading in the top voltmeter is the voltage drop across the battery. The reading in the lower voltmeter is the voltage across the resistor. As we discovered from the Loop Rule, the readings in these voltmeters must be the same; the voltage supplied by the battery is equal to the voltage drop across the resistor. As there are no junctions in the circuit (no current flows throw the voltmeters due to their nearly infinite resistance), each ammeter will also read an equal value for the current. In this simple circuit, all the current that flows out of the battery flows into the resistor, and all the current that flows out of the resistor flows back into the battery.

ANALYSIS OF CIRCUITS WITH RESISTORS

Circuits with several resistors can be computationally intensive. Problems frequently involve iteratively solving for the current and voltage of several devices in several steps. The typical approach to quantitative analysis is to determine the overall equivalent resistance of the circuit. Then given the battery voltage, the current supported by the battery can be calculated, as well as the power supplied by the battery. From the Junction Rule, Ohm's Law is then applied to any resistors in series with the battery, as the current supported by the battery flows through each of those resistors. The iterative process then requires using the Loop Rule to determine the voltage drop across the paths that contain parallel branches.

Example 4 The circuit below shows a 12 V battery connected to three resistors. The resistances include $R_1 = 4\ \Omega$, $R_2 = 3\ \Omega$, and $R_3 = 6\ \Omega$. Determine the current supported by the battery and the voltage drop across each resistor.

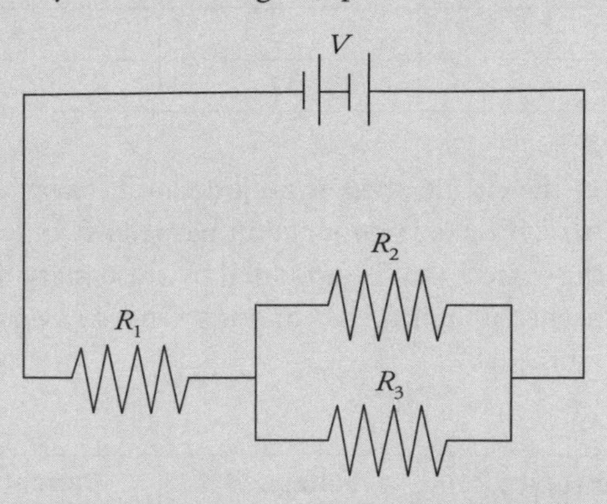

Solution. To begin, it is helpful to make a chart of the resistance, voltage, current, and power for each device. We need a row for each resistor and an additional row for the entire circuit. Any time that any two of the columns are known in a row, Ohm's Law and the power equation can be used to solve for the other two columns. We can immediately fill out any known resistance, current, voltage, or power values. Here, we have three resistances and one voltage as given.

	Resistance (Ω)	Voltage (V)	Current (A)	Power (W)
R_1	4			
R_2	3			
R_3	6			
Entire Circuit		12		

Next, we can use the concept of equivalent resistance to get the resistance of the entire circuit, which will give us two values in the bottom row.

To begin, we combine R_2 and R_3 to get an equivalent resistor $R_{2,3}$. Since R_2 and R_3 are in parallel,

$$\frac{1}{R_{2,3}} = \frac{1}{R_2} + \frac{1}{R_3} = \frac{1}{3} + \frac{1}{6} = \frac{2}{6} + \frac{1}{6} = \frac{3}{6} = \frac{1}{2}$$

$$R_{2,3} = 2\ \Omega$$

We now have R_1 in series with $R_{2,3}$. The equivalent resistance $R_{1,2,3} = R_1 + R_{2,3} = 4 + 2 = 6\ \Omega$. The total resistance of the circuit is $R_{total} = 6\ \Omega$. We add that to the bottom row of the table. Then, using Ohm's Law and the power equation, we can complete the bottom row of the table.

	Resistance (Ω)	Voltage (V)	Current (A)	Power (W)
R_1	4			
R_2	3			
R_3	6			
Entire Circuit	**6**	12	12/6 = 2	$12^2/6 = 24$

According to the image of the circuit, there is no junction between V and R_1. The corollary to the Junction Rule states that if there is no junction between two devices, the current through each must be equal. All the current that is supported by the battery must flow through R_1. This allows us to fill in the current through R_1. Then, using Ohm's Law and the power equation, we can complete the R_1 row of the table.

	Resistance (Ω)	Voltage (V)	Current (A)	Power (W)
R_1	4	$2 * 4 = 8$	**2**	$2^2 * 4 = 16$
R_2	3			
R_3	6			
Entire Circuit	6	12	2	24

After the Junction Rule is used to complete a row, the Loop Rule is typically used to complete the next step. We know that the total voltage drop through any complete loop must be equal to the voltage supplied by the battery. This circuit has two completed loops: $V = V_1 + V_2$ and $V = V_1 + V_3$. We know V is 12 V, and we know the voltage drop across R_1 is $V_1 = 8$ V, so the voltage drop across either R_2 or R_3 must be 4 V. This conclusion is consistent with the corollary to the Loop Rule that the voltage drops are equal across resistors that are in parallel. We can now complete the rest of the table.

	Resistance (Ω)	Voltage (V)	Current (A)	Power (W)
R_1	4	8	2	16
R_2	3	**4**	4/3 = 1.33	$4^2/3 = 5.33$
R_3	6	**4**	4/6 = 0.67	$4^2/6 = 2.67$
Entire Circuit	6	12	2	24

This example did not ask for the power dissipated by the circuit or by any component, but it is useful to have that column. As you will note, the resistances do not sum to the entire resistance. The voltages do not sum to the entire voltage. The currents do not sum to the entire current. However, the power will always sum to the total power. Here $24 = 16 + 5.33 + 2.67$ is true. This simple check allows you to verify the likelihood that your answer is correct and that you didn't make any algebra mistakes in calculating the voltages or currents.

Example 5 In this circuit, the voltage of the battery and the resistance of three of the resistors is known. An ammeter placed at P_2 reads 6 A.

(a) What is the value of the unknown resistor?

(b) What will a voltmeter read that is placed between P_1 and P_2?

(c) What percentage of the power dissipated by the circuit is dissipated by the 150 Ω resistor?

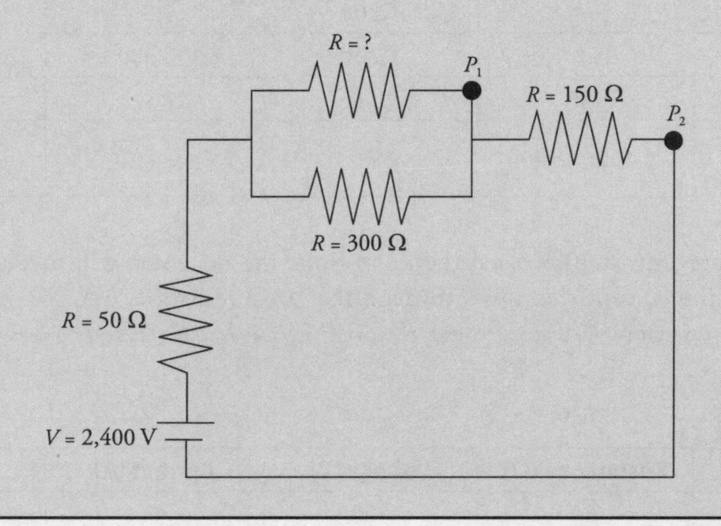

Solution. For the chart, we will need to label each resistor. Going in the direction of current flow from the battery, the 50 Ω resistor is R_1, the unknown resistor is R_2, the 300 Ω resistor is R_3, and the 150 Ω resistor is R_4. Fill out the given three resistances and one voltage.

	Resistance (Ω)	Voltage (V)	Current (A)	Power (W)
R_1	50			
R_2				
R_3	300			
R_4	150			
Entire Circuit		2,400		

Additionally, we are given the current at point P_2 is 6 A. Whenever there is no junction between two devices, the current through each must be equal, so any current through an ammeter at P_2 must flow through the battery, R_1, and R_4.

	Resistance (Ω)	Voltage (V)	Current (A)	Power (W)
R_1	50	50 * 6 = 300	**6**	6^2 * 50 = 1,800
R_2				
R_3	300			
R_4	150	150 * 6 = 900	**6**	6^2 * 150 = 5,400
Entire Circuit	2,400/6 = 400	2,400	**6**	2,400 * 6 = 14,400

Since the last step involved the Junction Rule, the next step will involve the Loop Rule. The voltage across R_2 (which we know is equal to the voltage across R_3) is $V - V_1 - V_4 = 2{,}400 - 900 - 300 = 1{,}200$ V.

	Resistance (Ω)	Voltage (V)	Current (A)	Power (W)
R_1	50	300	6	1,800
R_2		**1,200**		
R_3	300	**1,200**	1,200/300 = 4	$1{,}200^2/300 = 4{,}800$
R_4	150	900	6	5,400
Entire Circuit	2,400/6 = 400	2,400	6	14,400

Now, since the previous step involved the Loop Rule, the next step will involve the Junction Rule. We know the current that flows through the 50 Ω resistor is 6 A. We know this must all flow into the junction between R_2 and R_3. So $I_1 = I_2 + I_3$ and therefore $6 = I_2 + 4$. We can conclude that I_2 is 2 A.

	Resistance (Ω)	Voltage (V)	Current (A)	Power (W)
R_1	50	300	6	1,800
R_2	1,200/2 = 600	1,200	**2**	2 * 1,200 = 2,400
R_3	300	1,200	1,200/300 = 4	$1{,}200^2/300 = 4{,}800$
R_4	150	900	6	5,400
Entire Circuit	2,400/6 = 400	2,400	6	14,400

Finally, we can check to see if the sum of the power of each resistor correctly adds to the entire power: 1,800 + 2,400 + 4,800 + 5,400 = 14,400 is correct.

With the analysis complete, we can address the questions that were asked in the problem.

(a) The unknown resistance (R_2) is 600 Ω.

(b) A voltmeter placed between P_1 and P_2 will be in parallel with R_3 and will therefore have a reading of 1,200 V.

(c) The 150 Ω resistor dissipates 5,400 W of the entire 14,400 W, which is 37.5%.

EMF and Internal Resistance

The voltage supplied by an ideal battery is often, for historical reasons, referred to as EMF (electromotive force, ε). This is simply another name for voltage and electric potential difference, but it is almost always encountered in the context of "voltage supplied by the battery" (or in the context of Faraday's Law in the magnetism chapter).

A battery in a circuit can be modeled as an ideal EMF in series with a resistor, *r*. In this model, *r* is called the "internal resistance" of the battery and the voltage measured across the EMF source, and *r* is the "terminal voltage." This would be the measurement reading from a voltmeter placed across the terminals of the battery.

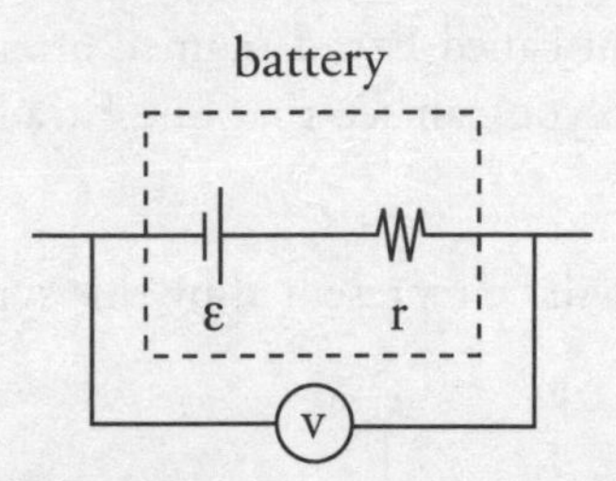

Example 6 If the 50 Ω resistor in the previous example were an internal resistor, what would the terminal voltage of the battery in that example equal?

Solution. The terminal voltage would be the voltage measured from one side of the battery across the internal resistor. Here, that would just be $V_{terminal} = V - V_1 = 2{,}400 - 300 = 2{,}100$ V.

CAPACITORS IN CIRCUITS

In the previous chapter, we looked at the electric field generated by parallel-plate capacitors. The electric field between the plates of a capacitor allows capacitors to be used in circuits as devices to store energy. They also have interesting properties related to the amount of time they take to charge, but those non-steady-state conditions are beyond the scope of the AP Physics 2 Exam, as their analysis requires calculus.

A capacitor stores energy in an electric field. A parallel-plate capacitor is comprised of two plates of area *A* and separation *d*. An insulator (called a dielectric in this context) may fill the area between the plates, or the space between them may be in vacuum.

The capacitance, which determines the amount of charge that can be stored on the plates for a given voltage, depends on the plate area, separation, and the dielectric constant, κ, of whatever material is between the plates (for vacuum, κ = 1).

$$C = \kappa\varepsilon_0 \frac{A}{d}$$

Equation Sheet

As current flows in a circuit with a capacitor, charges move onto one plate and off of the other plate. As a charge excess builds up on one plate, the electric field between the plates builds. The ratio of the charge on either plate to the voltage across the plates is the capacitance.

Equation Sheet

$$\Delta V = Q/C$$

Capacitance is measured in a unit called **Farad** (named after Michael Faraday). Rearranging the equation above to $C = Q/\Delta V$, you can see that one Farad is one Coulomb per Volt, 1F = 1C/V.

In a circuit diagram, a capacitor will be represented by this symbol:

Combinations of Capacitors and Equivalent Capacitance

Multiple capacitors may be in a circuit in which the equivalent capacitance may be calculated. Two capacitors that are arranged in parallel are equivalent to a single capacitor of area $A_p = A_1 + A_2$. Following a similar analysis to a pair of resistors, we see that

$$C_p = \kappa\varepsilon_0 \frac{A_1 + A_2}{d} = \kappa\varepsilon_0 \frac{A_1}{d} + \kappa\varepsilon_0 \frac{A_2}{d} = C_1 + C_2$$

This result holds for more than two capacitors as well, so long as they are all arranged in parallel.

Equation Sheet

$$C_p = \sum_i C_i$$

Similarly, the capacitors may be arranged in series. That arrangement is equivalent to a single capacitor of separation: $d_s = d_1 + d_2$. Following a similar analysis to a pair of resistors in parallel, we see that for capacitors in series

$$C_s = \kappa\varepsilon_0 \frac{A}{d_1 + d_2} \rightarrow \frac{1}{C_s} = \frac{(d_1 + d_2)}{\kappa\varepsilon_0 A} = \frac{d_1}{\kappa\varepsilon_0 A} + \frac{d_2}{\kappa\varepsilon_0 A} = \frac{1}{C_1} + \frac{1}{C_2}$$

This result holds for more than two capacitors as well, so long as they are all arranged in series.

Equation Sheet

$$\frac{1}{C_s} = \sum_i \frac{1}{C_i}$$

Example 7 Calculate the equivalent capacitance of a pair of capacitors that are in series when that combination is in parallel with a third capacitor. All the capacitors have capacitance C.

Solution. In the same way we combined part of the resistor circuits, we first combine only the portion of this mixed series and parallel capacitor arrangement in which the capacitors are in series. $\frac{1}{C_{1,2}} = \frac{1}{C_1} + \frac{1}{C_2}$ so $C_{1,2} = \frac{C_1 C_2}{C_1 + C_2}$. Because all the capacitors have capacitance C, $C_{1,2} = \frac{C^2}{2C} = \frac{C}{2}$. Then combining that equivalent capacitor with the third capacitor in parallel, $C_{1,2,3} = C_{1,2} + C_3 = \frac{C}{2} + C = \frac{3C}{2}$.

Charging and Discharging Capacitors

An uncharged capacitor may become charged by connecting it to a battery. The instant that connection is made, the uncharged capacitor has a potential difference of 0 V across it and current flows into the capacitor just like current would flow through a wire. For some amount of time, the capacitor is partially charged and the potential difference across the plates grows as more and more charge builds up on the plate. Eventually, the capacitor reaches a state at which the potential difference across the plates is equal to the voltage of the battery and the current flow onto the capacitor plates ceases. In this state, the capacitor acts like an open switch and no current flows onto or off of the plates.

A charged capacitor may be discharged by connecting the two plates to a resistor. The voltage between the two plates causes current to flow across the resistor. As the current flows, the amount of charge on the plates, Q, decreases and the voltage ΔV also decreases. Eventually, when the capacitor is fully discharged so that $Q = 0$ C, the voltage across the resistor is $\Delta V = 0$ V, from $\Delta V = Q/C$. Ohm's Law, $V = IR$, indicates that no more charge flows across the resistor.

Calculating the amount of time required for a capacitor to charge and discharge requires calculus. It is not part of the AP Physics 2 Exam. Instead, the exam focuses on capacitors in "steady state," in which enough time has passed for the capacitor to begin each problem in a state that is either completely discharged or completely charged.

Altering the Capacitance of a Capacitor

The capacitance is given by

Equation Sheet

$$C = \kappa \varepsilon_0 \frac{A}{d}$$

Three things may be done to alter the capacitance of a capacitor: a dielectric may be inserted (altering κ), the area of the plates may change (changing A), or the spacing between the plates may change (which changes d).

There are two scenarios for making these changes to a charged capacitor in a circuit:

1. The battery may remain connected, causing the potential difference to remain constant.
2. The battery may be disconnected, causing the charge on each plate to remain constant.

Which of these scenarios unfolds will influence an analysis of the capacitor in terms of its charge on each plate, potential difference, stored electric field, and stored energy.

Let's look at what happens when the capacitance increases as a result of adding a dielectric (κ increases).

Capacitance is increased with a battery connected. *V* is constant and *Q* changes.	
V	Constant
$Q = CV$	Increases
$E = V/d$	Constant
$U_C = \frac{1}{2}QV$	Increases

Capacitance is increased with a battery disconnected. *Q* is constant and *V* changes.	
$V = Q/C$	Decreases
Q	Constant
$E = V/d$	Decreases
$U_C = \frac{1}{2}QV$	Decreases

If the capacitance were decreased instead of increased, say by removing a dielectric from a charged capacitor (κ decreases), the same analysis would apply. The results would be as follows.

Capacitance is decreased with a battery connected. *V* is constant and *Q* changes.	
V	Constant
$Q = CV$	Decreases
$E = V/d$	Constant
$U_C = \frac{1}{2}QV$	Decreases

Capacitance is decreased with a battery disconnected. *Q* is constant and *V* changes.	
$V = Q/C$	Increases
Q	Constant
$E = V/d$	Increases
$U_C = \frac{1}{2}QV$	Increases

RC CIRCUITS WITH CAPACITORS IN STEADY STATE

The two possible steady state conditions for a capacitor in a circuit occur when the capacitor is discharged or when it is fully charged.

When a capacitor is discharged, the voltage between the plates is 0 V. As a result, the discharged capacitor *acts like a wire.* Then, from any junction where one parallel path has the discharged capacitor and the other has a resistor, all of the current will flow into the capacitor and none will flow into the resistor. Essentially, the discharged capacitor short-circuits any resistors arranged in parallel with the capacitor.

The situation is reversed when a capacitor is fully charged. The voltage between the plates is the maximum amount that can be supplied by the battery. As a result, the fully charged capacitor *acts like a broken wire.* Therefore, from any junction where one parallel path has the fully charged capacitor and the other has a resistor, all of the current will flow into the resistor and none will flow into the capacitor. Essentially, the fully charged capacitor short-circuits any resistors arranged in series with the capacitor.

When we analyze circuits with capacitors, we still use a chart of the resistance, voltage, current, and power. However, because our analysis is limited to one of two particular instants in time (the discharged capacitor or the fully charged capacitor), both the "current" and the "power" columns are not applicable to the capacitor, as both of those quantities explicitly depend on Δt.

Example 8 The image shows an RC circuit with an uncharged capacitor. Determine the following:

(a) the current to the capacitor when the switch has just been closed

(b) the voltage across the capacitor after the switch has been closed for a long time

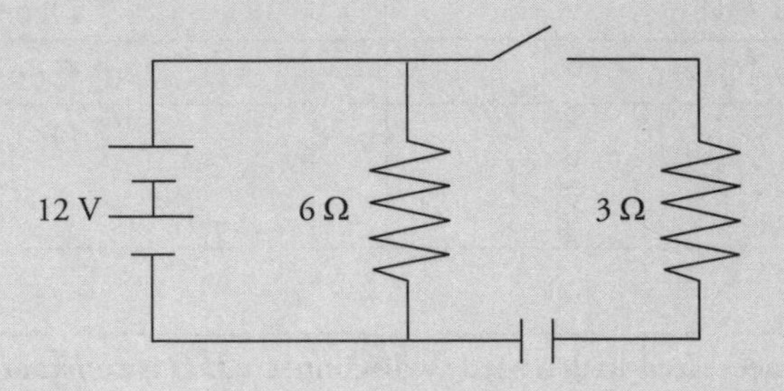

Solution.

(a) Immediately after the switch is closed, the uncharged capacitor acts like a wire.

	Resistance (Ω)	Voltage (V)	Current (A)	Power (W)
R_1	6			
R_2	3			
C	**0**	**0**	N/A	N/A
Entire Circuit		12		

We can find the resistance of the entire circuit by noting the 6 Ω and 3 Ω resistors are in parallel, so the equivalent resistance is $\frac{1}{R} = \frac{1}{3} + \frac{1}{6}$ so that $R = 2\ \Omega$.

	Resistance (Ω)	Voltage (V)	Current (A)	Power (W)
R_1	6			
R_2	3			
C	0	0	N/A	N/A
Entire Circuit	**2**	12	12/2 = 6	$12^2/2 = 72$

Now that we have used the current, we look to the Loop Rule. Since the 6 Ω and 3 Ω resistors are both in parallel with the battery, they all have equal voltages.

	Resistance (Ω)	Voltage (V)	Current (A)	Power (W)
R_1	6	**12**	12/6 = 2	$12^2/6 = 24$
R_2	3	**12**	12/3 = 4	$12^2/3 = 48$
C	0	0	N/A	N/A
Entire Circuit	2	12	6	72

Finally, we return to the Junction Rule and see that the capacitor is in series with the 3 Ω resistor, so the current flowing to the capacitor must be 4 A.

Note that as with wires, we cannot apply Ohm's Law to the capacitor because we would be dividing by 0 when calculating the current.

(b) After the switch is closed for a long time, the capacitor has filled. The capacitor acts like a broken wire. Now, R_2 is in series with the fully charged capacitor, so it has been shorted out of the circuit. Therefore, when we complete the table, we place a value of 0 Ω in for R_2 instead of 3 Ω. The resistance of the resistor is still 3 Ω, but because it is shorted out of the circuit, its effect is 0 Ω.

	Resistance (Ω)	**Voltage (V)**	**Current (A)**	**Power (W)**
R_1	6			
R_2	**0**			
C	∞		N/A	N/A
Entire Circuit		12		

The equivalent resistance of the circuit is just 6 Ω.

	Resistance (Ω)	**Voltage (V)**	**Current (A)**	**Power (W)**
R_1	6			
R_2	0			
C	∞		N/A	N/A
Entire Circuit	**6**	12	12/6 = 2	$12^2/6 = 24$

All of the current supported by the battery flows through R_1 and none through R_2. Because no current flows across R_2, there must be 0 V voltage between the two ends of the resistor, and it must be dissipating 0 W of power.

	Resistance (Ω)	**Voltage (V)**	**Current (A)**	**Power (W)**
R_1	6	6 * 2 = 12	2	$12^2/6 = 24$
R_2	0	**0**	**0**	**0**
C	∞		N/A	N/A
Entire Circuit	**6**	12	2	24

We know that each parallel branch has to have the same voltage. Since the current, voltage, and resistance of R_2 are all 0, then the voltage across the filled capacitor must be 12 V.

Example 9 Determine the current through and the voltage across each resistor and the battery in the following circuit when
(a) the switch has just been closed
(b) the switch has been closed for a long period of time

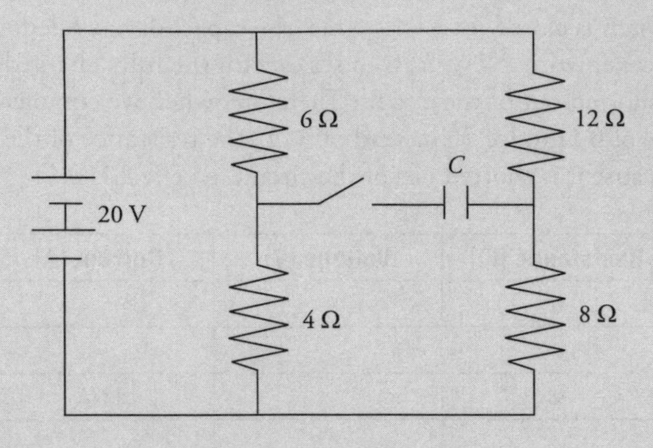

Solution.

(a) If we imagine the capacitor as a wire, the 6 Ω and 12 Ω resistors are in parallel, with an equivalent resistance of 4 Ω. The 4 Ω and 8 Ω resistors are in parallel, with an equivalent resistance of $\frac{8}{3}$ Ω. The equivalent 4 Ω resistor and $\frac{8}{3}$ Ω resistor are in series. The total resistance of the circuit is $\frac{20}{3}$ Ω.

	Resistance (Ω)	Voltage (V)	Current (A)	Power (W)
R_1	6			
R_2	12			
R_3	4			
R_4	8			
C	**0**	**0**	N/A	N/A
Entire Circuit	**20/3**	20	20/6.67 = 3	$20^2/6.67 = 60$

Now, we again have a mixture of parallel and series. We could continue our methodical approach and find the equivalent resistance combination of R_1 and R_2 since we know the current through that combination is 3 A. That would allow us to calculate the voltage across each of those resistors. However, since we know the 3 A of total current divides between the 6 Ω and 12 Ω resistors, and since those are in parallel, they must have the same voltage, so $I_1 * 6 = I_2 * 12$, and we can conclude that the current in R_1 is 2 A while the current in R_2 is 1 A. Similarly, the 3 A of total current divides between the 4 Ω and 8 Ω resistors. Since those are in parallel, they must have the same voltage, so $I_3 * 4 = I_4 * 8$, and we can conclude that the current in R_3 is 2 A while the current in R_4 is 1 A.

	Resistance (Ω)	Voltage (V)	Current (A)	Power (W)
R_1	6	6 * 2 = 12	**2**	$2^2 * 6 = 24$
R_2	12	12 * 1 = 12	**1**	$1^2 * 12 = 12$
R_3	4	4 * 2 = 8	**2**	$2^2 * 4 = 16$
R_4	8	8 * 1 = 8	**1**	$1^2 * 8 = 8$
C	**0**	**0**	N/A	N/A
Entire Circuit	20/3	20	3	60

(b) When the switch has been closed for a long time, it acts like a broken wire. This makes the 6 Ω and 4 Ω resistors in series, acting as a single 10 Ω resistor, and that combination is in parallel with the combination of the 12 Ω and 8 Ω resistors in series, acting as a single 20 Ω resistor. The equivalent resistance of the entire circuit is $\frac{1}{R}=\frac{1}{10}+\frac{1}{20}$, so $R = 6.67$.

	Resistance (Ω)	Voltage (V)	Current (A)	Power (W)
R_1	6			
R_2	12			
R_3	4			
R_4	8			
C	∞		N/A	N/A
Entire Circuit	**6.67**	20	20/6.67 = 3	$20^2/6.67 = 60$

Now, we again have a mixture of parallel and series. We could continue our methodical approach and find the equivalent resistance combination of R_1 and R_3 since we know the voltage across that combination is 20 V. That would allow us to calculate the current through each of those resistors. However, since we know the 6 A of total current divides between the 10 Ω and 20 Ω paths, we know that 2 A goes through R_1 and R_3 while 1 A goes through R_2 and R_4.

	Resistance (Ω)	Voltage (V)	Current (A)	Power (W)
R_1	6	6 * 2 = 12	**2**	$12^2/6 = 24$
R_2	12	12 * 1 = 12	**1**	$12^2/2 = 12$
R_3	4	4 * 2 = 8	**2**	$8^2/4 = 16$
R_4	8	8 * 1 = 8	**1**	$8^2/8 = 8$
C	∞		N/A	N/A
Entire Circuit	6.67	20	20/6.67 = 3	$20^2/6.67 = 60$

Chapter 7 Review Questions

Solutions can be found in Chapter 11.

Section I: Multiple Choice

1. A wire made of brass and a wire made of silver have the same length, but the diameter of the brass wire is 4 times the diameter of the silver wire. The resistivity of brass is 5 times greater than the resistivity of silver. If R_B denotes the resistance of the brass wire and R_S denotes the resistance of the silver wire, which of the following is true?

   (A) $R_B = \frac{5}{16} R_S$
   (B) $R_B = \frac{4}{5} R_S$
   (C) $R_B = \frac{5}{4} R_S$
   (D) $R_B = \frac{5}{2} R_S$

2. For an ohmic conductor, doubling the voltage without changing the resistance will cause the current to

   (A) decrease by a factor of 4
   (B) decrease by a factor of 2
   (C) increase by a factor of 2
   (D) increase by a factor of 4

3. A capacitor, battery, and two resistors are to be arranged in a circuit. Which configurations allow there to be a current that is initially through one resistor and finally through the other resistor?

   (A) The capacitor should be in series with the battery, and the resistors should be in parallel.
   (B) The capacitor should be in parallel with the battery, and the resistors should be in series.
   (C) One resistor should be in series with the battery, and the other should be in parallel with the capacitor.
   (D) One resistor should be in parallel with the battery, and the other should be in series with the capacitor.

4. A student wants to determine the resistivity of copper. She has a voltmeter reading for a copper wire of known length. What other information will she need?

   (A) An ammeter reading
   (B) The diameter of the wire
   (C) Both an ammeter reading and the diameter of the wire
   (D) Neither an ammeter reading nor the diameter of the wire

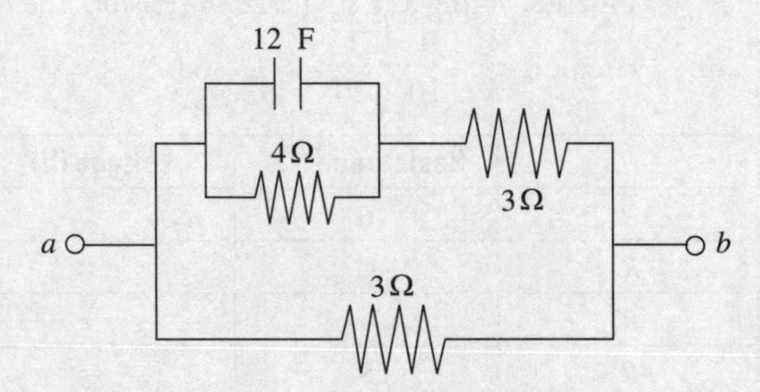

5. The circuit shown above has a constant voltage between points *a* and *b*. The voltage has been applied for a long time. What is the resistance of the circuit?

   (A) 0.47 Ω
   (B) 1.5 Ω
   (C) 2.1 Ω
   (D) 10 Ω

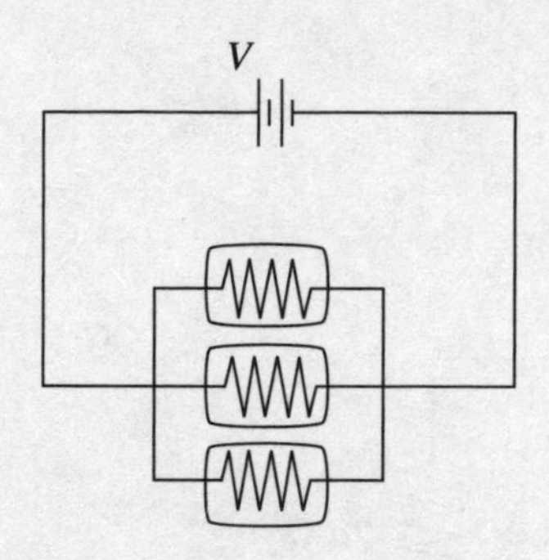

6. Three identical light bulbs are connected to a source of emf, as shown in the diagram above. What will happen if the middle bulb burns out?

(A) The light intensity of the other two bulbs will decrease (but they won't go out).
(B) The light intensity of the other two bulbs will increase.
(C) The light intensity of the other two bulbs will remain the same.
(D) More current will be supported by the source of emf.

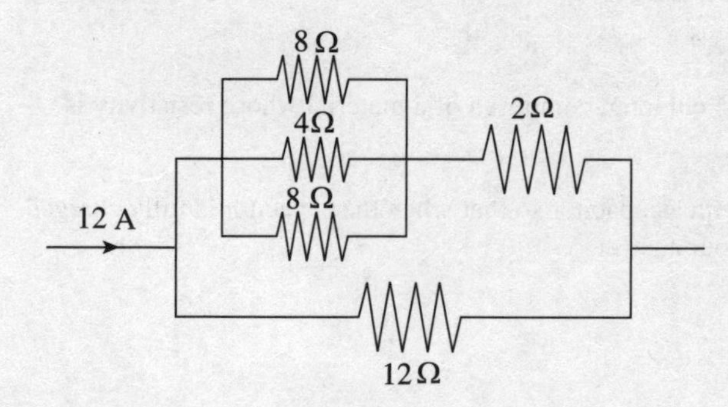

7. What is the voltage drop across the 12 Ω resistor in the portion of the circuit shown above?

(A) 24 V
(B) 36 V
(C) 48 V
(D 72 V

8. A simple DC circuit with a single ohmic resistor is set up. A graph is produced to show the voltage drop across the resistor versus current. For an ideal battery, the graph is directly proportional. If the battery has some internal resistance, how, if at all, would such a graph change?

(A) The graph would remain directly proportional.
(B) The graph would be linear, but would have a negative y-intercept.
(C) The graph would be linear, but would have a positive y-intercept.
(D) The graph would be nonlinear.

9. How much energy is dissipated as heat in 20 s by a 100 Ω resistor that carries a current of 0.5 A?

(A) 50 J
(B) 100 J
(C) 250 J
(D) 500 J

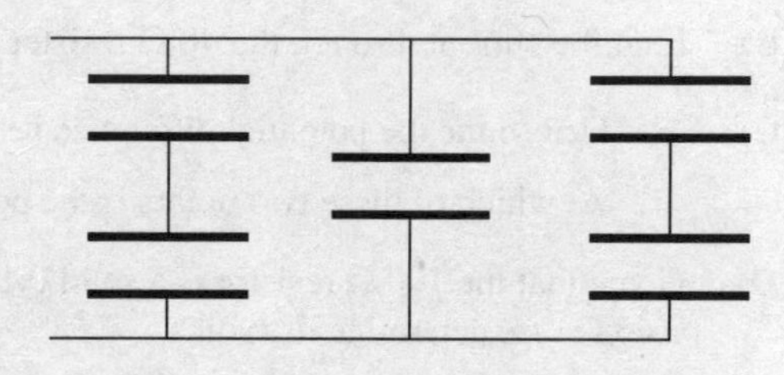

10. If each of the capacitors in the array shown above is C, what is the capacitance of the entire combination?

(A) $C/2$
(B) $2C/3$
(C) $5C/6$
(D) $2C$

Section II: Free Response

1. Consider the following circuit:

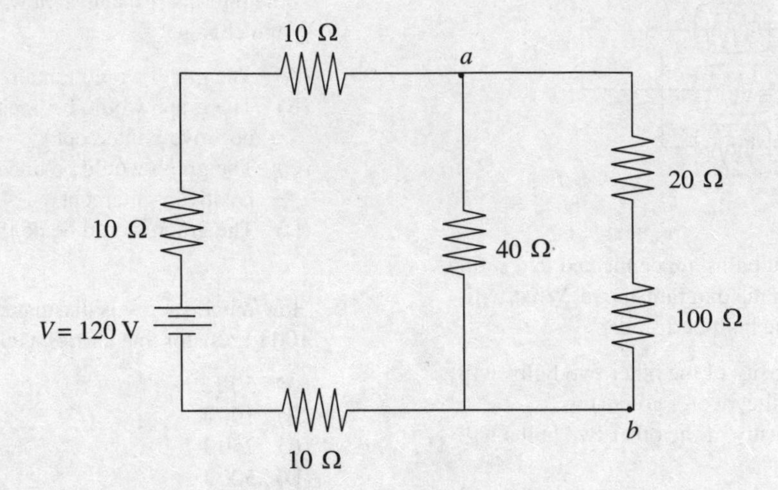

(A) At what rate does the battery deliver energy to the circuit?

(B) Find the current through the 40 Ω resistor.

(C) i. Determine the potential difference between points *a* and *b*.

ii. At which of these two points is the potential higher?

(D) Given that the 100 Ω resistor is a solid cylinder that's 4 cm long, composed of a material whose resistivity is 0.45 Ω • m, determine its radius.

(D) Which resistor or resistors, if any, could be replaced with a capacitor so that when the capacitor is fully charged, no current flow is supported by the battery? Explain your answer.

2. Consider the following circuit:

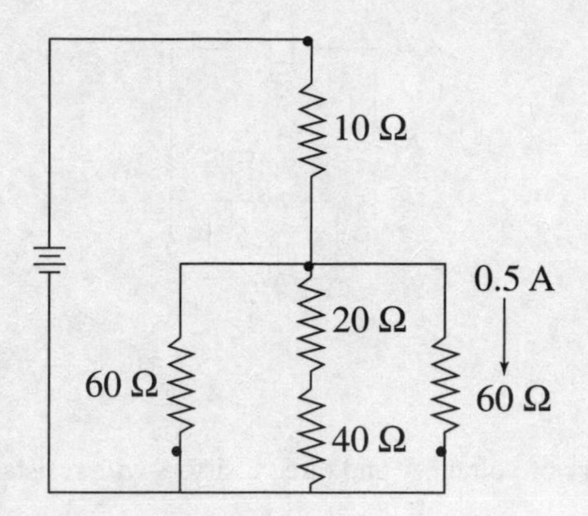

(A) What is the current through each resistor?

(B) What is the potential difference across each resistor?

(C) What is the equivalent resistance of the circuit?

(D) The rightmost 60 Ω resistor is replaced with a capacitor.

i. When the capacitor is uncharged and just beginning to fill, will the current in the leftmost 60 Ω resistor be greater than, equal to, or less than the 0.5 A it was before the capacitor was inserted into the circuit?

ii. When the capacitor is fully charged, will the current in the leftmost 60 Ω resistor be greater than, equal to, or less than the 0.5 A it was before the capacitor was inserted into the circuit?

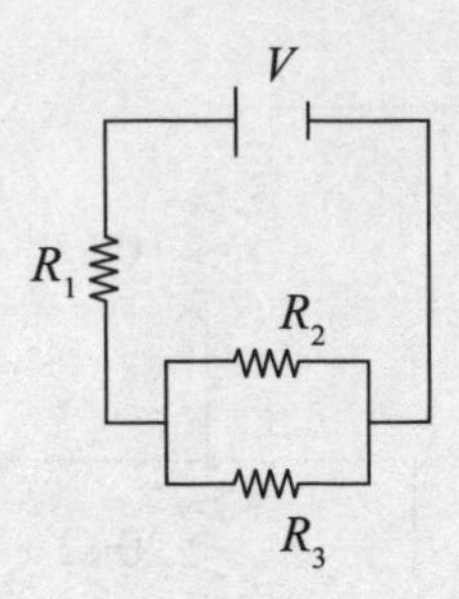

3. The above circuit contains a battery of voltage V and three resistors with resistances R_1, R_2, and R_3, respectively. As part of an experiment, a student has been given two measuring devices: a voltmeter and an ammeter. The first can be used to measure the changes in voltage of a circuit. The second can be used to measure the current flowing through a particular segment of wire. For answering the questions below, a voltmeter and ammeter look like ——(V)—— and ——(A)——, respectively, when drawn in a circuit diagram.

   (A) In terms of the known variables, what is the voltage lost in passing through the first resistor?

   (B) Draw a diagram showing how you would integrate the voltmeter to measure the voltage lost in the resistor labeled R_1. Explain the reasoning behind your decision.

   (C) Draw a diagram showing how you would integrate the ammeter to measure the current passing through the resistor labeled R_1. Explain the reasoning behind your decision.

   (D) What would be the ideal resistances for each device to have? Explain why each would be ideal for that device.

Summary

- The resistance of an object can be determined by $R = \frac{\rho \ell}{A}$, where ρ is the resistivity (a property of the material), ℓ is the length, and A is the cross-sectional area.
- The current is the rate at which charge is transferred and given by $I_{avg} = \frac{\Delta Q}{\Delta t}$.
- Many objects obey Ohm's Law, which is given by $V = IR$.
- The electrical power in a circuit is given by $P = IV$ or $P = I^2R$ or $P = \frac{V^2}{R}$.

  This is the same power we've encountered in our discussion of energy $P = \frac{W}{t}$.
- In a series circuit / In a parallel circuit

In a series circuit	In a parallel circuit
$V_B = V_1 + V_2 + \dots$	$V_B = V_1 = V_2 = \dots$
$I_B = I_1 = I_2 = \dots$	$I_B = I_1 + I_2 + \dots$
$R_{eq} = R_1 + R_2 + \dots$	$1/R_{eq} = 1/R_1 + 1/R_2 + \dots$
$R_s = \sum_i R_i$	$\frac{1}{R_p} = \sum_i \frac{1}{R_i}$

- Kirchhoff's Loop Rule tells us that the sum of the potential differences in any closed loop in a circuit must be zero.
- Kirchhoff's Junction Rule (Node Rule) tells us the total current that enters a junction must equal the total current that leaves the junction.
- To find the combined capacitance of capacitors in parallel (side by side), simply add their capacitances:

$$C_p = C_1 + C_2 + \dots \text{ or } C_p = \sum_i C_i$$

- To find the capacitance of capacitors in series (one after another), add their inverses:

$$\frac{1}{C_s} = \frac{1}{C_1} + \frac{1}{C_2} + \dots \text{ or } \frac{1}{C_s} = \sum_i \frac{1}{C_i}$$

- When a capacitor is in a circuit and is completely discharged, it will act like a closed switch (equivalent to a resistor with $R = 0\ \Omega$). When the capacitor is fully charged, it acts like an open switch (equivalent to an infinite resistor).

Chapter 8
Magnetism and Electromagnetic Induction

This chapter deals with CED Unit 5.1: Magnetic Systems.

INTRODUCTION

In a previous chapter, we learned that electric charges are the sources of electric fields and that other charges experience an electric force in those fields. The charges generating the field were assumed to be at rest, because if they weren't, then another force field would have been generated in addition to the electric field. Electric charges *that move* are the sources of **magnetic fields**, and other charges that move can experience a magnetic force in these fields.

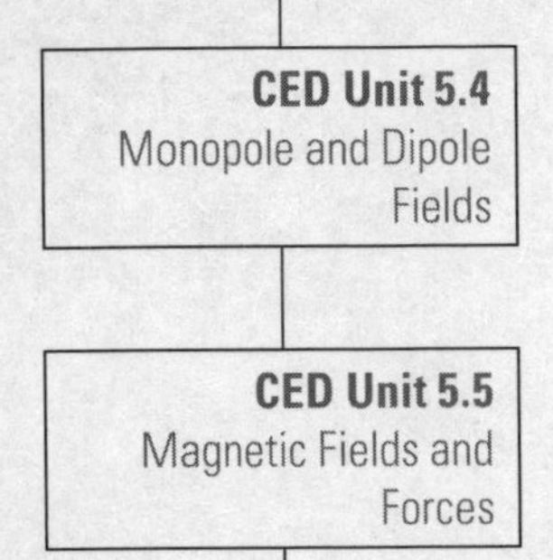

MAGNETIC FIELDS

Similar to our discussion about electric fields, the space surrounding a magnet is permeated by a magnetic field. The direction of the magnetic field is defined as pointing out of the north end of a magnet and into the south end of a magnet, as illustrated below.

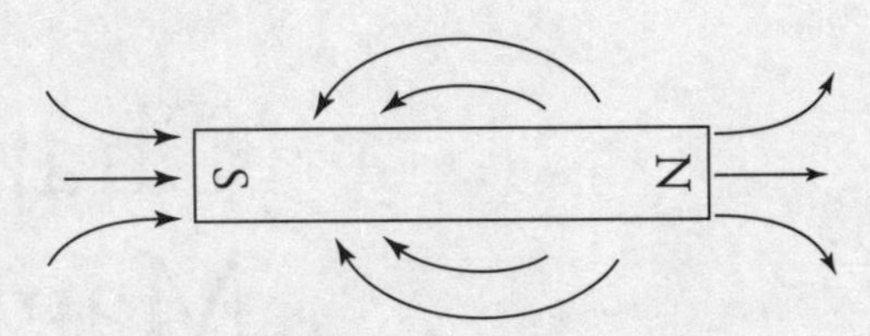

Magnetic fields have always been found to have both a north and a south pole, defined so that the field comes out of the north pole and enters back into the magnet at the south pole. As we saw in Chapter 5, electric fields could be created from a single positive or single negative charge. You will also see that a pair of charges, one positive and one negative, generated an electric dipole. Magnetic fields always have both a north and a south pole.

When two magnets get near each other, the magnetic fields interfere with each other and can be drawn as follows. Note that, for simplicity's sake, only the field lines closest to the poles are shown.

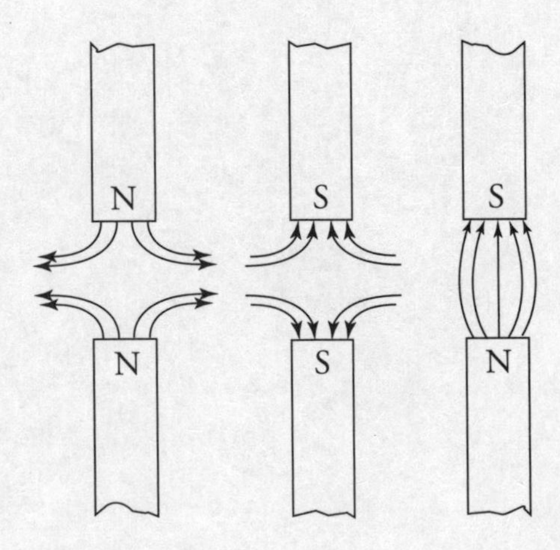

Notice there is a curve to the previous fields. We call the field uniform if the field lines are parallel and of equal strength. It is easy to recognize a uniform magnetic field to the right, left, top of the page, or bottom of the page. But you will also see a field going into or out of the page. A field into the page looks as if there were a north pole of a magnet above the page pointing down at the south pole of a magnet that is below the page. It is represented by an area with X's going into the page. A field coming out of the page looks as if there were a north pole of a magnet below the page pointing up at the south pole of a magnet that is above the page. It is represented by an area with dots (•) coming out of the page.

INTO	OUT OF
X X X X	• • • •
X X X X	• • • •
X X X X	• • • •
X X X X	• • • •

BAR MAGNETS

A permanent bar magnet creates a magnetic field that closely resembles the magnetic field produced by a circular loop of current-carrying wire:

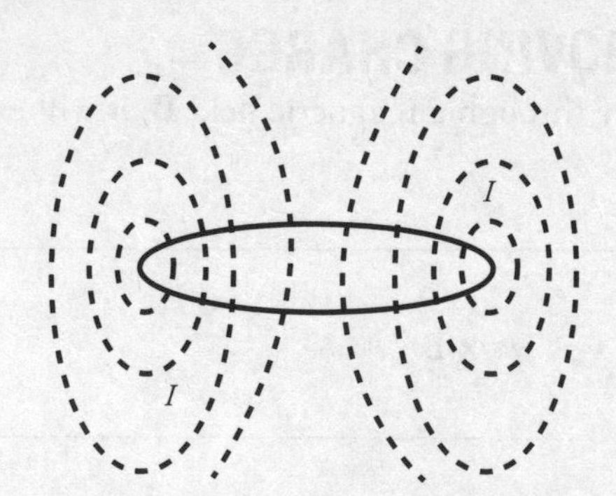

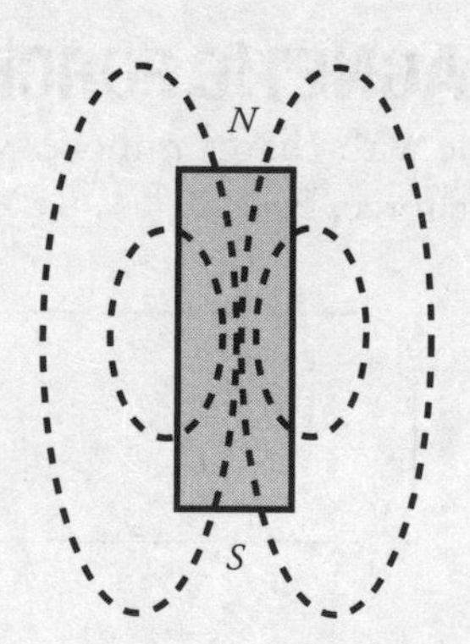

CED Unit 5.2
Magnetic Permeability and Magnetic Dipole Moment

By convention, the magnetic field lines emanate from the end of the magnet designated the north pole (N) and then curl around and reenter the magnet at the end designated the south pole (S). The magnetic field decreases in strength as distance from the magnet increases. The magnetic field created by a permanent bar magnet is due to the electrons, which have an intrinsic spin as they orbit the nuclei. They are literally charges in motion, which is the ultimate source of any and all magnetic fields. If a piece of iron is placed in an external magnetic field (for example, one created by a current-carrying solenoid), the individual magnetic dipole moments of the electrons will be forced to more or less line up. Because iron is ferromagnetic, these now-aligned magnetic dipole moments tend to retain this configuration, thus permanently magnetizing the bar and causing it to produce its own magnetic field. All materials have some **magnetic permeability**, μ, which determines how great a magnetic field an object will develop when placed in an external field.

As with electric charges, like magnetic poles repel each other, while opposite magnetic poles attract each other.

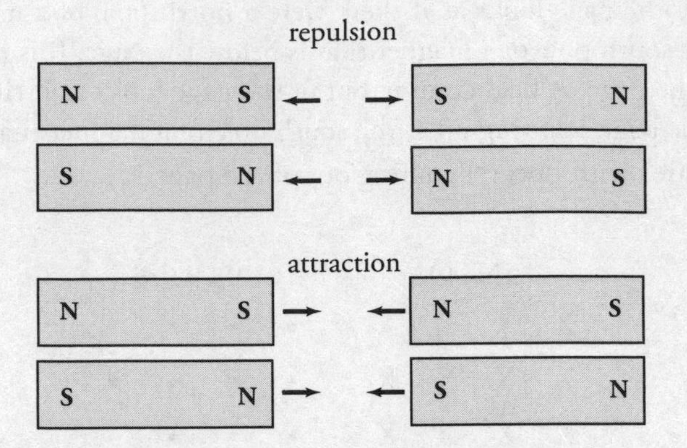

CED Unit 5.6
Magnetic Forces

CED Unit 5.7
Forces Review

However, while you can have a positive electric charge all by itself, you can't have a single magnetic pole all by itself: remember, the existence of a lone magnetic pole has never been confirmed. That is, there are no magnetic monopoles; magnetic poles always exist in pairs. If you break a bar magnet into two pieces, it does not produce one piece with just an N and another with just an S; it produces two separate, complete magnets, each with an N-S pair.

THE MAGNETIC FORCE ON A MOVING CHARGE

If a particle with charge q moves with velocity $\mathbf{v}$ through a magnetic field $\mathbf{B}$, it will experience a magnetic force, $\mathbf{F}_B$:

Equation Sheet

$$\mathbf{F}_B = q\mathbf{v} \times \mathbf{B}$$

with magnitude:

Equation Sheet

$$F_B = |q|\, vB \sin\theta$$

where θ is the angle between $\mathbf{v}$ and $\mathbf{B}$. From this equation, we can see that if the charge is at rest, then $v = 0$ immediately gives us $F_B = 0$. This tells us that magnetic forces act only on moving charges. Also, if $\mathbf{v}$ is parallel (or antiparallel) to $\mathbf{B}$, then $F_B = 0$ since, in either of these cases, $\sin\theta = 0$. So, only charges that cut across the magnetic field lines will experience a magnetic force. Furthermore, the magnetic force is maximized when $\mathbf{v}$ is perpendicular to $\mathbf{B}$, since if $\theta = 90°$, then $\sin\theta$ is equal to 1, its maximum value.

The direction of $\mathbf{F}_B$ is always perpendicular to both **v** and **B** and depends on the sign of the charge q and the direction of $\mathbf{v} \times \mathbf{B}$ (which can be found by using the right-hand rule).

> If q is *positive*, use your *right* hand and the *right*-hand rule.
>
> If q is *negative*, you need your thumb to point in the direction of the product of q and v (which is the direction opposite of v), not just the direction of the velocity when you apply the right-hand rule.

0 or 180
If the velocity **v** and the magnetic field **B** are parallel or antiparallel, which means oriented so the angle between the vectors is 180 degrees, the magnetic force $\mathbf{F}_B = 0$.

Whenever you use the right-hand rule, keep your hand flat, and follow these steps:

1. Orient your hand so that your thumb points in the direction of the velocity **v**. If the charge is negative, turn your thumb by 180 degrees.
2. Without changing the direction of your thumb, rotate your hand to point your fingers in the direction of **B**.
3. The direction of $\mathbf{F}_B$ will then be perpendicular to your palm.

Think of your palm pushing with the force $\mathbf{F}_B$; the direction it pushes is the direction of $\mathbf{F}_B$.

Right-Hand Rule:

For determining the direction of the magnetic force, $\mathbf{F}_B$, on a *positive* charge

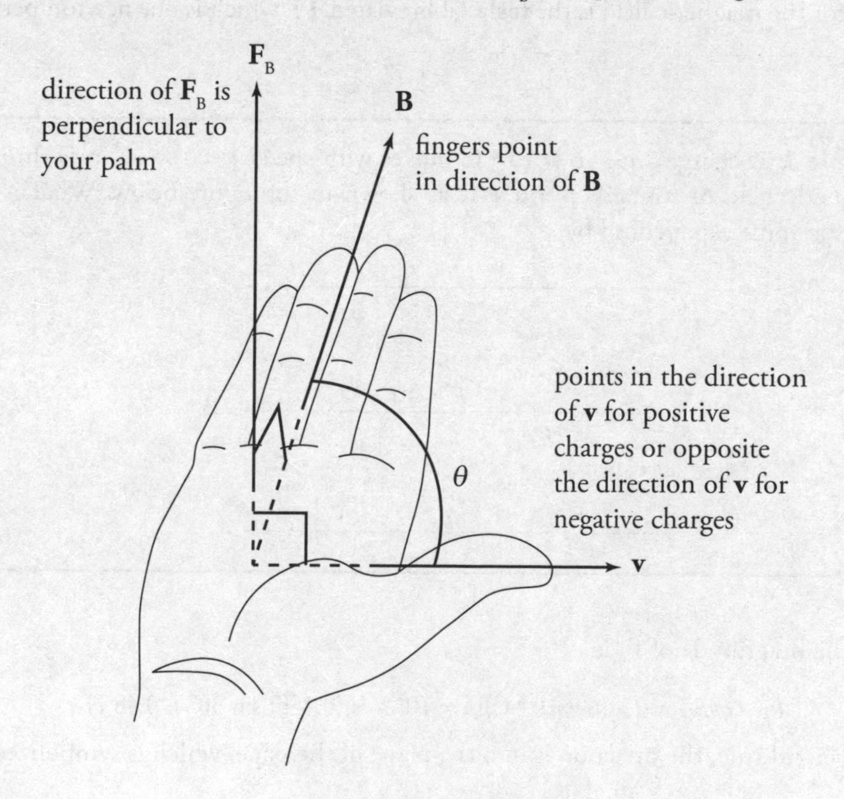

Note that there are fundamental differences between the electric force and magnetic force on a charge. First, a magnetic force acts on a charge only if the charge is moving; the electric force acts on a charge whether it moves or not. Second, the direction of the magnetic force is always perpendicular to the magnetic field, while the electric force is always parallel (or antiparallel) to the electric field.

Example 1 For each of the following charged particles moving through a magnetic field, determine the direction of the force acting on the charge.

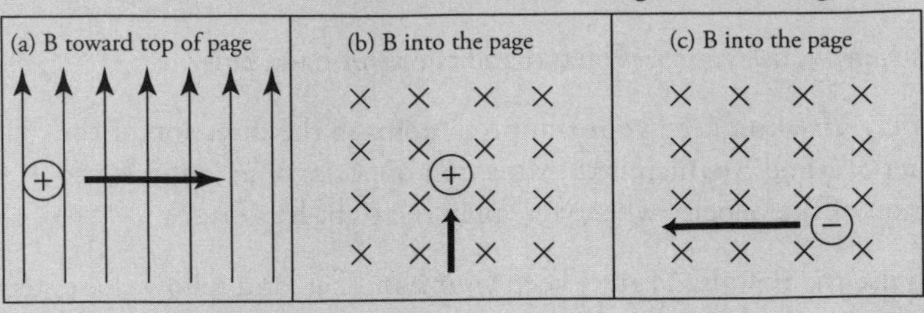

Solution.

(a) If you point your fingers to the top of the page and thumb to the right of the page, your palm should point out of the page. The force is out of the page.

(b) If you point your fingers into the page and thumb to the top of the page, your palm should point to the left of the page. The force is to the left of the page.

(c) If you point your fingers into the page and thumb to the right of the page (remember, when the charge is negative, your thumb points in the opposite direction from the velocity), your palm should point to the top of the page. The force points up toward the top of the page.

The SI unit for the magnetic field is the **tesla** (abbreviated **T**), which is one newton per ampere-meter.

Example 2 A charge $+q = +6 \times 10^{-6}$ C moves with speed $v = 4 \times 10^5$ m/s through a magnetic field of strength $B = 0.4$ T, as shown in the figure below. What is the magnetic force experienced by q?

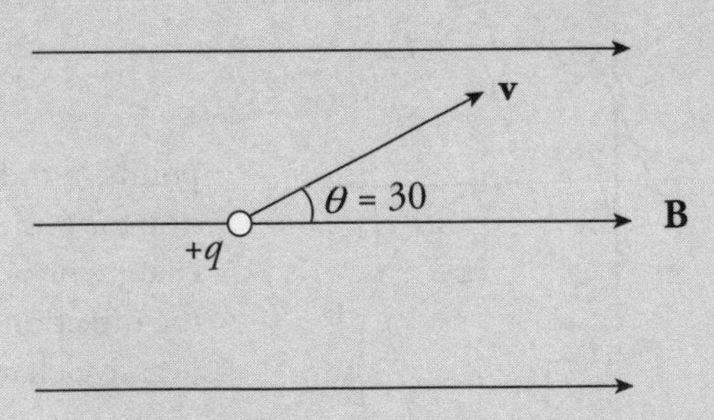

Solution. The magnitude of F_B is

$$F_B = qvB \sin \theta = (6 \times 10^{-6} \text{ C})(4 \times 10^5 \text{ m/s})(0.4 \text{ T}) \sin 30° = 0.48 \text{ N}$$

By the right-hand rule, the direction is into the plane of the page, which is symbolized by ×.

Example 3 A particle of mass m and charge $+q$ is projected with velocity $\mathbf{v}$ (in the plane of the page) into a uniform magnetic field $\mathbf{B}$ that points into the page. How will the particle move?

Solution. Since $\mathbf{v}$ is perpendicular to $\mathbf{B}$, the particle will feel a magnetic force of strength qvB, which will be directed perpendicular to $\mathbf{v}$ (and to $\mathbf{B}$) as shown:

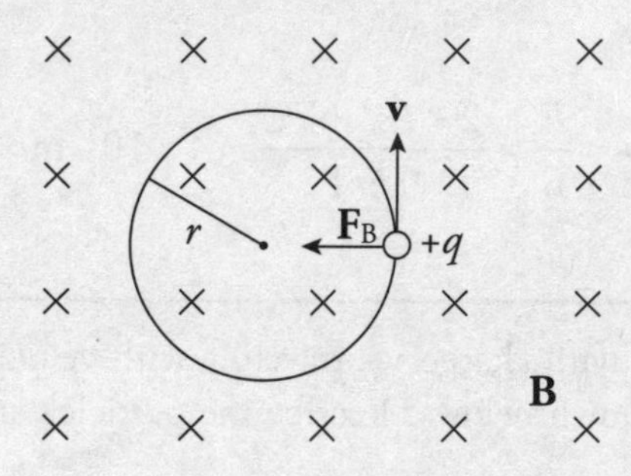

Since $\mathbf{F}_B$ is always perpendicular to $\mathbf{v}$, the particle will undergo uniform circular motion; $\mathbf{F}_B$ will provide the centripetal force. Notice that, because $\mathbf{F}_B$ is always perpendicular to $\mathbf{v}$, the magnitude of $\mathbf{v}$ will not change, just its direction. *Magnetic forces alone cannot change the speed of a charged particle; they can change only its direction of motion.* The radius of the particle's circular path is found from the equation $F_B = F_C$:

$$qvB = \frac{mv^2}{r} \Rightarrow r = \frac{mv}{qB}$$

Magnetic Fields

$\mathbf{F}_B$ is always perpendicular to both $\mathbf{v}$ and $\mathbf{B}$. Magnetic forces cannot change the speed of an object, only its direction. The magnetic field does no work on any charges.

Example 4 A particle of charge $-q$ is shot into a region that contains an electric field, **E**, crossed with a perpendicular magnetic field, **B**. If $E = 2 \times 10^4$ N/C and $B = 0.5$ T, what must be the speed of the particle if it is to cross this region without being deflected?

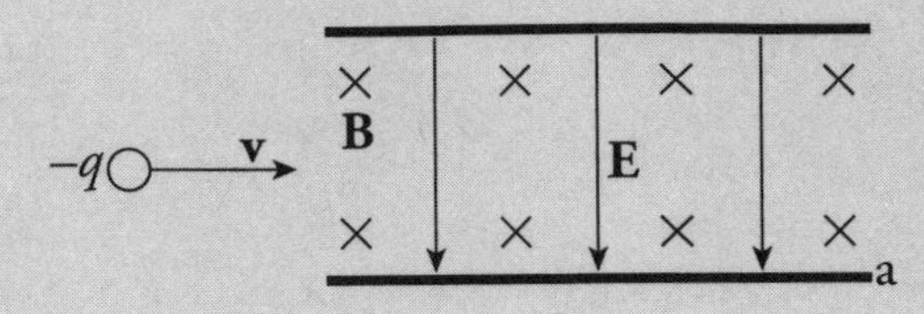

Solution. If the particle is to pass through undeflected, the electric force it feels has to be canceled by the magnetic force. In the diagram above, the electric force on the particle is directed upward (since the charge is negative and E is downward), and the magnetic force is directed downward by the right-hand rule. So F_E and F_B point in opposite directions, and in order for their magnitudes to balance, qE must equal qvB, so v must equal E/B, which in this case gives

$$v = \frac{E}{B} = \frac{2 \times 10^4 \text{ N/C}}{0.5 \text{ T}} = 4 \times 10^4 \text{ m/s}$$

Example 5 A particle with charge $+q$, traveling with velocity **v**, enters a uniform magnetic field **B**, as shown below. Describe the particle's subsequent motion.

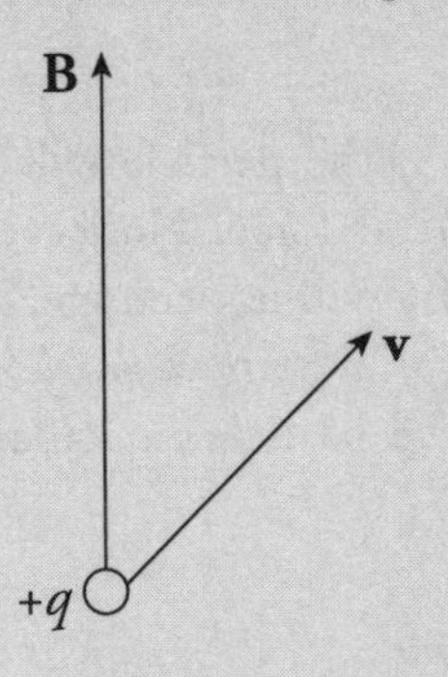

Solution. If the particle's velocity were parallel to **B**, then it would be unaffected by **B**. If **v** were perpendicular to **B**, then it would undergo uniform circular motion (as we saw in Example 2). In this case, **v** is neither purely parallel nor perpendicular to **B**. It has a component ($\mathbf{v}_1$) that's parallel to **B** and a component ($\mathbf{v}_2$) that's perpendicular to **B**.

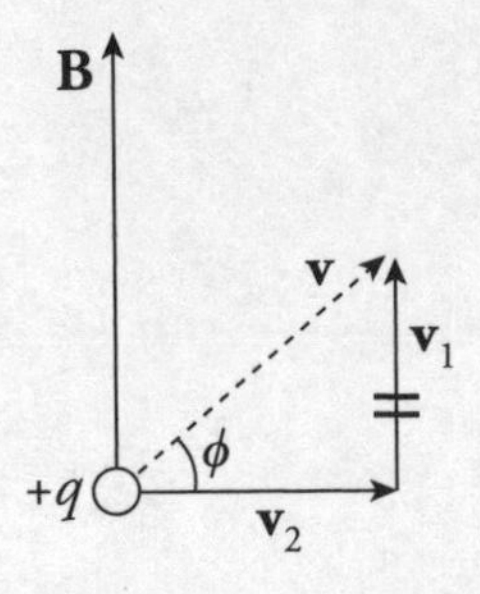

Component $\mathbf{v}_1$ will not be changed by $\mathbf{B}$, so the particle will continue upward in the direction of $\mathbf{B}$. However, the presence of $\mathbf{v}_2$ will create circular motion. The superposition of these two types of motion will cause the particle's trajectory to be a *helix*; it will spin in circular motion while traveling upward with the speed $v_1 = v \sin \phi$:

THE MAGNETIC FORCE ON A CURRENT-CARRYING WIRE

Since magnetic fields affect moving charges, they should also affect current-carrying wires. After all, a wire that contains a current contains charges that move. Remember, a current is the flow of positive charges.

Let a wire of length ℓ be immersed in magnetic field $\mathbf{B}$. If the wire carries a current I, then the force on the wire is

$$\mathbf{F}_B = I\boldsymbol{\ell} \times \mathbf{B}$$

Equation Sheet

with magnitude

$$F_B = \mathrm{B}I\ell \sin \theta$$

Equation Sheet

where θ is the angle between ℓ and $\mathbf{B}$. Here, the direction of ℓ is the direction of the current, I. The direction of $\mathbf{F}_B$ can be found using the right-hand rule and by letting your thumb point in the direction in which the current flows. Remember, conventional current is the flow of positive charges.

Another Way of Seeing It

Current I is charge over time (q/t) and the length of a wire I is a distance d. Hence, $I\ell = q(d/t)$. This gives us the same formula as before: $F_B = qv\mathrm{B}\sin \theta = I\ell\mathrm{B}\sin \theta$.

Example 6 A U-shaped wire of mass m is lowered into a magnetic field **B** that points out of the plane of the page. How much current I must pass through the wire in order to cause the net force on the wire to be zero?

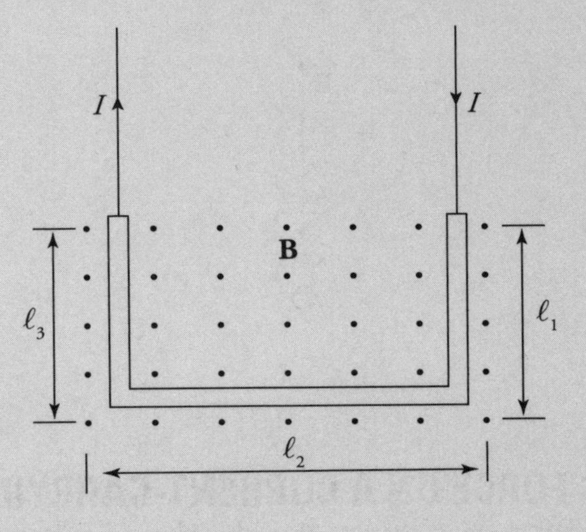

Solution. The total magnetic force on the wire is equal to the sum of the magnetic forces on each of the three sections of wire. The force on the first section (the right, vertical one), $\mathbf{F}_{B1}$, is directed to the left (applying the right-hand rule), and the force on the third piece (the left, vertical one), $\mathbf{F}_{B3}$, is directed to the right. Since these pieces are the same length, these two oppositely directed forces have the same magnitude, $I\ell_1B = I\ell_3B$, and they cancel. So the net magnetic force on the wire is the magnetic force on the middle piece. Since I points to the left and **B** is out of the page, the right-hand rule tells us the force is upward.

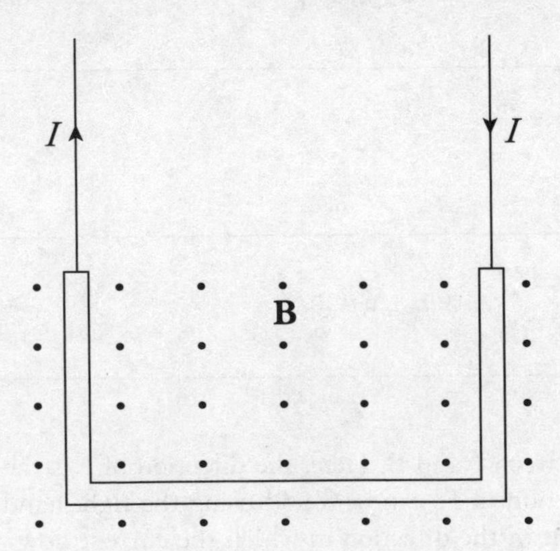

Since the magnetic force on the wire is $I\ell_2B$, directed upward, the amount of current must create an upward magnetic force that exactly balances the downward gravitational force on the wire. Because the total mass of the wire is m, the resultant force (magnetic + gravitational) will be zero if

$$I\ell_2B = mg \quad \Rightarrow \quad I = \frac{mg}{\ell_2B}$$

Example 7 A rectangular loop of wire that carries a current I is placed in a uniform magnetic field, **B**, as shown in the diagram below and is free to rotate. What torque does it experience?

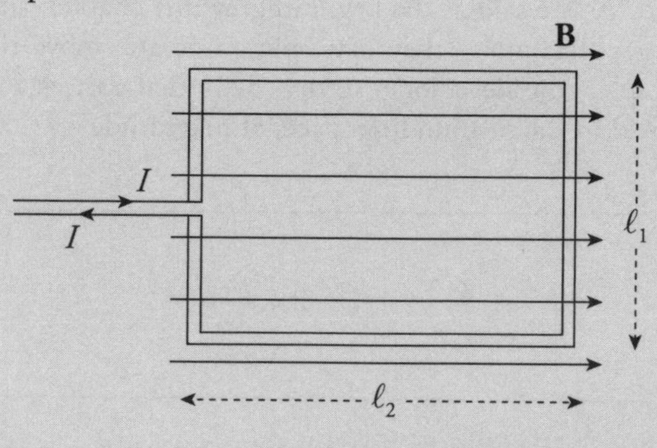

Solution. Ignoring the tiny gap in the vertical left-hand wire, we have two wires of length ℓ_1 and two of length ℓ_2. There is no magnetic force on either of the sides of the loop of length ℓ_2 because the current in the top side is parallel to **B** and the current in the bottom side is antiparallel to **B**. The magnetic force on the right-hand side points out of the plane of the page, while the magnetic force on the left-hand side points into the plane of the page.

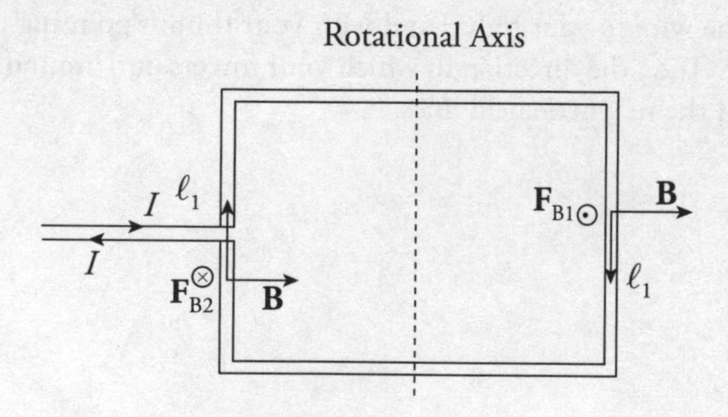

Each of these two forces exerts a torque that tends to turn the loop in such a way that the right-hand side rises out of the plane of the page, while the left-hand side rotates into the page. Relative to the axis shown above (which cuts the loop in half), the torque of $\mathbf{F}_{B1}$ is

$$\tau_1 = rF_{B1} \sin \theta = (\tfrac{1}{2}\ell_2)\,(I\,\ell_1 B) \sin 90° = \tfrac{1}{2} I\ell_1\ell_2 B$$

and the torque of $\mathbf{F}_{B2}$ is

$$\tau_2 = rF_{B2} \sin \theta = (\tfrac{1}{2}\ell_2)\,(I\ell_1 B) \sin 90° = \tfrac{1}{2} I\ell_1\ell_2 B$$

Since both these torques rotate the loop in the same direction, the net torque on the loop is

$$\tau_1 + \tau_2 = I\ell_1\ell_2 B$$

MAGNETIC FIELDS CREATED BY CURRENT-CARRYING WIRES

In the previous section, we examined the force that a current-carrying wire experiences when it is subjected to an external magnetic field. The source of the magnetic field in the previous section was unspecified. As we said at the beginning of this chapter, the sources of magnetic fields are electric charges that move; they may spin, circulate, move through space, or flow down a wire. For example, consider a long, straight wire that carries a current *I*. The current generates a magnetic field in the surrounding space, of magnitude

Equation Sheet

$$B = \frac{\mu_0}{2\pi}\frac{I}{r}$$

Magnetic Fields vs. Electric Fields
Unlike electric fields that start at a point charge and end at another point charge, magnetic fields are unending loops. Electric charges can be positive or negative and exist by themselves. There is no such thing as a monopole for magnets.

where r is the distance from the wire. The symbol μ_0 denotes a fundamental constant called the permeability of free space. Its value is

$$\mu_0 = 4\pi \times 10^{-7}\ \text{N/A}^2 = 4\pi \times 10^{-7}\ \text{T} \cdot \text{m/A}$$

The magnetic field lines are actually circles whose centers are on the wire. The direction of these circles is determined by a variation of the right-hand rule. Imagine grabbing the wire in your right hand with your thumb pointing in the direction of the current. Then the direction in which your fingers curl around the wire gives the direction of the magnetic field lines.

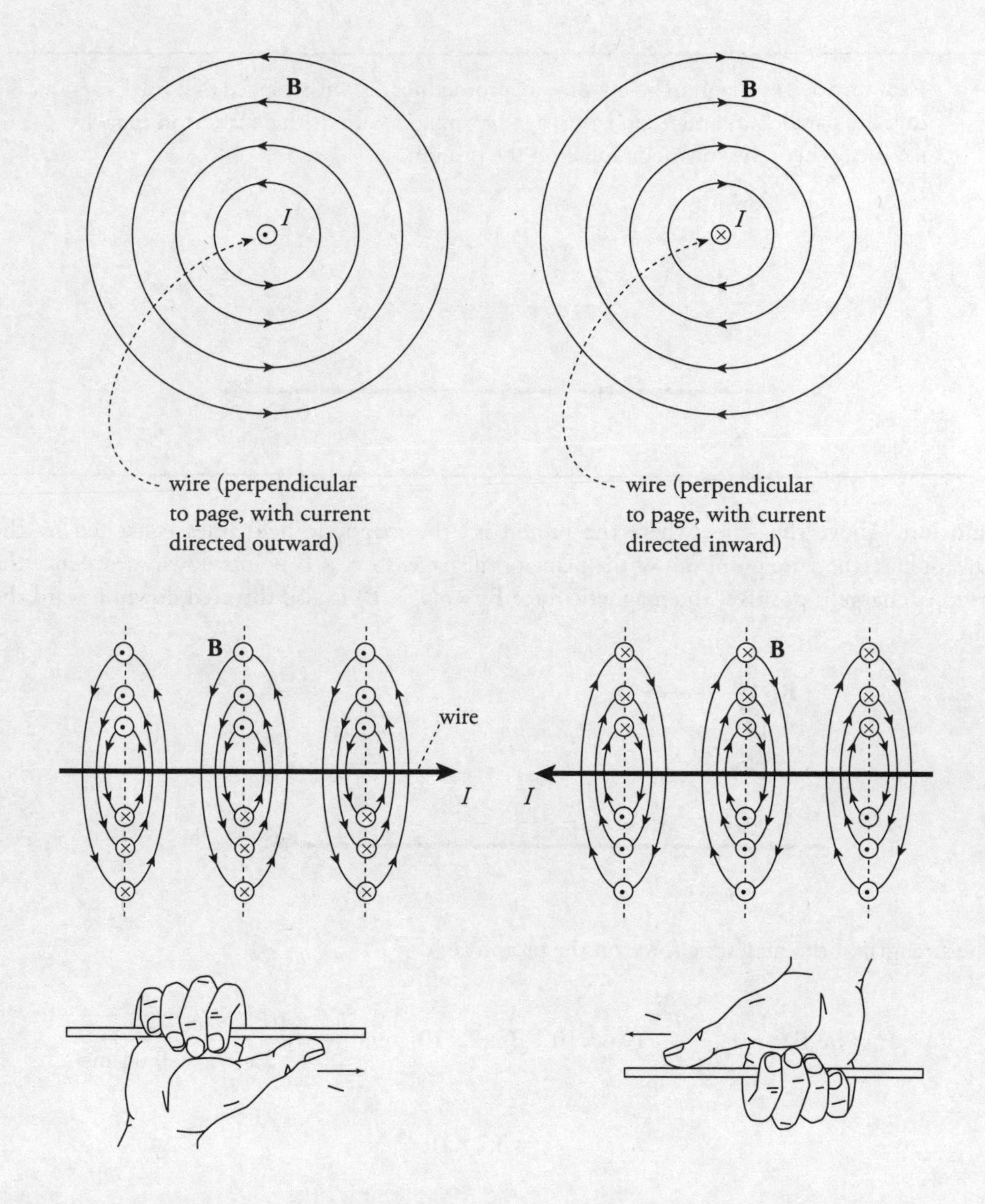

Right-Hand Rule for the Magnetic Field Created by a Current-Carrying Wire:

1. Put your thumb in the direction of the current or in the direction of a positive traveling charge.
2. Grab the wire/path.
3. As the fingers curl around your thumb, it represents the magnetic field going around the wire/path.

Example 8 The diagram below shows a proton moving with a speed of 2×10^5 m/s, initially parallel to, and 4 cm from, a long, straight wire. If the current in the wire is 20 A, what's the magnetic force on the proton?

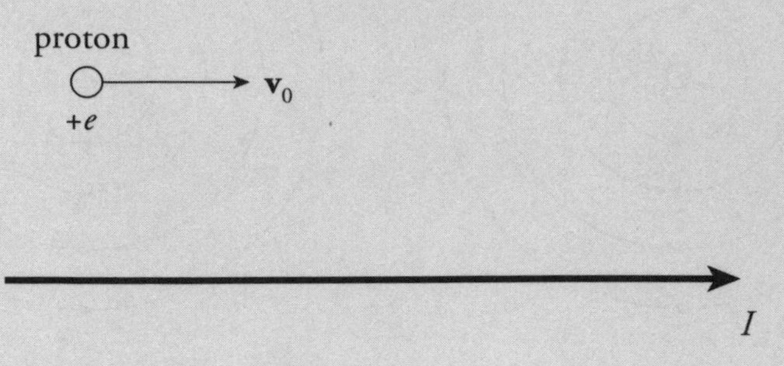

Solution. Above the wire (where the proton is), the magnetic field lines generated by the current-carrying wire point out of the plane of the page, so $\mathbf{v}_0 \times \mathbf{B}$ points downward. Since the proton's charge is positive, the magnetic force $\mathbf{F}_B = q(\mathbf{v}_0 \times \mathbf{B})$ is also directed down, toward the wire.

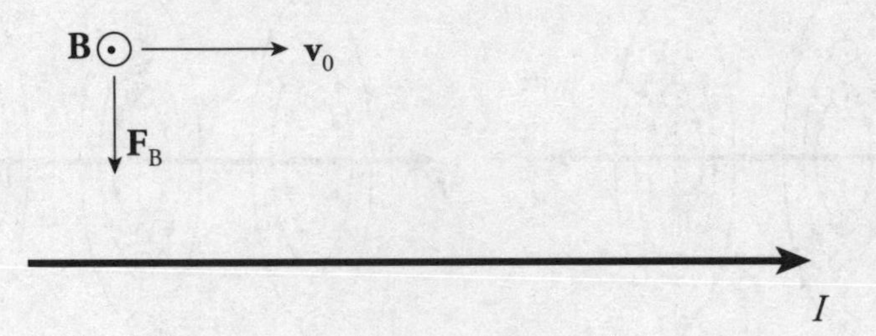

The strength of the magnetic force on the proton is

$$F_B = qv_0 B = ev_0 \frac{\mu_0}{2\pi}\frac{I}{r} = (1.6\times10^{-19}\text{ C})(2\times10^5\text{ m/s})\frac{4\pi\times10^{-7}\text{ N/A}^2}{2\pi}\frac{20\text{ A}}{0.04\text{ m}}$$

$$= 3.2\times10^{-18}\text{ N}$$

Example 9 The diagram below shows a pair of long, straight, parallel wires, separated by a small distance, r. If currents I_1 and I_2 are established in the wires, what is the magnetic force per unit length they exert on each other?

Solution. To find the force on Wire 2, consider the current in Wire 1 as the source of the magnetic field. Below Wire 1, the magnetic field lines generated by Wire 1 point into the plane of the page. Therefore, the force on Wire 2, as given by the equation $\mathbf{F}_{B2} = I_2(\ell_2 \times \mathbf{B}_1)$, points upward.

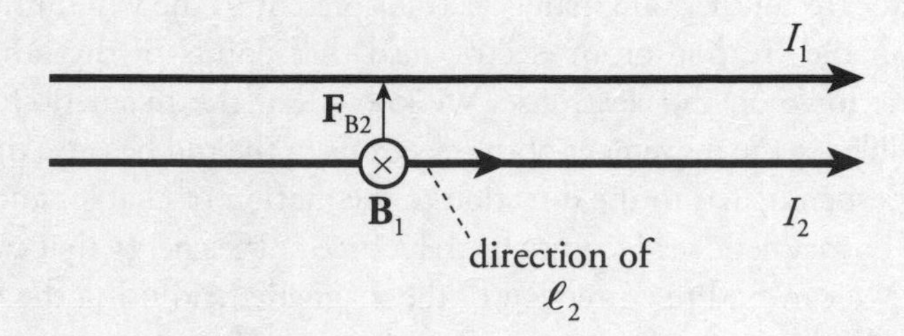

The magnitude of the magnetic force per unit length felt by Wire 2, due to the magnetic field generated by Wire 1, is found this way:

$$F_{B2} = I_2\ell_2 B_1 = I_2\ell_2 \frac{\mu_0}{2\pi}\frac{I_1}{r} \quad \Rightarrow \quad \frac{F_{B2}}{\ell_2} = \frac{\mu_0}{2\pi}\frac{I_1 I_2}{r}$$

By Newton's Third Law, this is the same force that Wire 1 feels due to the magnetic field generated by Wire 2. The force is attractive because the currents point in the same direction; if one of the currents were reversed, then the force between the wires would be repulsive.

SOLENOIDS CREATE UNIFORM FIELDS

The parallel-plate capacitor that was discussed in the chapter on electric fields appears frequently on the AP Physics 2 Exam because it generates a constant electric field in the region between the plates. Knowing that a current-carrying wire creates a magnetic field, if you look back at the section on bar magnets, you see that the magnetic field produced by a bar magnet closely resembles that produced by a single closed loop of current-carrying wire.

A device called a solenoid is constructed by a series of coaxial wires through which a continuous current flows. The magnetic fields from each of the individual loops of wire add together, and within the region inside the solenoid, the magnetic field becomes uniform. If the solenoid is very long in comparison to its diameter and the coils are tightly packed together, the field inside a solenoid created in a laboratory is very nearly uniform in both direction and strength.

Unlike a bar magnet, a solenoid is an electromagnet. When there is a current flowing through the wires of the solenoid, the solenoid creates a magnetic field. When there is no current flowing through the wire, there is not a magnetic field. The bar magnet, on the other hand, always produces a magnetic field.

MOTIONAL EMF

CED Unit 5.8
Magnetic Flux

The simple act of moving a conducting rod in the presence of an external magnetic field *creates an electric field* within the rod. The figure below shows a conducting wire of length ℓ moving with constant velocity v in the plane of the page through a uniform magnetic field **B** that's perpendicular to the page. The presence of a moving conductor (recall there are many electrons present in the wire) inside a magnetic field results in the creation of an electric field that points in the same direction as the magnetic force on the electrons. (We know that the magnetic force is not directly responsible for the movement of the electrons in the rod because the magnetic force is *always* perpendicular to the direction of the motion of charges and therefore does no work. The magnetic field cannot be the source of the energy that causes the charges to move.) We can find the direction of the magnetic field using the right-hand rule the instant the charges begin to move. The charges are negative and moving to the right, so $q\mathbf{v}$ points to the left (thumb of your right hand). **B** points into the page (fingers of your right hand). Your palm shows the direction of the magnetic force is downward. Therefore, the motion of the rod through the external magnetic field generates an electric field within the rod that causes electrons to move toward the bottom of the rod. This *induced electric field* points downward.

Vice-Versa
Earlier, we learned that moving charges generate magnetic fields. If we think about this backward, magnetic fields then can generate moving charges or current.

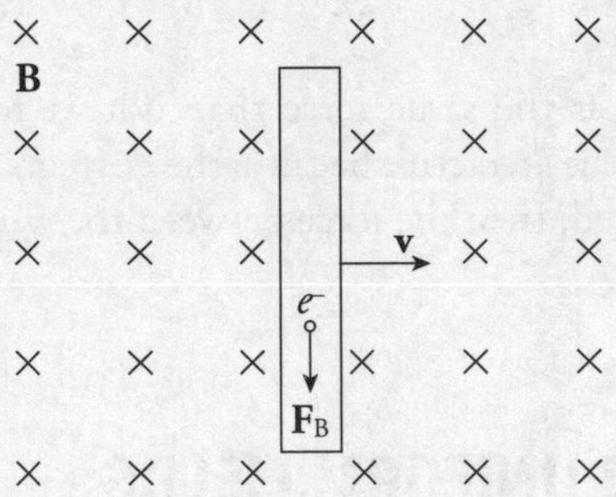

Right-Hand Rules
You will have to use a combination of your right-hand rules in order to find the direction of current. Just pay attention to orientation.

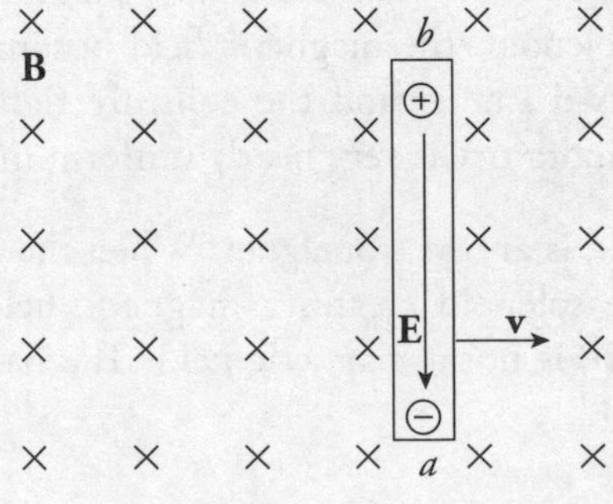

As long as the rod continues to move at velocity **v**, the electric field will be maintained within the rod. Electric fields are generated by charge separations, and therefore there will be a surfeit of negative charges on the bottom of the rod and a deficit of electrons at the top of the rod.

A charge q in the wire feels two forces: an electric force, $\mathbf{F}_E = q\mathbf{E}$, and a magnetic force, $F_B = qvB\sin\theta = qvB$, because $\theta = 90°$.

If q is negative, $\mathbf{F}_E$ is upward and $\mathbf{F}_B$ is downward; if q is positive, $\mathbf{F}_E$ is downward, and $\mathbf{F}_B$ is upward. So, in both cases, the forces act in opposite directions. Once the magnitude of $\mathbf{F}_E$ equals the magnitude of $\mathbf{F}_B$, the charges in the wire are in electromagnetic equilibrium. This occurs when $qE = qvB$; that is, when $E = vB$.

The presence of the electric field requires a potential difference between the ends of the rod. Since negative charge accumulates at the lower end (which we'll call point *a*) and positive charge accumulates at the upper end (point *b*), point *b* is at a higher electric potential. The potential difference V_{ba} is equal to $E\ell$ and, since $E = vB$, the potential difference can be written as $vB\ell$.

Now, imagine that the rod is sliding along a pair of conducting rails connected at the left by a stationary bar. The sliding rod now completes a rectangular circuit, and the potential difference V_{ba} causes current to flow.

Induced Current

An induced current can be created three different ways:

1. changing the area of the loop of wire in a stationary magnetic field
2. changing the magnetic field strength through a stationary circuit
3. changing the angle between the magnetic field and the wire loop

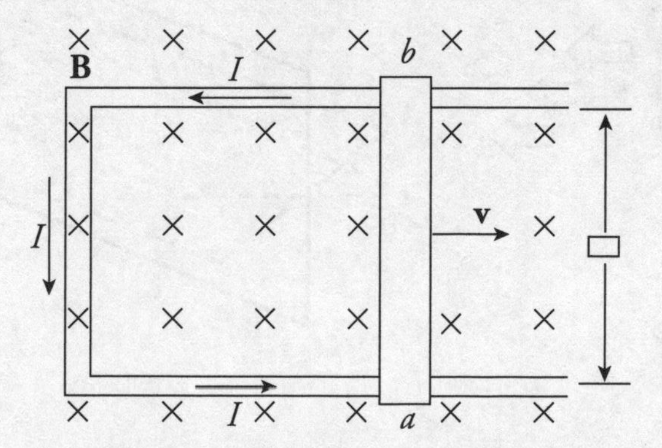

Each of these three changes causes a change in the flux, as we will see in the next section. The motion of the sliding rod through the magnetic field creates an electromotive force, called **motional emf**:

$$\mathcal{E} = B\ell v$$

Equation Sheet

The existence of a current in the sliding rod causes the magnetic field to exert a force on it. Using the formula $F_B = BI\ell$, the fact that ℓ points upward (in the direction of the current) and

B is into the page, tells us that the direction of $\mathbf{F}_B$ on the rod is to the left. An external agent must provide this same amount of force to the right to maintain the rod's constant velocity and keep the current flowing. The power that the external agent must supply is $P = Fv = I\ell Bv$, and the electrical power delivered to the circuit is $P = IV_{ba} = I\mathcal{E} = IvB\ell$. Notice that these two expressions are identical. The energy provided by the external agent is transformed first into electrical energy and then thermal energy as the conductors making up the circuit dissipate heat.

This relationship between current in a coil of wire and magnetic fields sets the basis for Faraday's Law.

FARADAY'S LAW OF ELECTROMAGNETIC INDUCTION

Electromotive force can be created by the motion of a conducting wire through a magnetic field, but there is another way to generate an emf from a magnetic field.

Faraday discovered that a current is induced when the magnetic flux passing through the coil or loop of wire changes. Magnetic flux helps us explain what that amount of field passing into the loop means.

Imagine holding a loop of wire in front of a fan as shown:

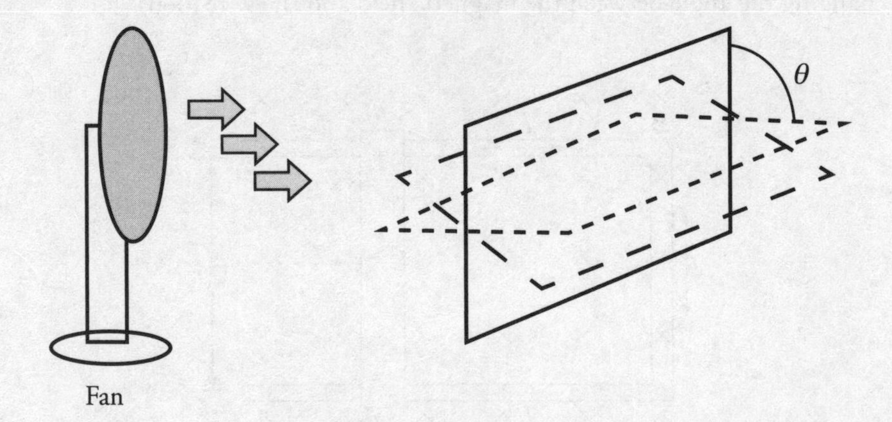

The amount of air that flows through the loop depends on the area of the loop as well as its orientation (tilt angle θ).

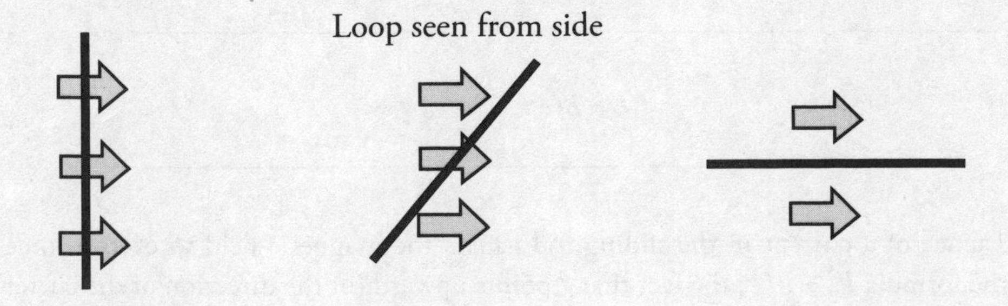

The most effective airflow occurs when the loop is completely perpendicular, as in the situation to the left. The least effective occurs when the airflow and loop are in the situation to the right.

We can apply this idea to a magnetic field passing through a loop. The magnetic flux, Φ_B, through an area A is equal to the product of **A** and the magnetic field parallel to the area vector. The area vector points normal to the surface.

$$\Phi_B = \mathbf{B} \bullet \mathbf{A} = BA \cos \theta$$

Equation Sheet

Magnetic flux measures the density of magnetic field lines that cross through an area. (Note that the direction of A is taken to be perpendicular to the plane of the loop.)

Magnetic Flux Units
The SI unit for magnetic flux is called the weber (Wb), which is equivalent to one Tesla meter-squared (T·m^2).

Example 10 The figure below shows two views of a circular loop of radius 3 cm placed within a uniform magnetic field, **B** (magnitude 0.2 T).

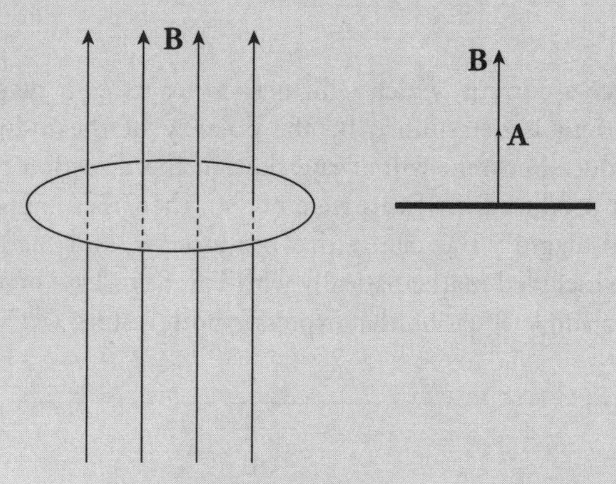

(a) What's the magnetic flux through the loop?
(b) What would be the magnetic flux through the loop if the loop were rotated 45°?
(c) What would be the magnetic flux through the loop if the loop were rotated 90°?

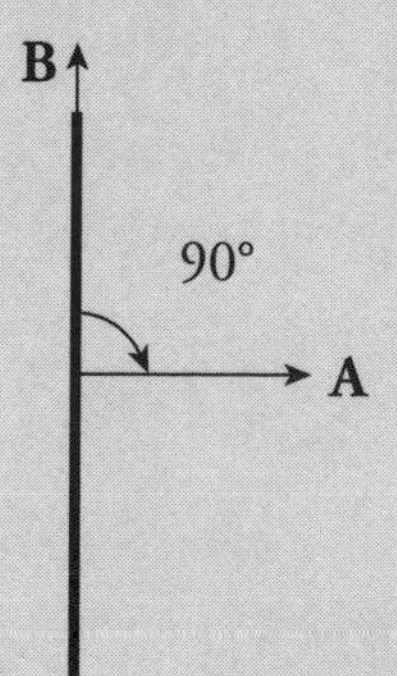

Solution.

(a) Since **B** is parallel to **A**, the magnetic flux is equal to BA:

$$\Phi_B = BA = B \cdot \pi r^2 = (0.2 \text{ T}) \cdot \pi(0.03 \text{ m})^2 = 5.7 \times 10^{-4} \text{ T·m}^2$$

The SI unit for magnetic flux, the tesla meter^2, is called a **weber** (abbreviated **Wb**). So $\Phi_B = 5.7 \times 10^{-4}$ Wb.

(b) Since the angle between **B** and **A** is 45°, the magnetic flux through the loop is

$$\Phi_B = BA \cos 45° = B \cdot \pi r^2 \cos 45° = (0.2 \text{ T}) \cdot \pi(0.03 \text{ m})^2 \cos 45° = 4.0 \times 10^{-4} \text{ Wb}$$

(c) If the angle between **B** and **A** is 90°, the magnetic flux through the loop is zero, since cos 90° = 0.

The concept of magnetic flux is crucial because changes in magnetic flux induce emf. According to **Faraday's Law of Electromagnetic Induction**, the magnitude of the emf induced in a circuit is equal to the rate of change of the magnetic flux through the circuit. This can be written mathematically in the form

$$\left|\mathcal{E}_{avg}\right| = \left|\frac{\Delta\Phi_B}{\Delta t}\right|$$

This induced emf can produce a current, which will then create its own magnetic field. The direction of the induced current is determined by the polarity of the induced emf and is given by **Lenz's Law**: the induced current will always flow in the direction that opposes the change in magnetic flux that produced it. If this were not so, then the magnetic flux created by the induced current would magnify the change that produced it, and energy would not be conserved. Lenz's Law can be included mathematically with Faraday's Law by the introduction of a minus sign; this leads to a single equation that expresses both results:

Equation Sheet

$$\mathcal{E}_{avg} = -\frac{\Delta\Phi_B}{\Delta t}$$

Example 11 The circular loop of Example 10 rotates at a constant angular speed through 45° in 0.5 s.

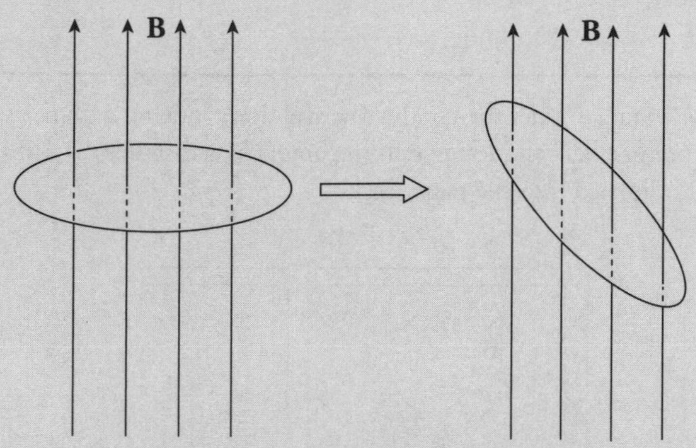

(a) What's the induced emf in the loop?

(b) In which direction will current be induced to flow?

Solution.

(a) As we found in Example 10, the magnetic flux through the loop changes when the loop rotates. Using the values we determined earlier, Faraday's Law gives

$$\mathcal{E}_{\text{avg}} = -\frac{\Delta\Phi_B}{\Delta t} = -\frac{(4.0\times10^{-4}\text{ Wb})-(5.7\times10^{-4}\text{ Wb})}{0.5\text{ s}} = 3.4\times10^{-4}\text{ V}$$

(b) The original magnetic flux was 5.7×10^{-4} Wb upward, and it was decreased to 4.0×10^{-4} Wb. So the change in magnetic flux is -1.7×10^{-4} Wb upward, or, equivalently, $\Delta\Phi_B = 1.7 \times 10^{-4}$ Wb downward. To oppose this change, we would need to create some magnetic flux upward. The current would be induced in the counterclockwise direction (looking down on the loop), because the right-hand rule tells us that then the current would produce a magnetic field that would point up.

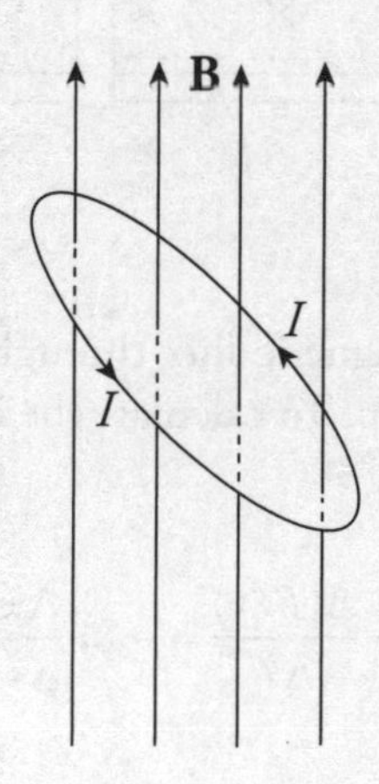

The current will flow only while the loop rotates, because emf is induced only when magnetic flux is changing. If the loop rotates 45° and then stops, the current will disappear.

Example 12 Again consider the conducting rod that's moving with constant velocity **v** along a pair of parallel conducting rails (separated by a distance ℓ), within a uniform magnetic field directed into the page, **B**:

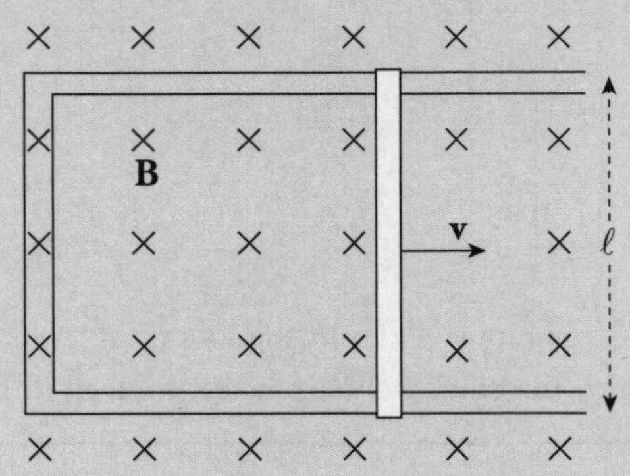

Find the induced emf and the direction of the induced current in the rectangular circuit.

Solution. The area of the rectangular loop is ℓx, where x is the distance from the left-hand bar to the moving rod:

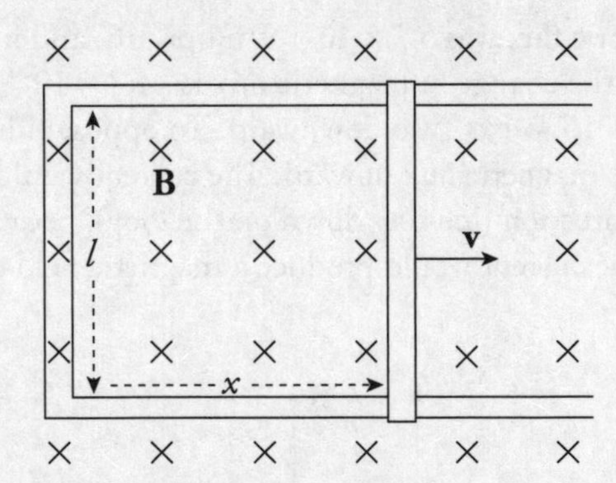

Because the area is changing, the magnetic flux through the loop is changing, which means that an emf will be induced in the loop. To calculate the induced emf, we first write $\Phi_B = BA = B\ell x$; then, since $\Delta x/\Delta t = v$, we get

$$\mathcal{E}_{avg} = -\frac{\Delta\Phi_B}{\Delta t} = -\frac{\Delta(B\ell x)}{\Delta t} = -B\ell\frac{\Delta x}{\Delta t} = -B\ell v$$

We can figure out the direction of the induced current from Lenz's Law. As the rod slides to the right, the magnetic flux into the page increases. How do we oppose an increasing into-the-page flux? By producing out-of-the-page flux. In order for the induced current to generate a magnetic field that points out of the plane of the page, the current must be directed counterclockwise (according to the right-hand rule).

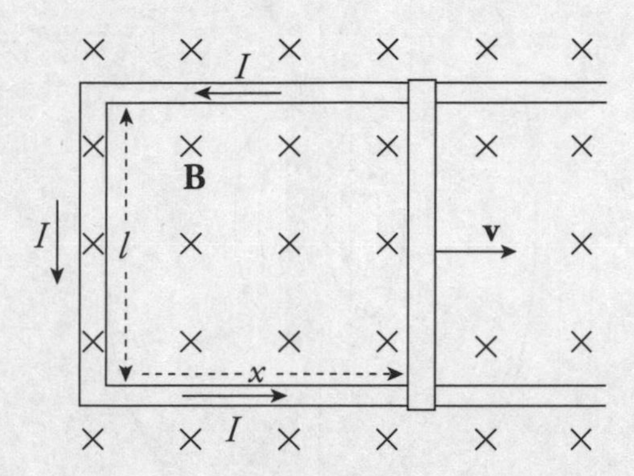

Note that the magnitude of the induced emf and the direction of the current agree with the results we derived earlier, in the section on motional emf.

This example also shows how a violation of Lenz's Law would lead directly to a violation of the Law of Conservation of Energy. The current in the sliding rod is directed upward, as given by Lenz's Law, so the conduction electrons are drifting downward. The force on these drifting electrons—and thus, the rod itself—is directed to the left, opposing the force that's pulling the rod to the right. If the current were directed downward, in violation of Lenz's Law, then the magnetic force on the rod would be to the right, causing the rod to accelerate to the right with ever-increasing speed and kinetic energy, without the input of an equal amount of energy from an external agent.

Example 13 A permanent magnet creates a magnetic field in the surrounding space. The end of the magnet at which the field lines emerge is designated the **north pole** (N), and the other end is the **south pole** (S):

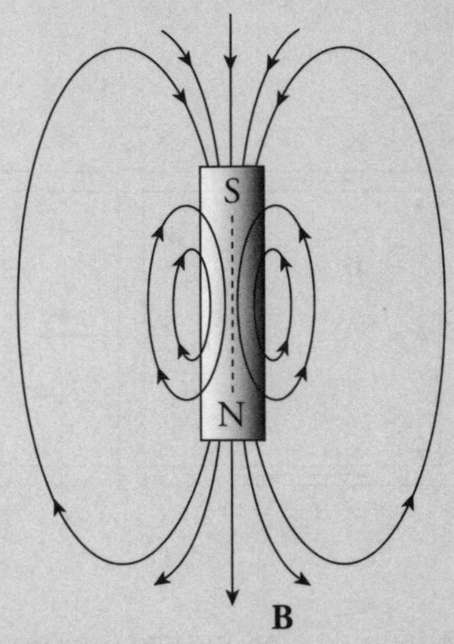

(a) The figure below shows a bar magnet moving down, through a circular loop of wire. What will be the direction of the induced current in the wire?

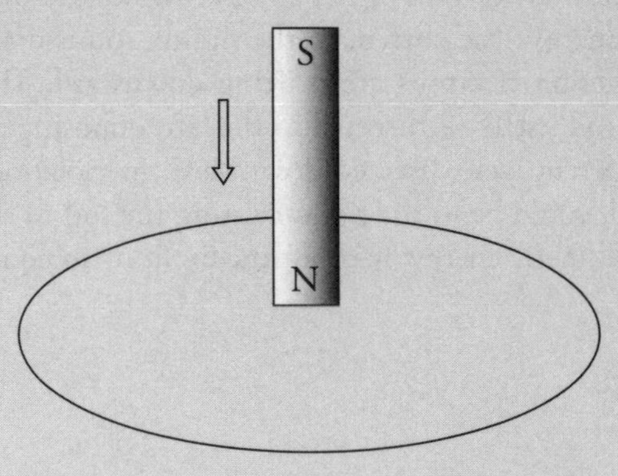

(b) What will be the direction of the induced current in the wire if the magnet is moved as shown in the following diagram?

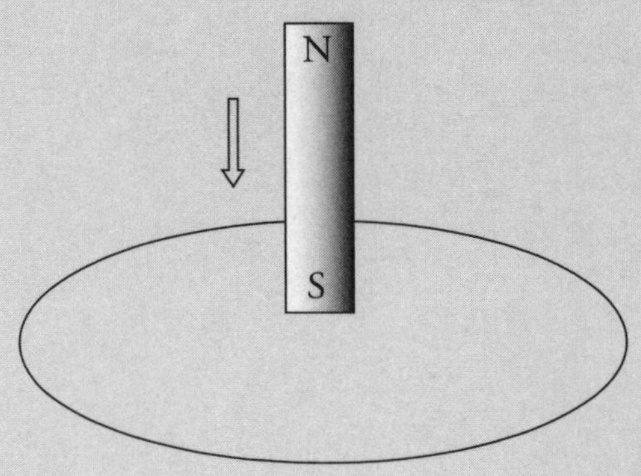

Solution.

(a) The magnetic flux down, through the loop, increases as the magnet is moved. By Lenz's Law, the induced emf will generate a current that opposes this change. How do we oppose a change of *more flux downward*? By creating flux *upward*. So, according to the right-hand rule, the induced current must flow counterclockwise (because this current will generate an upward-pointing magnetic field):

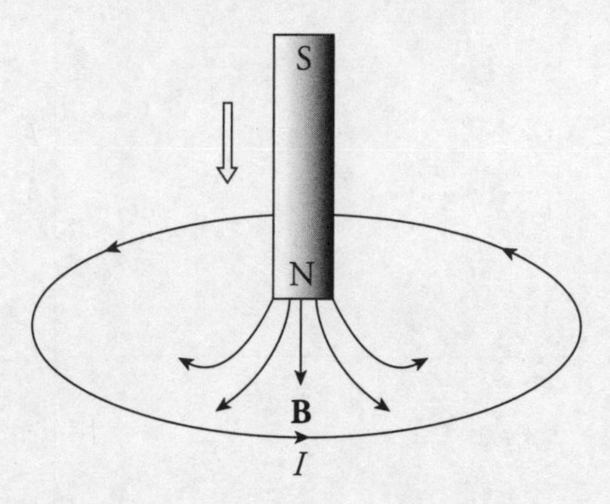

(b) In this case, the magnetic flux through the loop is upward. As the south pole moves closer to the loop, the magnetic field strength increases, so the magnetic flux through the loop increases upward. How do we oppose a change of *more flux upward*? By creating flux *downward*. Therefore, in accordance with the right-hand rule, the induced current will flow clockwise (because this current will generate a downward-pointing magnetic field):

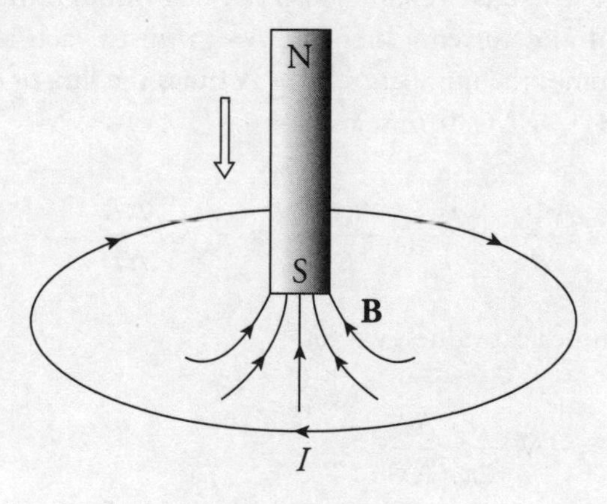

Example 14 A square loop of wire 2 cm on each side contains 5 tight turns and has a total resistance of 0.0002 Ω. It is placed 20 cm from a long, straight, current-carrying wire. If the current in the straight wire is increased at a steady rate from 20 A to 50 A in 2 s, determine the magnitude and direction of the current induced in the square loop. (Because the square loop is at such a great distance from the straight wire, assume that the magnetic field through the loop is uniform and equal to the magnetic field at its center.)

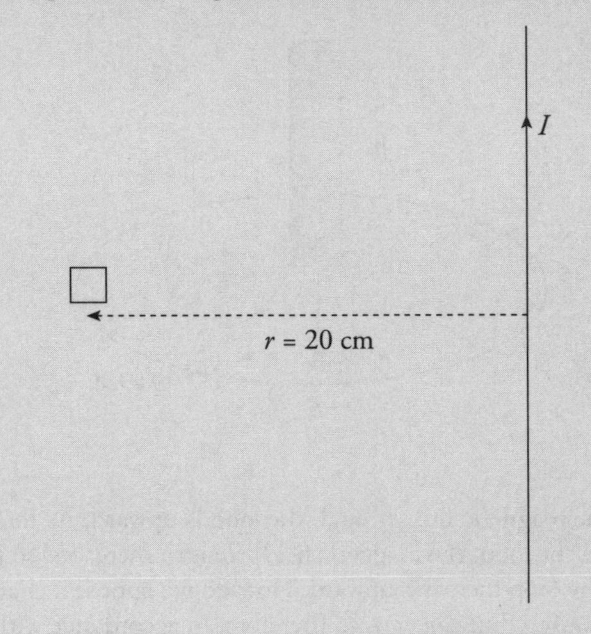

Solution. At the position of the square loop, the magnetic field due to the straight wire is directed out of the plane of the page, and its strength is given by the equation $B = (\mu_0/2\pi)(I/r)$. As the current in the straight wire increases, the magnetic flux through the turns of the square loop changes, inducing an emf and current. There are $N = 5$ turns; each loop contributes the same flux, so the total flux becomes the number of loops N times the flux of each individual loop, Φ_B. Faraday's Law becomes $\mathcal{E}_{avg} = -N(\Delta\Phi_B/\Delta t)$, and

$$\mathcal{E}_{avg} = -N\frac{\Delta\Phi_B}{\Delta t} = -N\frac{\Delta(BA)}{\Delta t} = -NA\frac{\Delta B}{\Delta t} = -NA\frac{\mu_0}{2\pi r}\frac{\Delta I}{\Delta t}$$

Substituting the given numerical values, we get

$$\begin{aligned}\mathcal{E}_{avg} &= -NA\frac{\mu_0}{2\pi r}\frac{\Delta I}{\Delta t}\\ &= -(5)(0.02\text{ m})^2\frac{4\pi\times10^{-7}\text{ T}\cdot\text{m/A}}{2\pi(0.20\text{ m})}\frac{(50\text{ A}-20\text{ A})}{2\text{ s}}\\ &= -3\times10^{-8}\text{ V}\end{aligned}$$

The magnetic flux through the loop is out of the page and increases as the current in the straight wire increases. To oppose an increasing out-of-the-page flux, the direction of the induced current should be clockwise, thereby generating an into-the-page magnetic field (and flux).

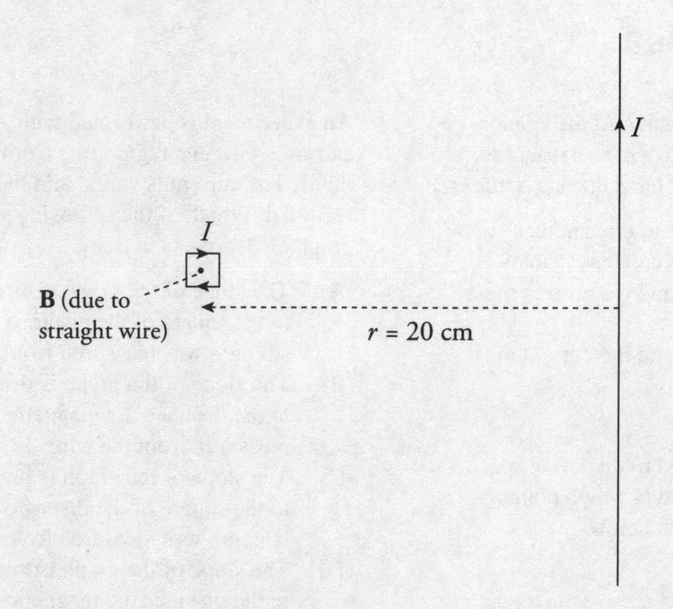

The value of the current in the loop will be

$$I = \frac{\mathcal{E}}{R} = \frac{3 \times 10^{-8}\ \text{V}}{0.0002\ \Omega} = 1.5 \times 10^{-4}\ \text{A}$$

Chapter 8 Review Questions

Solutions can be found in Chapter 11.

Section I: Multiple Choice

1. Two long wires carry a nonzero current I. At the location midway between the two wires, there will be a magnetic field strength of 0 T only if which of the following is true?

(A) The wires are perpendicular to one another.
(B) The wires must be coiled to create solenoids.
(C) The wires must be parallel and the current must flow in the same direction.
(D) The wires must be parallel and the current must flow in opposite directions.

2. A particle of charge $-q$ is launched to the left at speed **v** through a uniform magnetic field **B** which points into the second quadrant, as shown below.

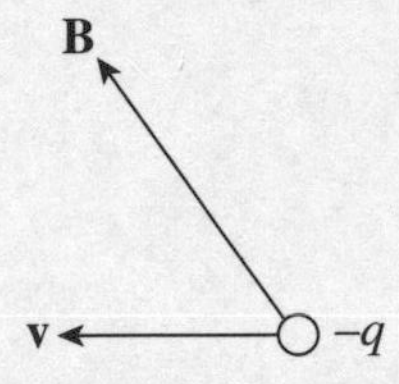

Another particle is then launched and experiences a magnetic force of the same magnitude and direction as the first particle. Which of the following could NOT be the conditions for the second particle?

(A) **B** is unchanged, but q is now positive and **v** is directed to the right.
(B) The charge is still negative, but **v** and **B** are rotated 90° into the plane of the paper so that **v** is directed along the $-z$-axis.
(C) The charge is still negative, but **v** and **B** are rotated 90° counterclockwise so that **v** is directed along the $-y$-axis.
(D) **v** is unchanged, but q is now positive and **B** is rotated 180° to point into the fourth quadrant.

3. An experiment is performed with a long current-carrying wire in a region free from any other magnetic fields. The current is varied and the field strength is recorded. Which of the following statements about a graph of **B** versus I is true?

(A) The slope of the graph is directly proportional to the square of the distance the magnetic field strength was measured from the wire.
(B) The slope of the graph is directly proportional to the distance the magnetic field strength was measured from the wire.
(C) The slope of the graph is inversely proportional to the square of the distance the magnetic field strength was measured from the wire.
(D) The slope of the graph is inversely proportional to the distance the magnetic field strength was measured from the wire.

4. Which of the following situations would result in the largest magnetic field?

(A) Measuring at a distance r from a wire carrying a current of I
(B) Measuring at a distance $2r$ from a wire carrying a current of $I/2$
(C) Measuring at a distance $r/2$ from a wire carrying a current of $2I$
(D) All three measurements would result in the same magnetic field.

5. In the figure below, what is the magnetic field at the point P, which is midway between the two wires?

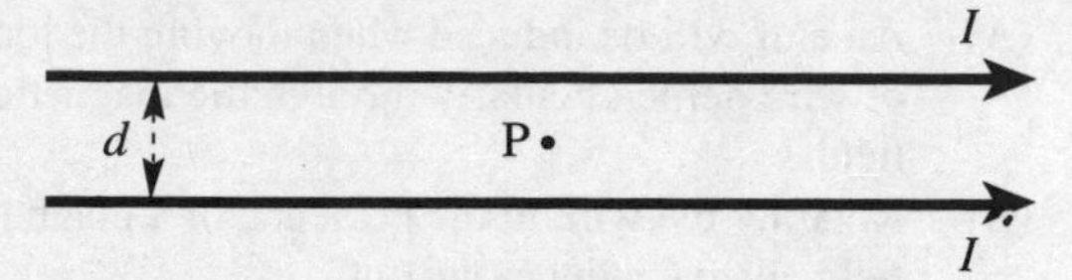

(A) $2\mu_0 I/(\pi d)$, into the plane of the page
(B) $\mu_0 I/(2\pi d)$, out of the plane of the page
(C) $\mu_0 I/(2\pi d)$, into the plane of the page
(D) Zero

6. Here is a section of a wire with a current moving to the right. Where is the magnetic field strongest and pointing INTO the page?

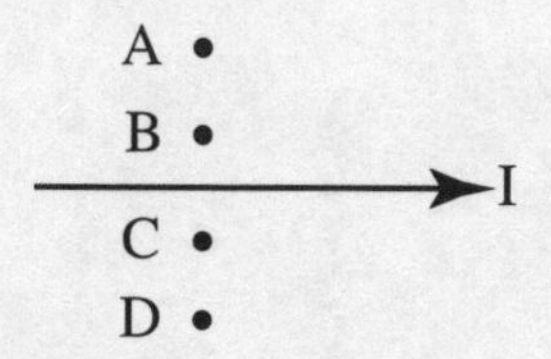

(A) A
(B) B
(C) C
(D) D

7. What is the direction of force acting on the current-carrying wire as shown below?

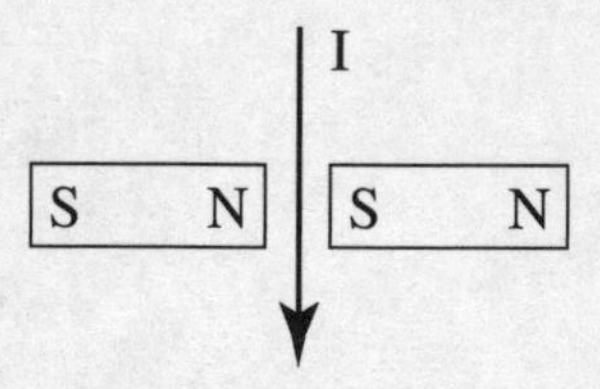

(A) To the bottom of the page
(B) Into the page
(C) Out of the page
(D) To the right of the page

8. A metal rod of length L is pulled upward with constant velocity **v** through a uniform magnetic field **B** that points into the plane of the page.

What is the potential difference between points a and b?

(A) 0

(B) $\frac{1}{2}vBL$, with point b at the higher potential

(C) vBL, with point a at the higher potential

(D) vBL, with point b at the higher potential

9. A conducting rod of length 0.2 m and resistance 10 ohms between its endpoints slides without friction along a U-shaped conductor in a uniform magnetic field B of magnitude 0.5 T perpendicular to the plane of the conductor, as shown in the diagram below.

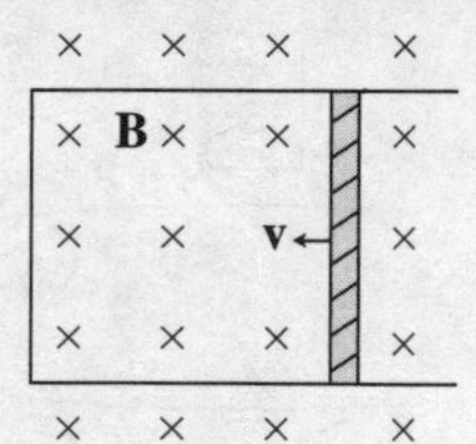

If the rod is moving with velocity **v** = 3 m/s to the left, what is the magnitude and direction of the current induced in the rod?

(A) 0.03A down the rod
(B) 0.03A up the rod
(C) 0.3A down the rod
(D) 0.3A up the rod

10. As shown in the figures below, a bar magnet is moved at a constant speed through a loop of wire. Figure A shows the bar magnet when it is at a position below the loop of wire. Figure B shows the loop of wire after the bar magnet has passed completely through the loop.

Figure A

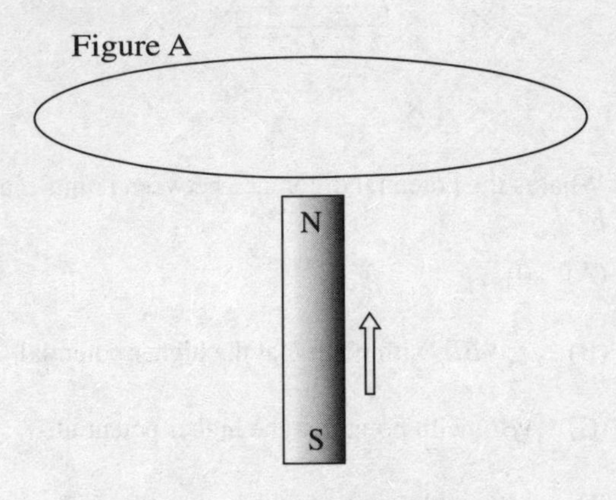

Figure B

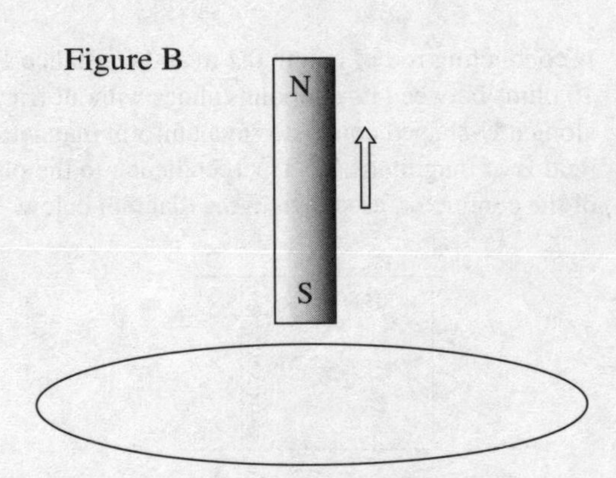

Which of the following best describes the direction or directions of the current induced in the loop when the loop is looked at from above? Note that when looking at the loop from above, the bar magnet will be moving toward the viewer.

(A) Always clockwise
(B) Always counterclockwise
(C) First clockwise, then counterclockwise
(D) First counterclockwise, then clockwise

11. Which of the following statements about induced emf is true?

(A) An emf will be induced when moving the loop of wire perpendicularly through the magnetic field.
(B) Rotating the wire in the presence of a magnetic field always induces an emf.
(C) Shrinking the size of a loop of wire in constant magnetic field will induce an emf in the wire.
(D) A time varying magnetic field is required to induce an emf.

Section II: Free Response

1. The diagram below shows a simple mass spectrograph. It consists of a source of ions (charged atoms) that are accelerated (essentially from rest) by the voltage V and enter a region containing a uniform magnetic field $\mathbf{B}$. The polarity of V may be reversed so that both positively charged ions (cations) and negatively charged ions (anions) can be accelerated. Once the ions enter the magnetic field, they follow a semicircular path and strike the front wall of the spectrograph, on which photographic plates are constructed to record the impact. Assume that the ions have mass m.

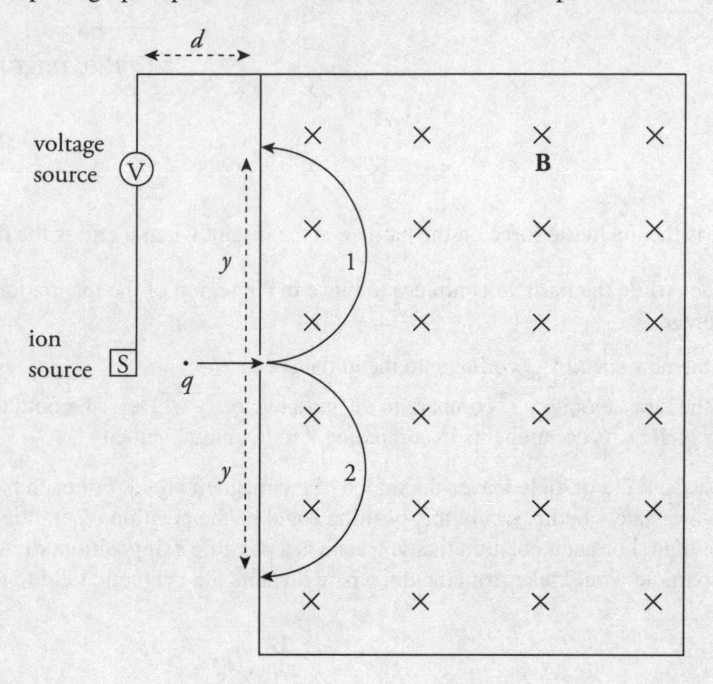

(A) What is the acceleration of an ion of charge q just before it enters the magnetic field?

(B) Find the speed with which an ion of charge q enters the magnetic field.

(C) Answer the following:

i. Which semicircular path, 1 or 2, would a cation follow?

ii. Which semicircular path, 1 or 2, would an anion follow?

(D) Determine the mass of a cation entering the apparatus in terms of y, q, $\mathbf{B}$, and V.

(E) Once a cation of charge q enters the magnetic field, how long does it take to strike the photographic plate?

(F) What is the work done by the magnetic force in the spectrograph on a cation of charge q?

2. A region of uniform magnetic field directed into the page (in the $-z$ direction) is generated. A positively charged particle traveling in the $+x$ direction at a speed v enters the region of the magnetic field, as shown below.

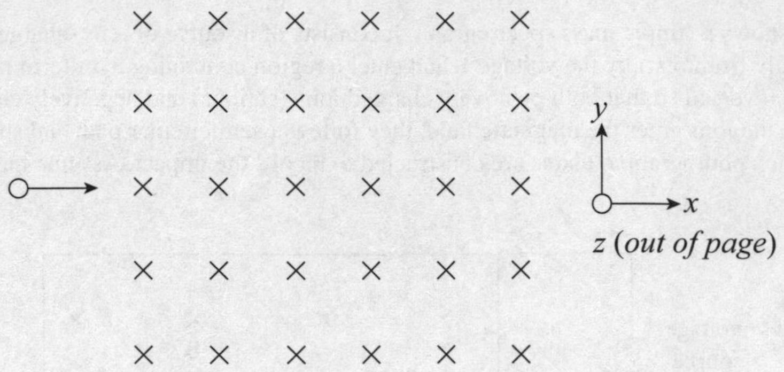

(A) What direction is the magnetic force on the particle at the instant when it enters the magnetic field region?

(B) Some time passes while the particle continues to move in the region of the magnetic field under the influence of the magnetic force.

i. How will the new speed v_{new} compare to the initial speed v?

ii. How will the new velocity $\overrightarrow{v_{new}}$ compare to the initial velocity $\vec{v}$? Describe both the magnitude of the velocity as well as its components in comparing it to the initial velocity.

(C) More time passes and the particle leaves the region of the magnetic field. For each position A, B, C, and D on the diagram below, state whether or not that position could be the position where the particle leaves the region of the magnetic field. For each position that you state is a possible exit position, draw the path on the diagram below that the particle would take from the dot where it enters the magnetic field to the letter where it exits.

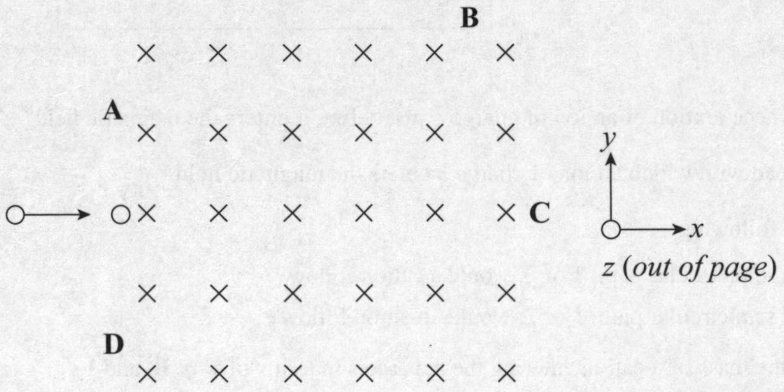

3. A rectangular wire is pulled through a uniform magnetic field of 2 T going into the page, as shown. The resistor has a resistance of 20 Ω.

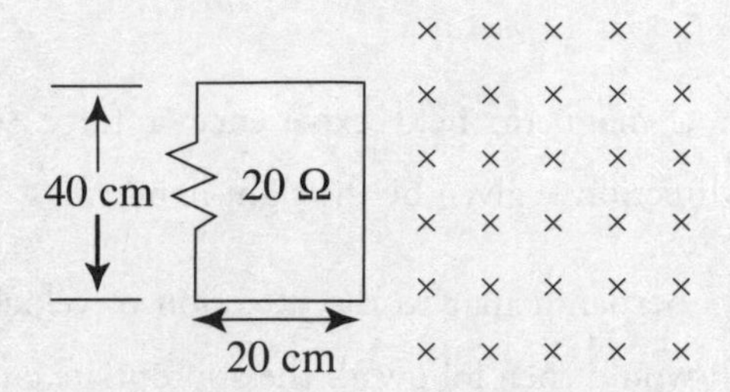

(A) What is the voltage across the resistor as the wire is pulled horizontally at a velocity of 1 m/s and it just enters the field?

(B) What is the current through the circuit in the above case and in what direction does it flow?

(C) The region containing the magnetic field is 1 meter long in the direction the loop is being pulled. If the loop is pulled at the constant velocity of 1 m/s, describe the current in the loop from a time before the loop encounters the left edge of the region of the magnetic field until the time that the left edge of the loop completely passes out of the magnetic field region. Indicate the values and the direction of the current.

(D) The loop of wire is rotated 90° clockwise so that the 20 cm side is vertical and the 40 cm side is horizontal at the top. Explain, without relying solely on equations, whether the loop would have to be pulled faster than 1 m/s, at exactly 1 m/s, or slower than 1 m/s for the current to have the same magnitude as in part (b).

Summary

- Charges moving through a magnetic field experience a force whose magnitude is given by $F_B = qvB\sin\theta$ and whose direction is given by the right-hand rule.
- Because the force is always perpendicular to the direction of velocity, the charge may experience uniform circular motion. It would then follow all the appropriate circular motion relationships and orbit in a radius given by $r = \frac{mv}{qB}$.
- Because wires have charges moving through them, a wire will experience a force if placed in a magnetic field. This is expressed by $F_B = BI\ell\sin\theta$. Also, a current-carrying wire will produce a magnetic field whose strength is given by $B = \frac{\mu_o}{2\pi}\frac{I}{r}$, where $\mu_0 = 4\pi \times 10^{-7}$ T·m/A, I is the current through the wire, and r is the radial distance from the wire.
- An electromotive force is produced as a conducting wire's position changes or as the magnetic field changes. This idea is summarized by $\varepsilon = B\ell v$.
- The flux (Φ_B) tells the amount of something (for this chapter, the magnetic field) that goes through a surface. The magnetic flux depends on the strength of the magnetic field, the surface area through which the field passes, and the angle between the two. This idea is summed up by $\Phi_B = BA\cos\theta$.
- Faraday's Law of Induction says that if a wire is formed in a loop, an electromotive force is produced if the magnetic flux through the loop changes with time. This idea can be summarized by $\varepsilon = -\frac{\Delta\Phi_B}{\Delta t}$.

Chapter 9
Geometric and Physical Optics

The beginning of this chapter deals with CED Unit 6.1: Waves in addressing "the wave nature of light." The latter half of this chapter deals with another description of light as a ray.

INTRODUCTION

Light (or visible light) makes up only a small part of the entire spectrum of electromagnetic waves, which ranges from radio waves to gamma rays. Most waves require a material medium for their transmission, but electromagnetic waves can propagate through empty space. Electromagnetic waves consist of time-varying electric and magnetic fields that oscillate perpendicular to each other and to the direction of propagation of the wave. Through a vacuum, all electromagnetic waves travel at a fixed speed:

$$c = 3.00 \times 10^8 \text{ m/s}$$

regardless of their frequency. Like all the waves, we can say $v = f\lambda$.

CED Unit 6.2
Electromagnetic Waves

CED Unit 6.3
Periodic Waves

ELECTROMAGNETIC WAVES

We saw previously that if we oscillate one end of a long rope, we generate a wave that travels down the rope and has the same frequency as that of the oscillation.

You can think of an electromagnetic wave in a similar way: an oscillating electric charge generates an electromagnetic (EM) wave, which is composed of oscillating electric and magnetic fields. These fields oscillate with the same frequency at which the electric charge that created the wave oscillated. The fields oscillate in phase with each other, perpendicular to each other and the direction of propagation. For this reason, electromagnetic waves are transverse waves. The direction in which the wave's electric field oscillates is called the direction of polarization of the wave.

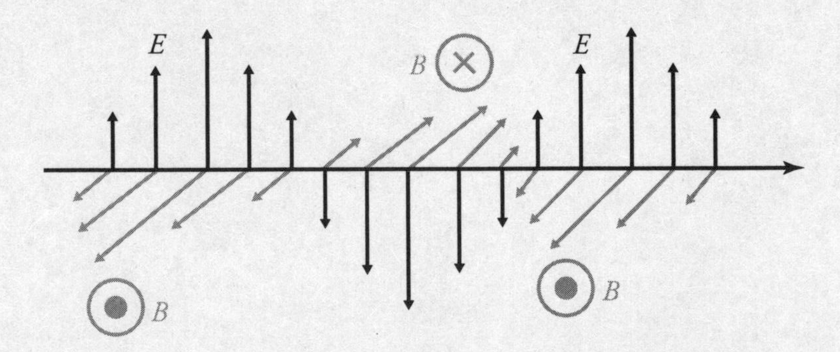

THE ELECTROMAGNETIC SPECTRUM

Electromagnetic waves can be categorized by their frequency (or wavelength); the full range of waves is called the **electromagnetic** (or **EM**) **spectrum**. Types of waves include **radio waves**, **microwaves**, **infrared**, **visible light**, **ultraviolet**, **X-rays**, and **Υ-rays** (**gamma rays**) and, although they've been delineated in the spectrum below, there's no universal agreement on all the boundaries, so many of these bands overlap. You should be familiar with the names of the major categories, and, in particular, memorize the order of the colors within the visible spectrum (which, as you can see, accounts for only a tiny sliver of the full EM spectrum). In order of increasing wave frequency, the colors are red, orange, yellow, green, blue, and violet, which is commonly remembered as ROYGBV ("roy-gee-biv"). The wavelengths of the colors in the visible spectrum are usually expressed in nanometers (nm). For example, electromagnetic waves whose wavelengths are between 577 nm and 597 nm are seen as yellow light.

Electromagnetic Wave Speed
All electromagnetic waves, regardless of frequency, travel through a vacuum at this speed. The most important equation for waves, $v = \lambda f$, is also true for electromagnetic waves. For EM waves traveling through a vacuum, $v = c$, so the equation becomes $\lambda f = c$.

Visible Spectrum
White light is not colorless. White light is actually composed of a combination of all of the colors of the visible spectrum.

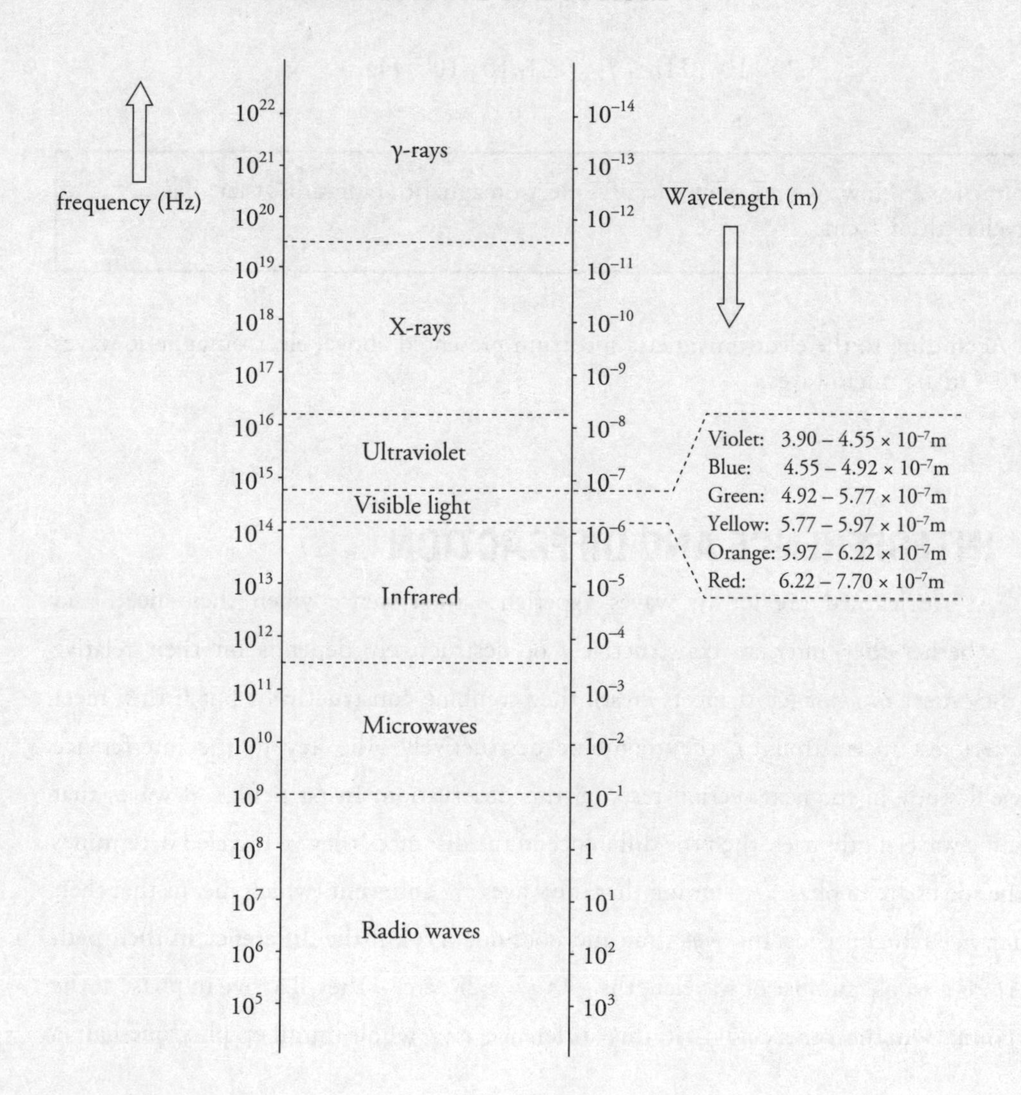

Example 1 What's the frequency range for green light?

Solution. According to the spectrum, light is green if its wavelength is between 4.92×10^{-7} m and 5.77×10^{-7} m. Using the equation $v = f\lambda$, we find that the upper end of this wavelength range corresponds to a frequency of

$$f_1 = \frac{v}{\lambda_1} = \frac{3.00 \times 10^8 \text{ m/s}}{5.77 \times 10^{-7} \text{ m}} = 5.20 \times 10^{14} \text{ Hz}$$

while the lower end corresponds to

$$f_2 = \frac{v}{\lambda_2} = \frac{3.00 \times 10^8 \text{ m/s}}{4.92 \times 10^{-7} \text{ m}} = 6.10 \times 10^{14} \text{ Hz}$$

So the frequency range for green light is

$$5.20 \times 10^{14} \text{ Hz} \le f_{\text{green}} \le 6.10 \times 10^{14} \text{ Hz}$$

Example 2 How would you classify electromagnetic radiation that has a wavelength of 1 cm?

Solution. According to the electromagnetic spectrum presented above, electromagnetic waves with $\lambda = 10^{-2}$ m are microwaves.

CED Unit 6.6
Interference and Diffraction

INTERFERENCE AND DIFFRACTION

As we learned previously, waves experience interference when they meet, and whether they interfere constructively or destructively depends on their relative phase. If they meet *in phase* (crest meets crest), they combine constructively, but if they meet *out of phase* (crest meets trough), they combine destructively. The key to the interference patterns we'll study in the next section rests on this observation. In particular, if waves that have the same wavelength meet, then the difference in the distances they've traveled determines whether the waves are in phase. Assuming that the waves are **coherent** (which means that their phase difference remains constant over time and does not vary), if the difference in their path lengths, $\Delta \ell$, is a whole number of wavelengths—0, $\pm\lambda$, $\pm 2\lambda$, etc.—they'll arrive in phase at the meeting point. On the other hand, if this difference is a whole number plus one-half a wavelength—$\pm\frac{1}{2}\lambda$, $\pm(1+\frac{1}{2})\lambda$, $\pm(2+\frac{1}{2})\lambda$, etc.—then they'll arrive exactly out of phase. That is,

$$\left.\begin{array}{ll}\text{constructive interference:} & \Delta\ell = m\lambda \\ \text{destructive interference:} & \Delta\ell = (m+\frac{1}{2})\lambda\end{array}\right\} \text{where } m = 0, 1, 2...$$

Equation Sheet

Young's Double-Slit Interference Experiment

The following figure shows incident light on a barrier that contains two narrow slits (perpendicular to the plane of the page), separated by a distance *d*. On the right is a screen whose distance from the barrier, *L*, is much greater than *d*. The question is, what will we see on the screen? You might expect that we'll just see two bright, narrow strips of light, directly opposite the slits in the barrier. As reasonable as this may sound, it doesn't take into account the wave nature of light.

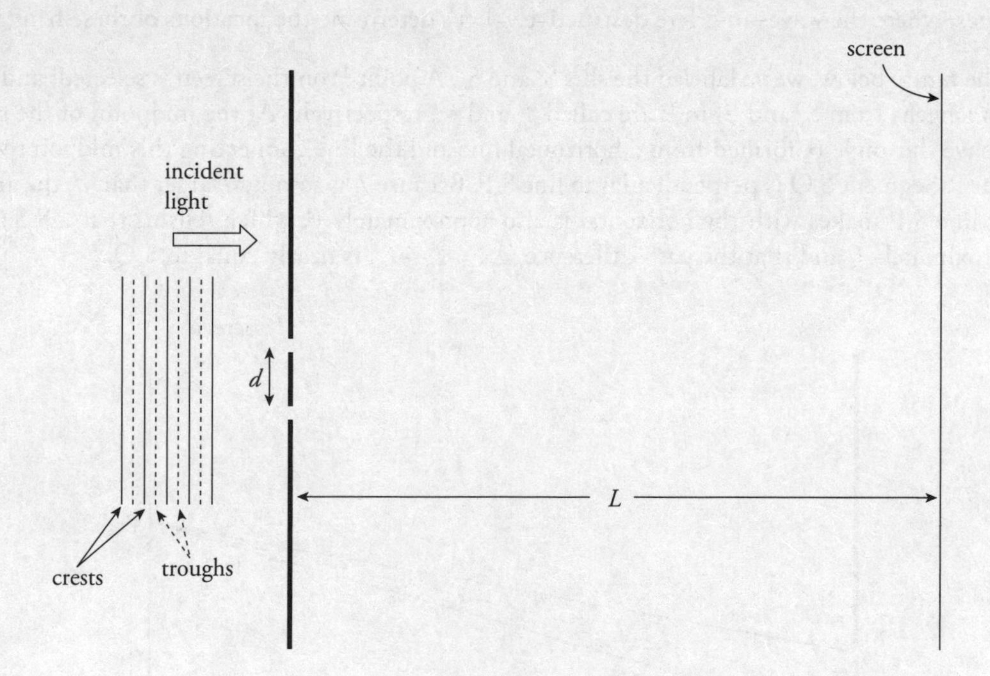

When a wave encounters an aperture whose width is comparable to its wavelength, the wave will fan out after it passes through. The alteration in the straight-line propagation of a wave when it encounters a barrier is called **diffraction**. In the setup above, the waves will diffract through the slits, and spread out and interfere as they travel toward the screen.

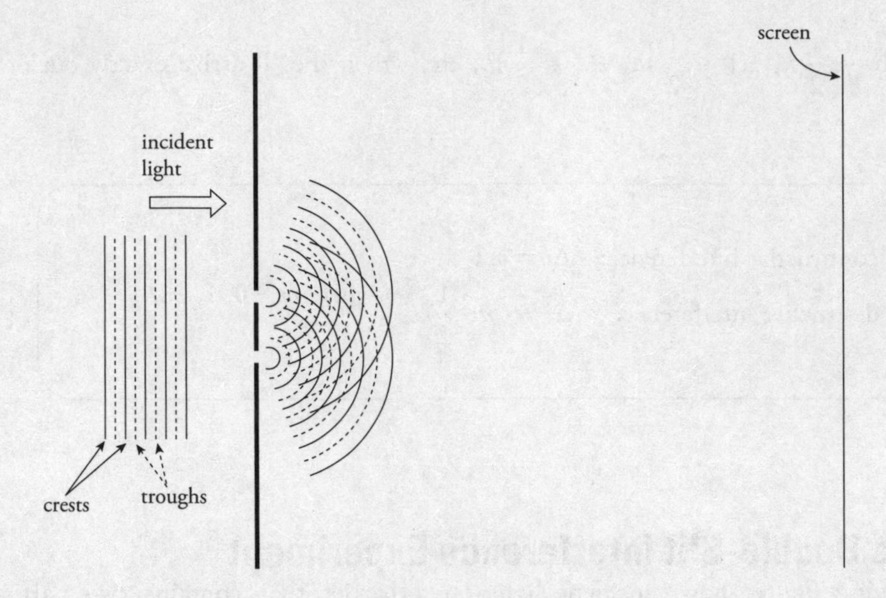

You can clearly see points of interference in the figure above. Look for solid lines intersecting other solid lines—these are points of constructive interference. Look for solid lines intersecting dashed lines—these are points of destructive interference.

The screen will show the results of this interference: there will be bright bands (bright **fringes**) centered at those points at which the waves interfere constructively, alternating with dark fringes, where the waves interfere destructively. Let's determine the locations of these fringes.

In the figure below, we've labeled the slits S_1 and S_2. A point P on the screen is selected, and the path lengths from S_1 and S_2 to P are called ℓ_1 and ℓ_2, respectively. At the midpoint of the slits, we have the angle θ, formed from a horizontal line and the line connecting this midpoint with point P. Segment S_1Q is perpendicular to line S_2P. Because L is so much larger than d, the angle that line S_2P makes with the horizontal is also approximately θ, which tells us that $\angle S_2S_1Q$ is approximately θ and that the path difference, $\Delta\ell = \ell_2 - \ell_1$, is nearly equal to S_2Q.

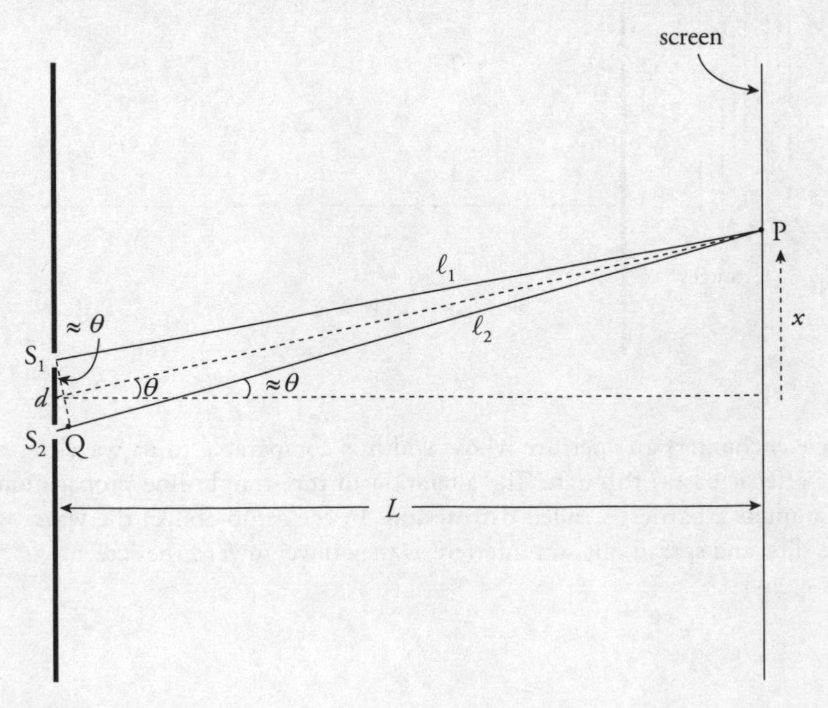

Because $S_2Q = d \sin \theta$, we get $\Delta \ell = d \sin \theta$. Now, using what we learned earlier about how constructive or destructive interference depends on $\Delta \ell$, we can write:

constructive interference (intensity maximum bright fringe on screen): $d \sin \theta = m\lambda$

destructive interference (intensity minimum dark fringe on screen): $d \sin \theta = (m + \frac{1}{2})\lambda$

where $m = 0, 1, 2...$

Equation Sheet

To locate the positions of, say, the bright fringes on the screen, we use the fact that $x = L \tan \theta$. If θ is small, then $\tan \theta \approx \sin \theta$, so we can write $x = L \sin \theta$ (we can tell this from the figure). Since $\sin \theta = m\lambda/d$ for bright fringes, we get

$$x_m = \frac{m\lambda L}{d}$$

Also, the intensity of the bright fringes decreases as m increases in magnitude. The bright fringe directly opposite the midpoint of the slits—the **central maximum**—will have the greatest intensity when $m = 0$. The bright fringes with $m = 1$ will have a lower intensity, those with $m = 2$ will be fainter still, and so on. If more than two slits are cut in the barrier, the interference pattern becomes sharper, and the distinction between dark and bright fringes becomes more pronounced. Barriers containing thousands of tiny slits per centimeter—called **diffraction gratings**—are used precisely for this purpose.

Example 3 For the experimental setup we've been studying, assume that $d = 1.5$ mm, $L = 6.0$ m, and that the light used has a wavelength of 589 nm.

(a) How far above the center of the screen will the second brightest maximum appear?

(b) How far below the center of the screen is the third dark fringe?

(c) What would happen to the interference pattern if the slits were moved closer together?

Solution.

(a) The central maximum corresponds to $m = 0$ ($x_0 = 0$). The first maximum above the central one is labeled x_1 (since $m = 1$). The other bright fringes on the screen are labeled accordingly.

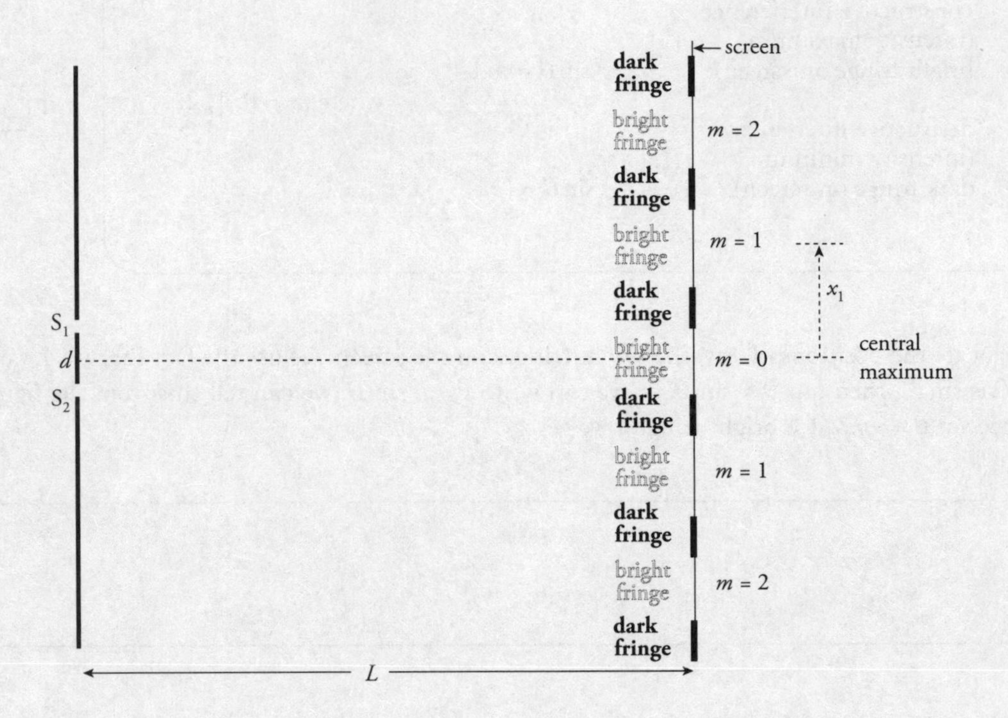

The value of x_1 is

$$x_1 = \frac{1 \cdot \lambda L}{d} = \frac{(589\times10^{-9}\ \text{m})(6.0\ \text{m})}{1.5\times10^{-3}\ \text{m}} = 2.4\times10^{-3}\ \text{m} = 2.4\ \text{mm}$$

(b) The first dark fringe occurs when the path difference is 0.5λ, the second dark fringe occurs when the path difference is 1.5λ, and the third dark fringe occurs when the path difference is 2.5λ, so

$$x_{\substack{\text{3rd minimum}\\ \text{below central max}}} = \frac{(2+\frac{1}{2})\lambda L}{d} = \frac{(2+\frac{1}{2})(589\times10^{-9}\ \text{m})(6.0\ \text{m})}{1.5\times10^{-3}\ \text{m}} = 5.9\times10^{-3}\text{m} = 5.9\ \text{mm}$$

(c) Since $x_m = m\lambda L/d$, a decrease in d would cause an increase in x_m. That is, the fringes would become larger; the interference pattern would be more spread out.

REFLECTION AND REFRACTION

CED Unit 6.4
Refraction, Reflection, and Absorption

Imagine a beam of light directed toward a smooth, transparent surface. When it hits this surface, some of its energy will be reflected off the surface and some will be transmitted into the new medium. Some of the transmitted light will be absorbed, which is the source of heating from radiation discussed in Thermodynamics, while some of the light will emerge from the new medium as a transmitted beam. We can figure out the directions of the reflected and transmitted beams by calculating the angles that the beams make with the normal to the interface. In the following figure, an incident beam strikes the boundary of another medium; it could be a beam of light in air striking a piece of glass. Notice all angles are measured from the normal.

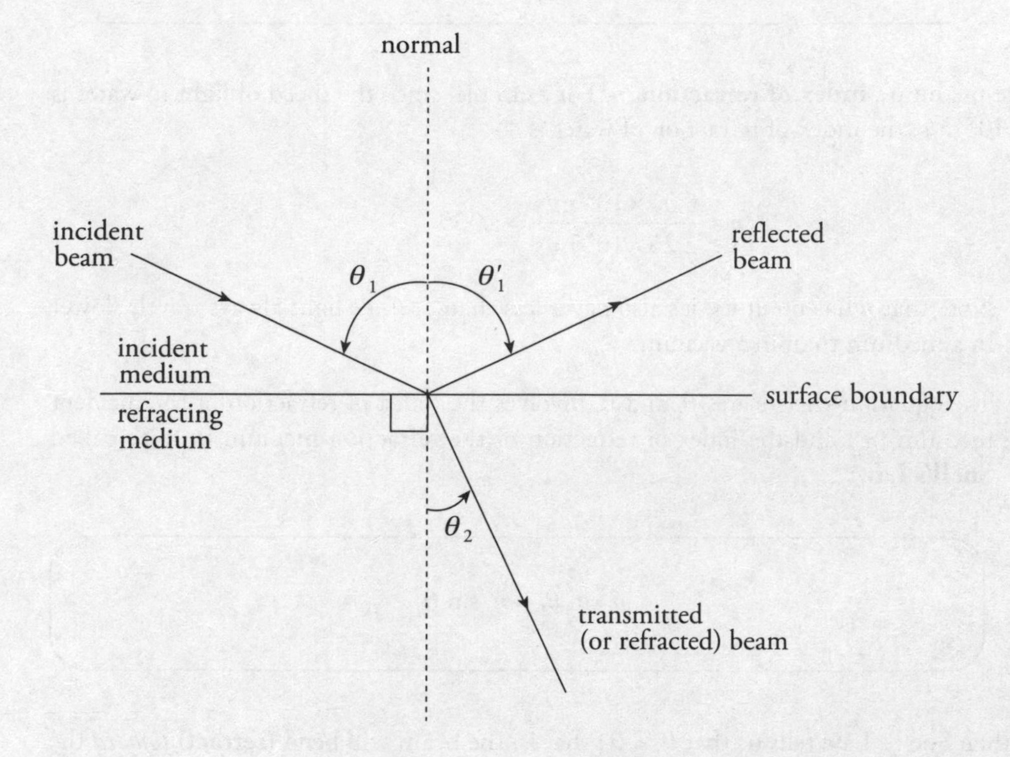

Tricky Problems!
Remember that the normal is always perpendicular to the surface. When a question asks for the angle of incidence or reflection, it wants the angle formed with the normal, not the angle formed with the surface. Make sure you know if you're solving for the angle of incidence, the angle of reflection, or the angle of refraction.

The angle that the **incident beam** makes with the normal is called the **angle of incidence**, or θ_1. The angle that the **reflected beam** makes with the normal is called the **angle of reflection**, θ'_1, and the angle that the **transmitted beam** makes with the normal is called the **angle of refraction**, θ_2. The incident, reflected, and transmitted beams of light all lie in the same plane.

The relationship between θ_1 and θ'_1 is pretty easy; it is called the **Law of Reflection:**

$$\theta_1 = \theta'_1$$

The Law of Reflection basically states the Angle of Reflection is equal to the Angle of Incidence. In order to describe how θ_1 and θ_2 are related, we first need to talk about a medium's index of refraction.

When light travels through empty space (vacuum), its speed is $c = 3.00 \times 10^8$ m/s; this is one of the fundamental constants of nature. But when light travels through a medium (such as water or glass), it's constantly being absorbed and re-emitted by the atoms that compose the material and, as a result, its apparent speed, v, is some fraction of c. The reciprocal of this fraction,

Equation Sheet

$$n = \frac{c}{v}$$

is called the medium's **index of refraction,** n. For example, since the speed of light in water is $v = 2.25 \times 10^8$ m/s, the index of refraction of water is

$$n = \frac{3.00 \times 10^8 \text{ m/s}}{2.25 \times 10^8 \text{ m/s}} = 1.33$$

***n* of air**
The index of refraction of air is 1.00029, which is so close to 1 that the norm is to consider it to just be 1.

Note that n has no units; it's also never less than 1, since light always travels slower in a medium than in a vacuum.

The equation that relates θ_1 and θ_2 involves the index of refraction of the incident medium (n_1) and the index of refraction of the refracting medium (n_2); it's called **Snell's Law**:

Equation Sheet

$$n_1 \sin \theta_1 = n_2 \sin \theta_2$$

If $n_2 > n_1$, then Snell's Law tells us that $\theta_2 < \theta_1$; that is, the beam will bend (**refract**) *toward* the normal as it enters the medium. On the other hand, if $n_2 < n_1$, then $\theta_2 > \theta_1$, and the beam will bend *away* from the normal.

Example 4 A beam of light in air is incident upon a piece of glass, striking the surface at an angle of 30°. If the index of refraction of the glass is 1.5, what are the angles of reflection and refraction?

Solution. If the light beam makes an angle of 30° with the surface, then it makes an angle of 60° with the normal; this is the angle of incidence. By the Law of Reflection, then, the angle of reflection is also 60°. We use Snell's Law to find the angle of refraction. The index of refraction of air is close to 1, so we can say that $n = 1$ for air.

$$\begin{aligned} n_1 \sin \theta_1 &= n_2 \sin \theta_2 \\ (1) \sin 60° &= 1.5 \sin \theta_2 \\ \sin \theta_2 &= 0.5774 \\ \theta_2 &= 35° \end{aligned}$$

Note that $\theta_2 < \theta_1$, as we would expect since the refracting medium (glass) has a greater index than does the incident medium (air).

Example 5 A fisherman drops a flashlight into a lake that's 10 m deep. The flashlight sinks to the bottom where its emerging light beam is directed almost vertically upward toward the surface of the lake, making a small angle (θ_1) with the normal. How deep will the flashlight appear to be to the fisherman? (Use the fact that tan θ is almost equal to sin θ if θ is small.)

Solution. Take a look at the figure below.

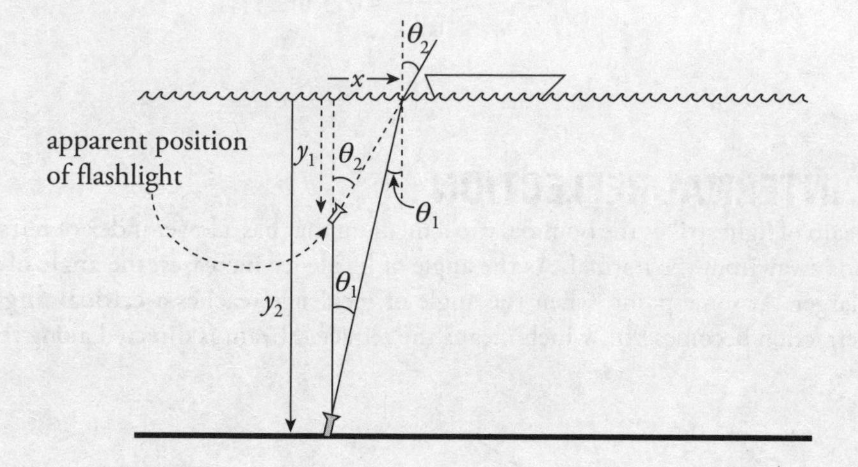

Since the refracting medium (air) has a lower index than the incident medium (water), the beam of light will bend away from the normal as it emerges from the water. As a result, the fisherman will think that the flashlight is at a depth of only y_1, rather than its actual depth of y_2 = 10 m. By simple trigonometry, we know that

$$\tan\theta_1 = \frac{x}{y_2} \quad \text{and} \quad \tan\theta_2 = \frac{x}{y_1}$$

So,

$$\frac{y_1}{y_2} = \frac{\tan\theta_1}{\tan\theta_2}$$

Snell's Law tells us that $n_1 \sin\theta_1 = \sin\theta_2$ (since $n_2 = n_{air} = 1$), so

$$\frac{\sin\theta_1}{\sin\theta_2} = \frac{1}{n_1}$$

Using the approximations $\sin \theta_1 \approx \tan \theta_1$ and $\sin \theta_2 \approx \tan \theta_2$, we can write

$$\frac{\tan \theta_1}{\tan \theta_2} \approx \frac{\sin \theta_1}{\sin \theta_2}$$

which means

$$\frac{y_1}{y_2} = \frac{1}{n_1}$$

Because $y_2 = 10$ m and $n_1 = n_{\text{water}} = 1.33$,

$$y_1 = \frac{y_2}{n_1} = \frac{10 \text{ m}}{1.33} = 7.5 \text{ m}$$

TOTAL INTERNAL REFLECTION

When a beam of light strikes the boundary to a medium that has a lower index of refraction, the beam bends away from the normal. As the angle of incidence increases, the angle of refraction becomes larger. At some point, when the angle of incidence reaches a **critical angle**, θ_c, the angle of refraction becomes 90°, which means the refracted beam is directed along the surface.

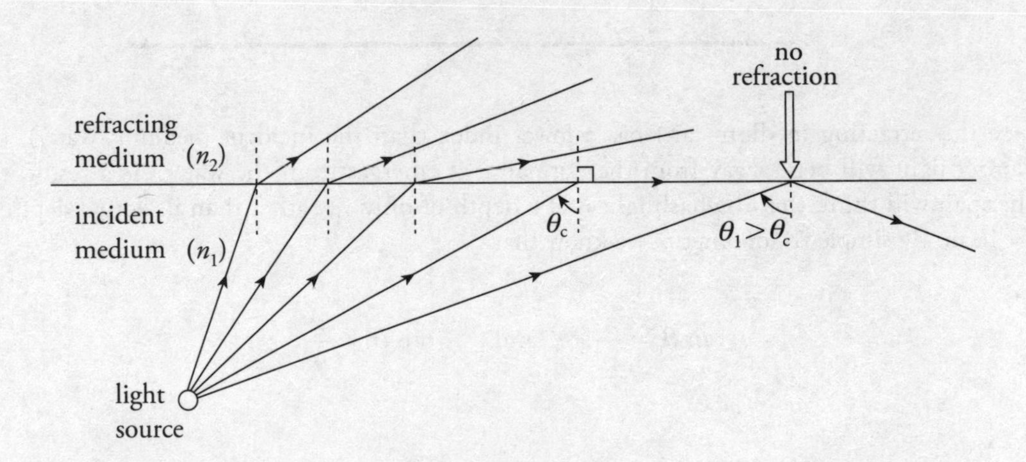

For angles of incidence that are greater than θ_c, there is *no* angle of refraction; the entire beam is reflected back into the original medium. This phenomenon is called **total internal reflection** (sometimes abbreviated **TIR**).

Subscripts

Always think of n_1 or θ_1 as the incidence ray, meaning where the light comes from. Treat n_2 or θ_2 as the resulting ray.

Total internal reflection occurs when:

1) $n_1 > n_2$

and

2) $\theta_1 > \theta_c$, where $\theta_c = \sin^{-1} (n_2/n_1)$

Notice that total internal reflection cannot occur if $n_1 < n_2$. This is because the largest output of sin θ is 1, so the largest input of $\sin^{-1}(x)$ is 1. If $n_1 > n_2$, then total internal reflection is a possibility; it will occur if the angle of incidence is large enough; that is, if it's greater than the critical angle, θ_c.

Example 6 What is the critical angle for total internal reflection between air and water? In which of these media must light be incident for total internal reflection to occur?

Solution. First, total internal reflection can occur only when the light is incident in the medium that has the greater refractive index and strikes the boundary to a medium that has a lower index. So, in this case, total internal reflection can occur only when the light source is in the water and the light is incident upon the water/air surface. The critical angle is found as follows:

$$\sin\theta_c = \frac{n_2}{n_1} \Rightarrow \sin\theta_c = \frac{n_{air}}{n_{water}} = \frac{1}{1.33} \Rightarrow \sin\theta_c = 0.75 \Rightarrow \theta_c = 49°$$

Total internal reflection will occur if the light from the water strikes the water/air boundary at an angle of incidence greater than 49°.

Example 7 How close must the fisherman be to the flashlight in Example 5 in order to see the light it emits?

Solution. In order for the fisherman to see the light, the light must be transmitted into the air from the water; that is, it cannot undergo total internal reflection. The figure below shows that, within a circle of radius x, light from the flashlight will emerge from the water. Outside this circle, the angle of incidence is greater than the critical angle, and the light would be reflected back into the water, rendering it undetectable by the fisherman above.

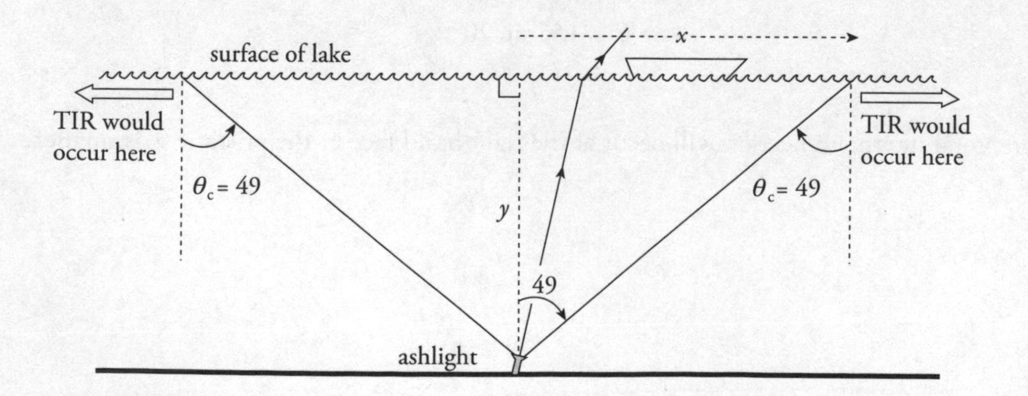

Remember to Check
For optics problems, you need to always remember to check if the refracted light ray undergoes total internal reflection. If it undergoes total internal reflection, there will be no refracted ray. Hence, if a trick problem asks for an angle of refraction for a situation that undergoes total internal reflection, there is no angle of refraction.

Because the critical angle for total internal reflection at a water/air interface is 49° (as we found in the preceding example), we can solve for *x*:

$$\tan 49° = \frac{x}{y} \quad \Rightarrow \quad x = y\tan 49° = (10 \text{ m})\tan 49° = 11.5 \text{ m}$$

Example 8 The refractive index for the glass prism shown below is 1.55. In order for a beam of light to experience total internal reflection at the right-hand face of the prism, the angle θ_1 must be smaller than what value?

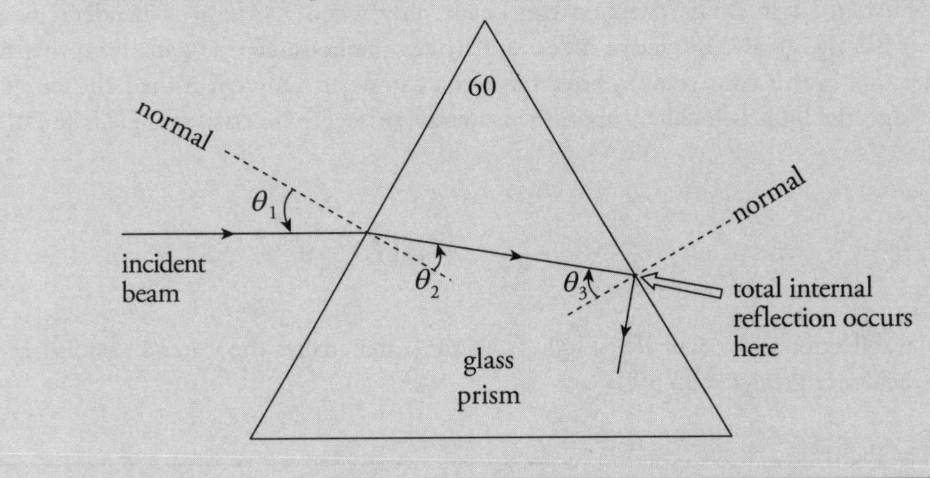

Solution. Total internal reflection will occur at the glass/air boundary if θ_3 is greater than the critical angle, θ_c, which we can calculate this way:

$$\sin \theta_c = \frac{n_{air}}{n_{glass}} = \frac{1}{1.55} \quad \Rightarrow \quad \theta_c = 40°$$

Because $\theta_3 = 60° - \theta_2$, total internal reflection will take place if θ_2 is smaller than 20°. Now, by Snell's Law, $\theta_2 = 20°$ if

$$\begin{aligned} n_{air} \sin \theta_1 &= n_{glass} \sin \theta_2 \\ \sin \theta_1 &= 1.55 \sin 20° \\ \theta_1 &= 32° \end{aligned}$$

Therefore, total internal reflection will occur at the right-hand face of the prism if θ_1 is smaller than 32°.

MIRRORS

A **mirror** is an optical device that forms an image by reflecting light. We've all looked into a mirror and seen images of nearby objects, and the purpose of this section is to analyze these images mathematically. We begin with a plane mirror, which is flat, and is the simplest type of mirror. Then we'll examine curved mirrors; we'll have to use geometrical methods or algebraic equations to analyze the patterns of reflection from these.

The rest of this chapter deals with CED Unit 6.5 Images from Lenses and Mirrors. Here, the wave nature of light is no longer of interest and light is treated simply as a ray.

Plane Mirrors

The figure below shows an object (denoted by a vertical, bold arrow) in front of a flat mirror. Light that's reflected off the object strikes the mirror and is reflected back to our eyes. The directions of the rays reflected off the mirror determine where we perceive the image to be.

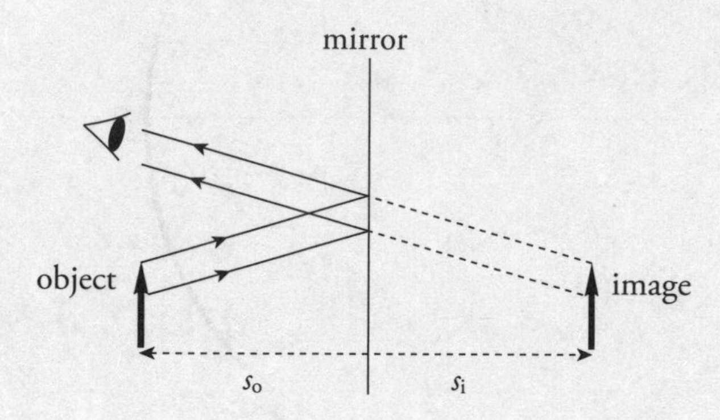

There are four questions we'll answer about the image formed by a mirror:

(1) Where is the image?

(2) Is the image real, or is it virtual?

(3) Is the image upright, or is it inverted?

(4) What is the height of the image (compared with that of the object)?

When we look at ourselves in a mirror, it seems like our image is *behind* the mirror, and if we take a step back, our image also takes a step back. The Law of Reflection can be used to show that the image seems as far behind the mirror as the object is in front of the mirror. This answers question (1).

An image is said to be **real** if light rays actually focus on the image. A real image can be projected onto a screen. For a flat mirror, light rays bounce off the front of the mirror; so, of course, no light focuses behind it. Therefore, the images produced by a flat mirror are not real; they are **virtual**. This answers question (2).

When we look into a flat mirror, our image isn't upside down; flat mirrors produce upright images, and question (3) is answered.

Finally, the image formed by a flat mirror is neither magnified nor diminished (minified) relative to the size of the object. This answers question (4).

Spherical Mirrors

A **spherical mirror** is a mirror that's curved in such a way that its surface forms part of a sphere.

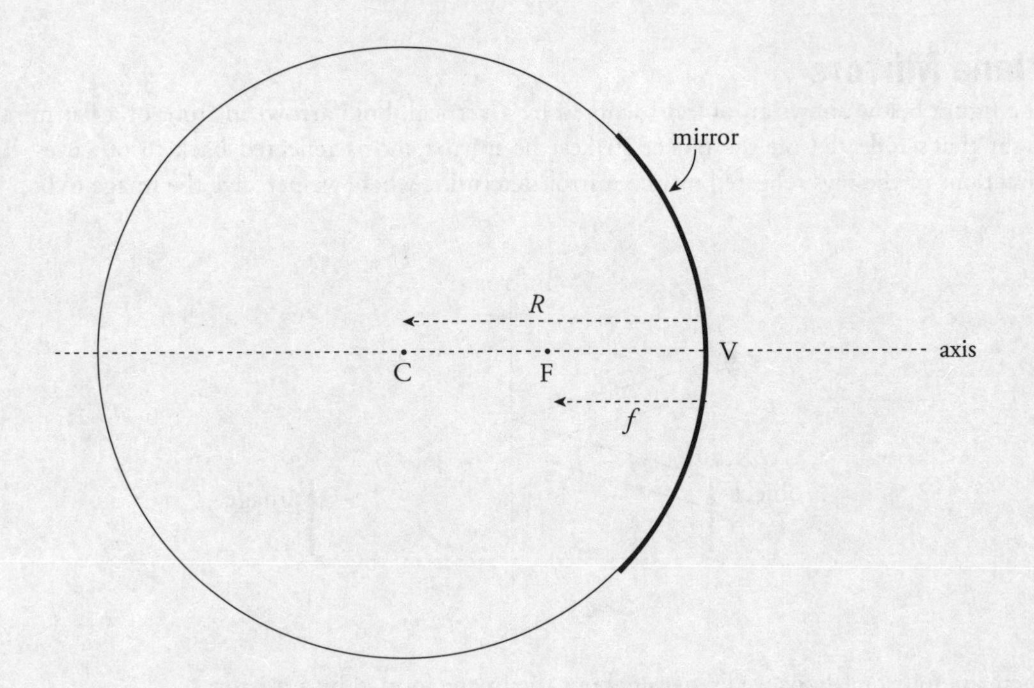

The center of this imaginary sphere is the mirror's **center of curvature**, and the radius of the sphere is called the mirror's **radius of curvature**, R. Halfway between the mirror and the center of curvature, C, is the **focus** (or **focal point**), F. The intersection of the mirror's optic **axis** (its axis of symmetry) with the mirror itself is called the **vertex**, V, and the distance from V to F is called the **focal length**, f, equal to one-half of the radius of curvature:

$$f = \frac{R}{2}$$

If the mirror had a parabolic cross-section, then any ray parallel to the axis would be reflected by the mirror through the focal point. Spherical mirrors do this for incident light rays near the axis (**paraxial rays**) because, in the region of the mirror that's close to the axis, the shapes of a parabolic mirror and a spherical mirror are nearly identical.

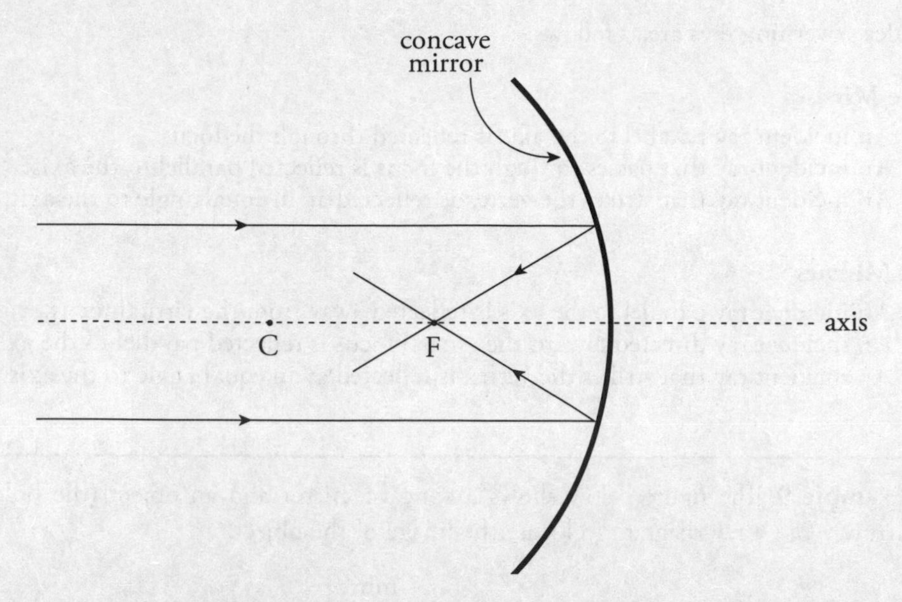

The previous two figures illustrate a **concave mirror**, a mirror whose reflective side is *caved in* toward the center of curvature. The following figure illustrates the **convex mirror**, which has a reflective side curving away from the center of curvature. We will call F the **virtual focus** in this case.

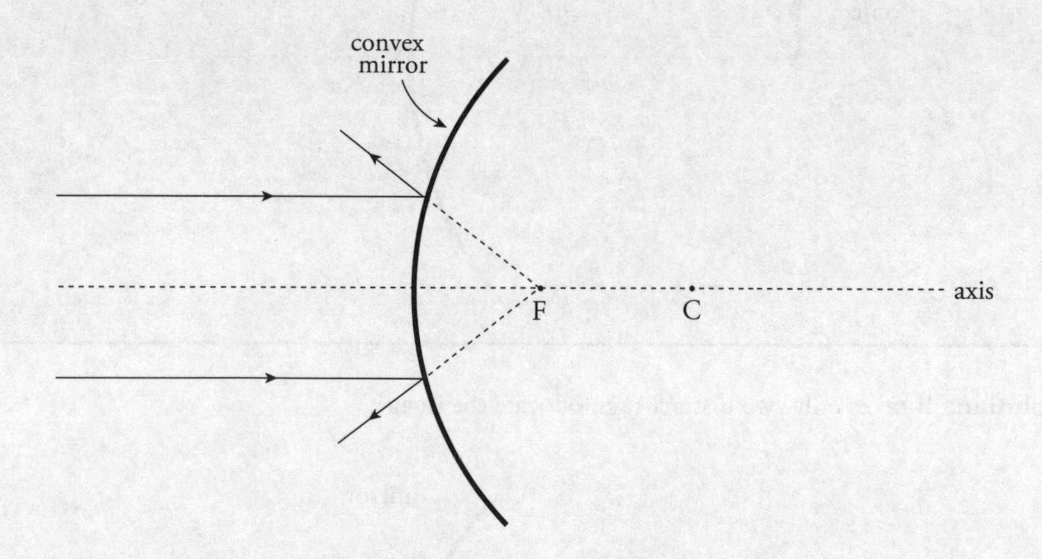

RAY TRACING FOR MIRRORS

One method of answering the four questions previously listed concerning the image formed by a mirror involves a geometric approach called **ray tracing**. Representative rays of light are sketched in a diagram that depicts the object and the mirror; the point at which the reflected rays intersect (or appear to intersect) is the location of the image.

Some rules governing rays are as follows:

Concave Mirrors

- An incident ray parallel to the axis is reflected through the focus.
- An incident ray that passes through the focus is reflected parallel to the axis.
- An incident ray that strikes the vertex is reflected at an equal angle to the axis.

Convex Mirrors

- An incident ray parallel to the axis is reflected away from the virtual focus.
- An incident ray directed toward the virtual focus is reflected parallel to the axis.
- An incident ray that strikes the vertex is reflected at an equal angle to the axis.

Example 9 The figure below shows a concave mirror and an object (the bold arrow). Use a ray diagram to locate the image of the object.

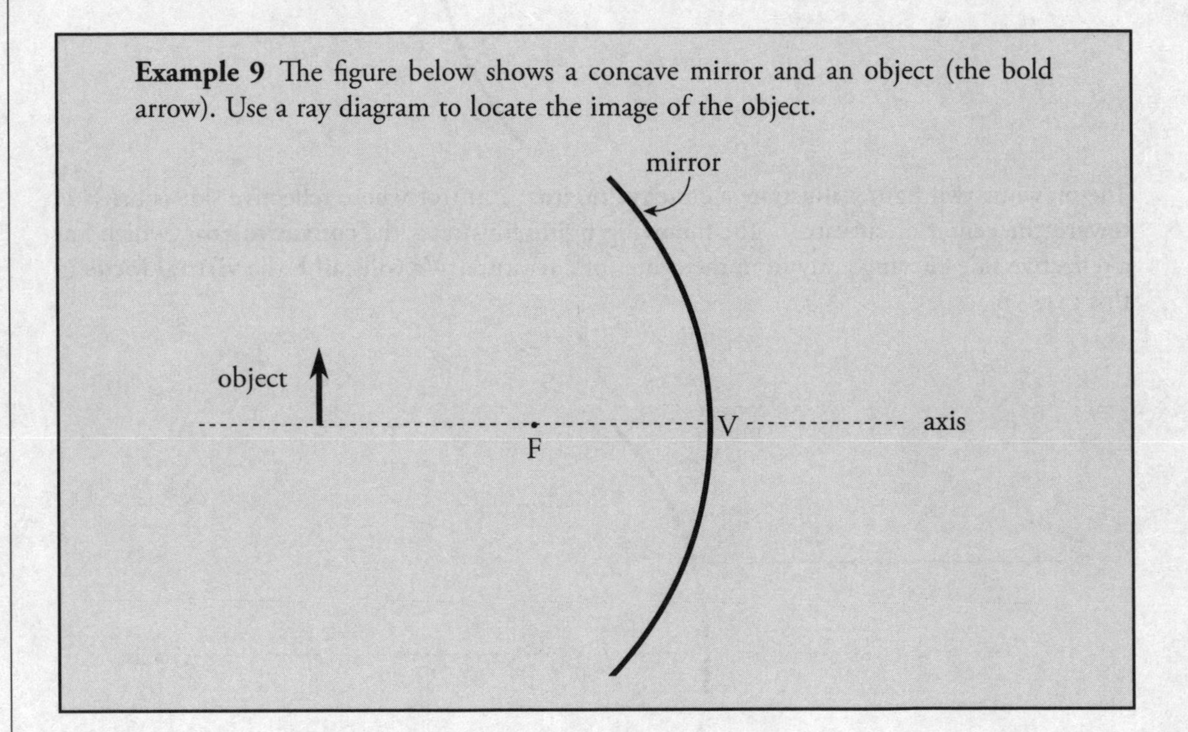

Solution. It takes only two distinct rays to locate the image:

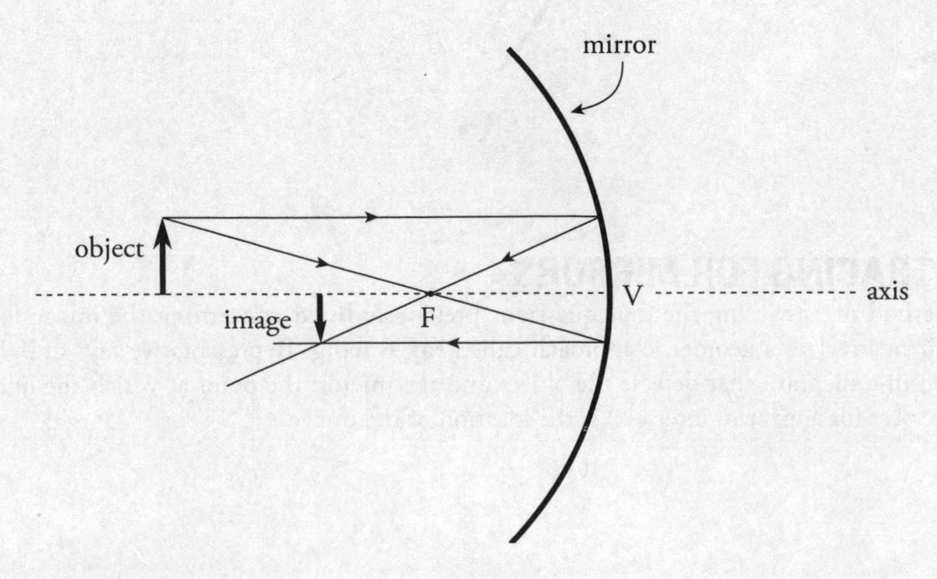

Example 10 The figure below shows a convex mirror and an object (the arrow). Use a ray diagram to locate the image of the object.

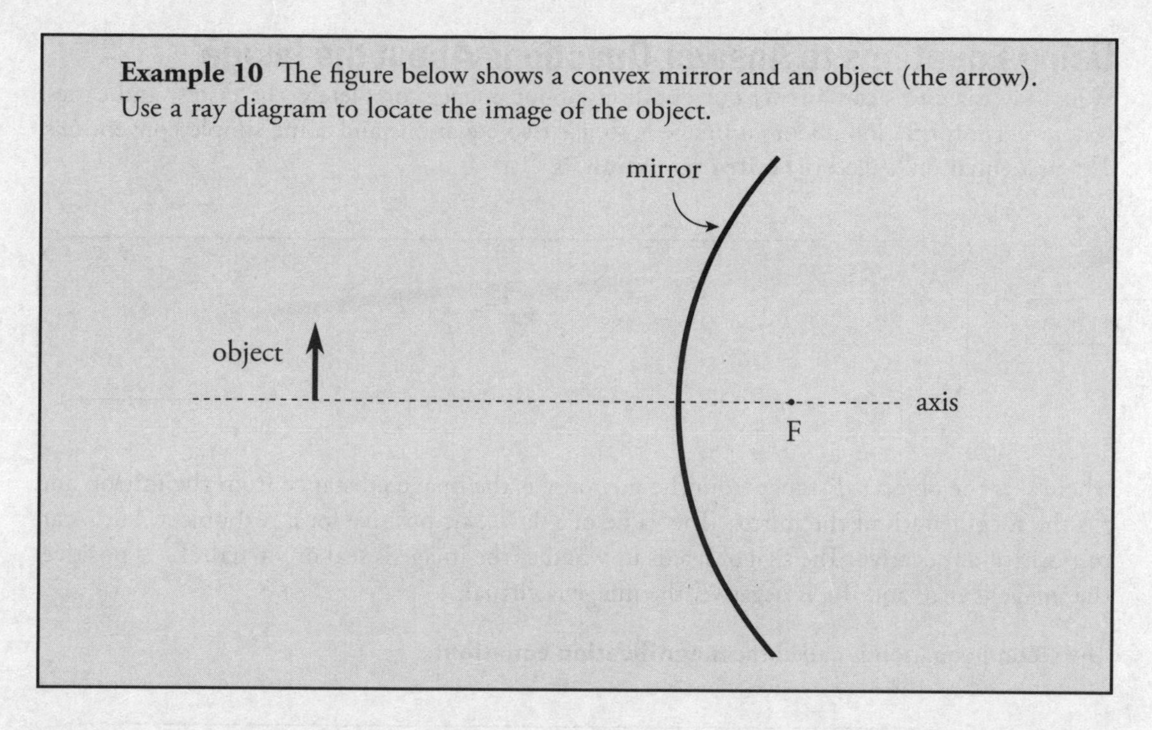

Solution.

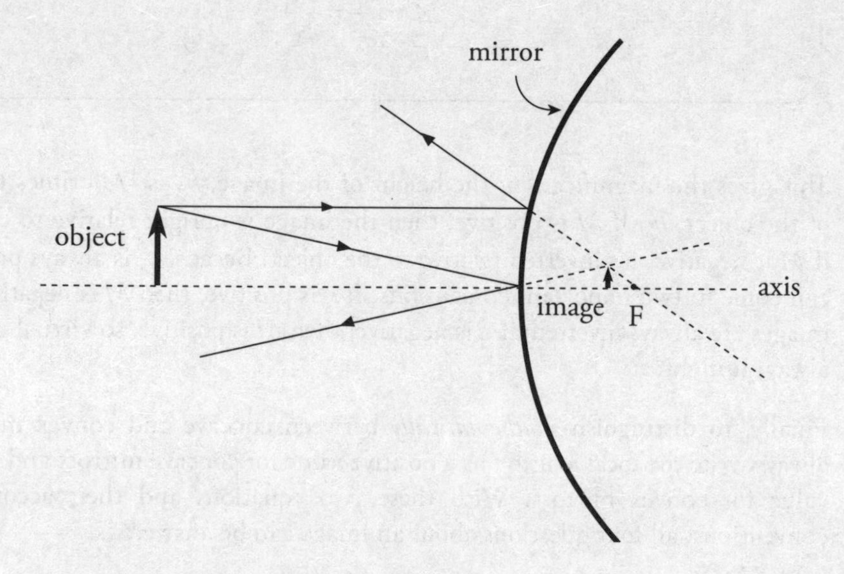

The ray diagrams of the preceding examples can be used to determine the location, orientation, and size of the image. The nature of the image—that is, whether it's real or virtual—can be determined by seeing on which side of the mirror the image is formed. If the image is formed on the same side of the mirror as the object, then the image is real, but if the image is formed on the opposite side of the mirror, it's virtual. Here's another way to look at this: if you had to trace lines back to form an image, that image is virtual. Therefore, the image in Example 9 is real, and the image in Example 10 is virtual.

Using Equations to Answer Questions About the Image

While ray diagrams can answer our questions about images completely, the fastest and easiest way to get information about an image is to use two equations and some simple conventions. The first equation, called the **mirror equation**, is

Equation Sheet

$$\frac{1}{s_o} + \frac{1}{s_i} = \frac{1}{f}$$

where s_o is the object's distance from the mirror, s_i is the image's distance from the mirror, and f is the focal length of the mirror. The value of s_o is *always* positive for a real object, but s_i can be positive or negative. The sign of s_i tells us whether the image is real or virtual: if s_i is positive, the image is real; and if s_i is negative, the image is virtual.

The second equation is called the **magnification equation**:

Equation Sheet

$$M = \frac{h_i}{h_o} = \frac{-s_i}{s_o}$$

Which Is the Positive Side?

We will always treat the object's position s_o as being the positive side. For mirrors, if the image is produced on the same side as the object, it is considered a positive distance. If an image is produced opposite the side of the object, it is considered a negative distance. Concave mirrors will always have a positive focal length (meaning the focal point is on the same side of the object). Convex mirrors will always have a negative focal length (meaning the focal point is on the opposite side of the object).

This gives the magnification; the height of the image, h_i, is $|M|$ times the height of the object, h_o. If M is positive, then the image is upright relative to the object; if M is negative, it's inverted relative to the object. Because s_o is always positive, we can come to two important conclusions. If s_i is positive, then M is negative, so real images are always inverted; if s_i is negative, then M is positive, so virtual images are always upright.

Finally, to distinguish *mathematically* between concave and convex mirrors, we always write the focal length f as a positive value for concave mirrors and a negative value for convex mirrors. With these two equations and their accompanying conventions, all four questions about an image can be answered.

MIRRORS

converging—concave ⇔ f positive
diverging—convex ⇔ f negative

$$\frac{1}{s_o} + \frac{1}{s_i} = \frac{1}{f} \begin{cases} s_o \text{ always positive (real object)} \\ s_i \text{ positive} \Rightarrow \text{image is real (located on the } \textit{same} \text{ side of mirror as object)} \\ s_i \text{ negative} \Rightarrow \text{image is virtual (located on the } \textit{opposite} \text{ side of mirror from object)} \end{cases}$$

$$M = \frac{h_i}{h_o} = \frac{-s_i}{s_o} \begin{cases} \text{given } h_o \text{ is positive} \\ h_i \text{ and } M \text{ positive} \Rightarrow \text{image is upright} \\ h_i \text{ and } M \text{ negative} \Rightarrow \text{image is inverted} \end{cases}$$

Real or Virtual?
Remember that all real images are inverted and all virtual images are upright.

Example 11 An object of height 4 cm is placed 30 cm in front of a concave mirror whose focal length is 10 cm.

(a) Where's the image?
(b) Is it real or virtual?
(c) Is it upright or inverted?
(d) What's the height of the image?

Solution.

(a) With s_o = 30 cm and f = 10 cm, the mirror equation gives

$$\frac{1}{s_o} + \frac{1}{s_i} = \frac{1}{f} \Rightarrow \frac{1}{30 \text{ cm}} + \frac{1}{s_i} = \frac{1}{10 \text{ cm}} \Rightarrow \frac{1}{s_i} = \frac{1}{15 \text{ cm}} \Rightarrow s_i = 15 \text{ cm}$$

The image is located 15 cm in front of the mirror.

(b) Because s_i is positive, the image is real.

(c) Real images are inverted.

(d) $$\frac{h_i}{h_o} = \frac{-s_i}{s_o} \Rightarrow h_i = \frac{-s_i h_o}{s_o} \Rightarrow h_i = \frac{(-15 \text{ cm})(4 \text{ cm})}{30 \text{ cm}} = -2 \text{ cm}$$

The –2 cm also confirms the image is inverted.

Example 12 An object of height 4 cm is placed 20 cm in front of a convex mirror whose focal length is –30 cm.

(a) Where's the image?

(b) Is it real or virtual?

(c) Is it upright or inverted?

(d) What's the height of the image?

Solution.

(a) With s_o = 20 cm and f = –30 cm, the mirror equation gives us

$$\frac{1}{s_o} + \frac{1}{s_i} = \frac{1}{f} \Rightarrow \frac{1}{20 \text{ cm}} + \frac{1}{s_i} = \frac{1}{-30 \text{ cm}} \Rightarrow \frac{1}{s_i} = -\frac{1}{12 \text{ cm}} \Rightarrow s_i = -12 \text{ cm}$$

So, the image is located 12 cm behind the mirror.

(b) Because s_i is negative, the image is virtual.

(c) Virtual images are upright.

(d) $$\frac{h_i}{h_o} = \frac{-s_i}{s_o} \Rightarrow h_i = \frac{-s_i h_o}{s_o} \Rightarrow h_i = \frac{-(-12 \text{ cm})(4 \text{ cm})}{20 \text{ cm}} = 2.4 \text{ cm}$$

The +2.4 cm also confirms the image is upright.

What Produces What?

Only concave mirrors can produce both real images (if $s_o > f$) and virtual images (if $s_o < f$). Convex mirrors and plane mirrors can produce only virtual images.

Example 13 Show how the statements made earlier about plane mirrors can be derived from the mirror and magnification equations.

Solution. A plane mirror can be considered a spherical mirror with an infinite radius of curvature (and an infinite focal length). If $f = \infty$, then $1/f = 0$, so the mirror equation becomes

$$\frac{1}{s_o} + \frac{1}{s_i} = 0 \Rightarrow s_i = -s_o$$

So, the image is as far behind the mirror as the object is in front. Also, since s_o is always positive, s_i is negative, so the image is virtual. The magnification is

$$M = -\frac{s_i}{s_o} = -\frac{-s_o}{s_o} = 1$$

and the image is upright and has the same height as the object. The mirror and magnification equations confirm our description of images formed by plane mirrors.

Example 14 Show why convex mirrors can form only virtual images.

Solution. Because f is negative and s_o is positive, the mirror equation

$$\frac{1}{s_o}+\frac{1}{s_i}=\frac{1}{f}$$

immediately tells us that s_i cannot be positive (if it were, the left-hand side would be the sum of two positive numbers, while the right-hand side would be negative). Since s_i must be negative, the image must be virtual.

Example 15 An object placed 60 cm in front of a spherical mirror forms a real image at a distance of 30 cm from the mirror.

(a) Is the mirror concave or convex?

(b) What's the mirror's focal length?

(c) Is the image taller or shorter than the object?

Solution.

(a) The fact that the image is real tells us that the mirror cannot be convex, since convex mirrors form only virtual images. The mirror is concave.

(b) With s_o = 60 cm and s_i = 30 cm (s_i is positive since the image is real), the mirror equation tells us that

$$\frac{1}{s_o}+\frac{1}{s_i}=\frac{1}{f} \Rightarrow \frac{1}{60\text{ cm}}+\frac{1}{30\text{ cm}}=\frac{1}{f} \Rightarrow \frac{1}{f}=\frac{1}{20\text{ cm}} \Rightarrow f=20\text{ cm}$$

Note that f is positive, which confirms that the mirror is concave.

(c) The magnification is

$$M=-\frac{s_i}{s_o}=-\frac{30\text{ cm}}{60\text{ cm}}=-\frac{1}{2}$$

Since the absolute value of M is less than 1, the mirror makes the object look smaller (minifies the height of the object). The image is only half as tall as the object and is inverted, since M is negative.

So Many Equations!
The purpose of a mirror is to reflect an image. The purpose of a lens is to refract an image (or see through it). Fortunately, the equations for mirrors and lenses will be the same with a minor difference on the focal length.

Example 16 A concave mirror with a focal length of 25 cm is used to create a real image that has twice the height of the object. How far is the image from the mirror?

Solution. Since h_i (the height of the image) is twice h_o (the height of the object), the value of the magnification is either +2 or –2. To figure out which, we just notice that the image is real; real images are inverted, so the magnification, M, must be negative. Therefore, $M = -2$, so

$$-\frac{s_i}{s_o} = -2 \;\Rightarrow\; s_o = \frac{1}{2}s_i$$

Substituting this into the mirror equation gives us

$$\frac{1}{s_o} + \frac{1}{s_i} = \frac{1}{f} \;\Rightarrow\; \frac{1}{\frac{1}{2}s_i} + \frac{1}{s_i} = \frac{1}{f} \;\Rightarrow\; \frac{3}{s_i} = \frac{1}{f} \;\Rightarrow\; s_i = 3f = 3(25\text{ cm}) = 75\text{ cm}$$

THIN LENSES

Mirrors vs. Lenses
The equations are exactly the same. The only difference is we switch the locations of the focus. A positive focal length lens will have a focus on the opposite side of the lens from the object. If the focal point is located on the same side of the object, then it is a negative focal length.

A lens is an optical device that forms an image by *refracting* light. We'll now talk about the equations and conventions that are used to analyze images formed by the two major categories of lenses: converging and diverging.

A **converging lens**—like the bi-convex one shown below—converges parallel paraxial rays of light to a focal point on the far side. (This lens is *bi-convex*; both of its faces are convex. All converging lenses have at least one convex face.) Because parallel light rays actually focus at F, this point is called a **real focus**. Its distance from the lens is the focal length, *f*.

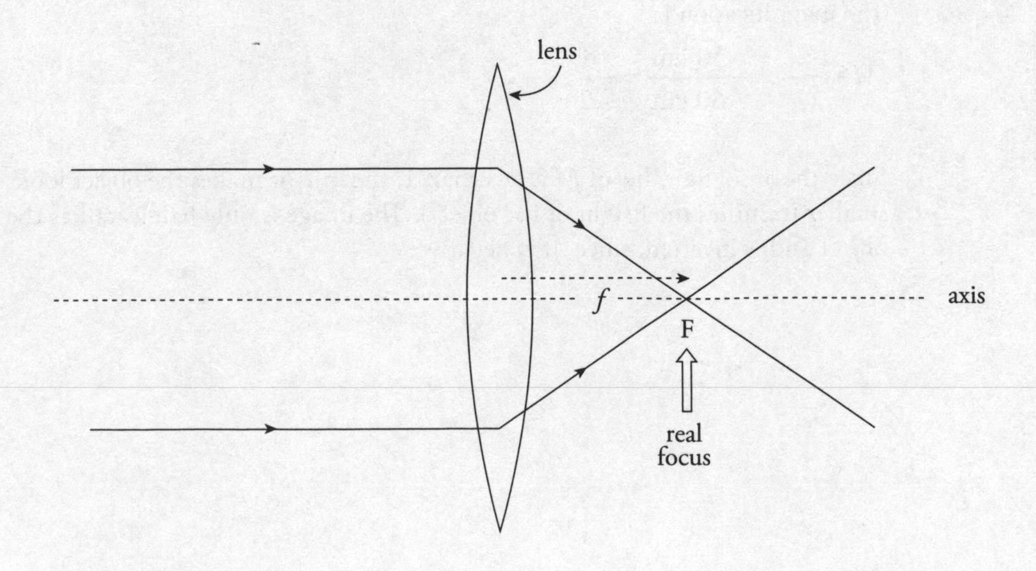

A **diverging lens**—like the *bi-concave* one shown below—causes parallel paraxial rays of light to diverge away from a **virtual focus**, F, on the same side as the incident rays. (All diverging lenses have at least one concave face.)

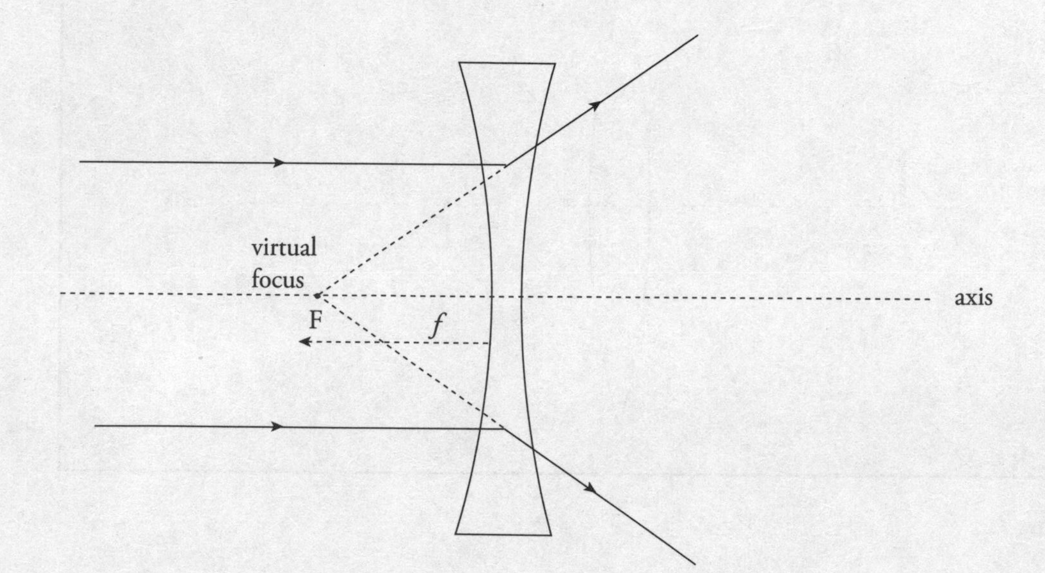

Don't Mix Them Up
Converging Lens = Bi-Convex Lens
Diverging Lens = Bi-Concave Lens

RAY TRACING FOR LENSES

Just as is the case with mirrors, representative rays of light can be sketched in a diagram along with the object and the lens; the point at which the reflected rays intersect (or appear to intersect) is the location of the image. The rules that govern these rays are as follows:

Converging Lenses

- An incident ray parallel to the axis is refracted through the real focus.
- An incident ray that passes through the focus is refracted parallel to the axis.
- Incident rays pass undeflected through the optical center, O (the central point within the lens where the axis intersects the lens).

Diverging Lenses

- An incident ray parallel to the axis is refracted away from the virtual focus.
- An incident ray heading toward the far focus is refracted parallel to the axis.
- Incident rays pass undeflected through the optical center, O.

Lenses actually have two focal points: a near focal point (on the same side of the lens as the object) and a far focal point (on the opposite side of the lens as the object).

Example 17 The figure below shows a converging lens and an object (denoted by the bold arrow). Use a ray diagram to locate the image of the object.

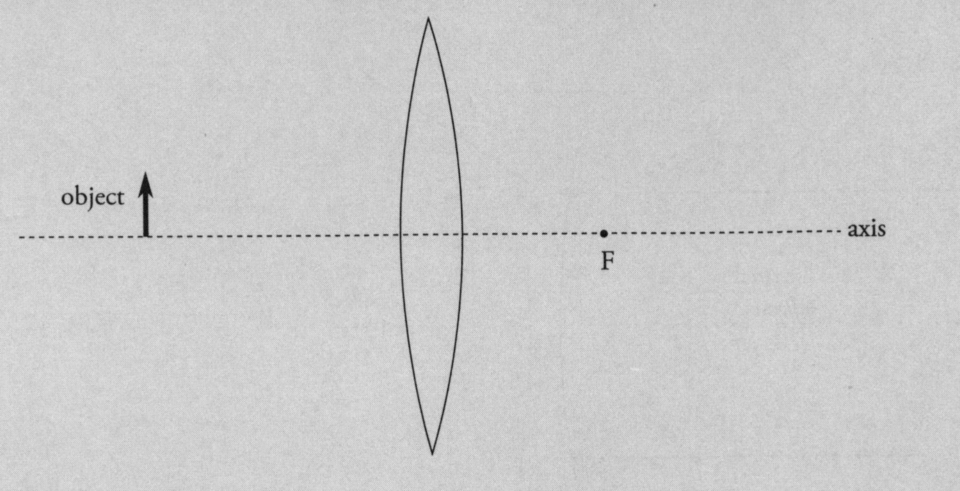

Solution.

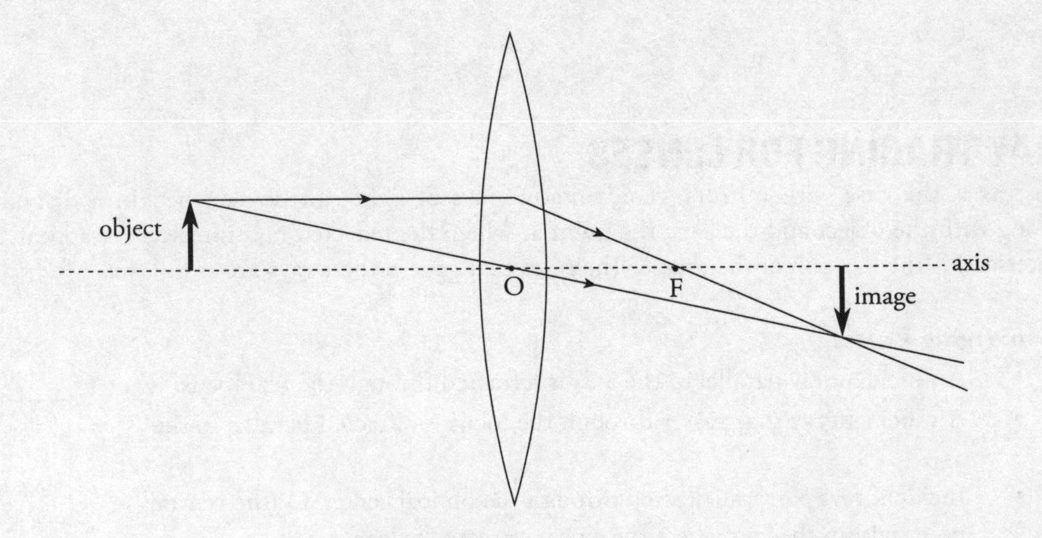

Example 18 The figure below shows a diverging lens and an object. Use a ray diagram to locate the image of the object.

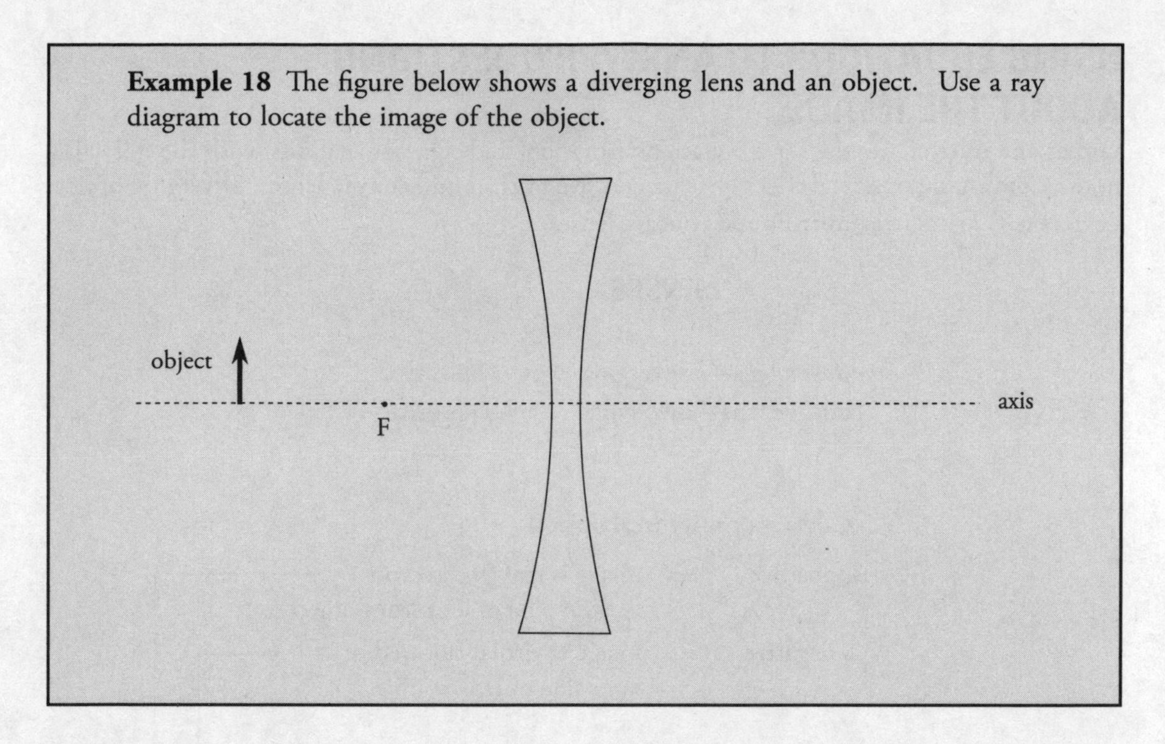

Solution.

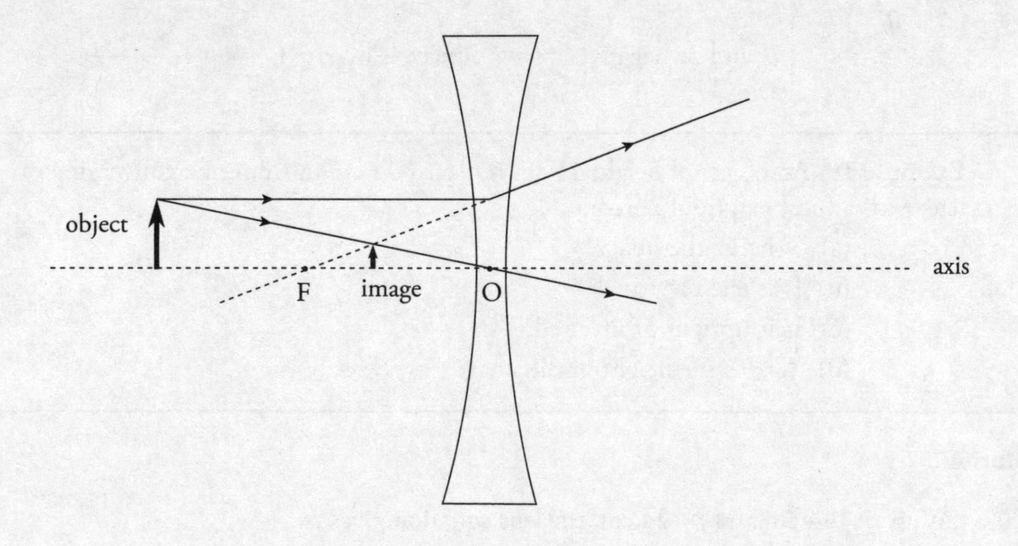

The nature of the image—that is, whether it's real or virtual—is determined by the side of the lens upon which the image is formed. If the image is formed on the side of the lens that's opposite the object, then the image is real, and if the image is formed on the same side of the lens as the object, then it's virtual. Another way of looking at this is if you had to trace lines back to form an image, that image is virtual. Therefore, the image in Example 17 is real, while the image in Example 18 is virtual.

USING EQUATIONS TO ANSWER QUESTIONS ABOUT THE IMAGE

Lenses and mirrors use the same equations, notation, and sign conventions, with the following note. Converging optical devices ($+f$) are con<u>cave</u> mirrors and con<u>vex</u> lenses. Diverging optical devices ($-f$) are con<u>vex</u> mirrors and con<u>cave</u> lenses.

LENSES

convex lens—converging $\Leftrightarrow$ f positive
concave lens—diverging $\Leftrightarrow$ f negative

$$\frac{1}{s_o} + \frac{1}{s_i} = \frac{1}{f}$$

- s_o always positive (real object)
- s_i positive $\Rightarrow$ image is real (located on *opposite* side of lens from object) → image is inverted
- s_i negative $\Rightarrow$ image is virtual (located on *same* side of lens as object) → image is upright

$$M = \frac{h_i}{h_o} = \frac{-s_i}{s_o}$$

- given h_o is positive
- h_i and M positive $\Rightarrow$ image is upright
- h_i and M negative $\Rightarrow$ image is inverted

Example 19 An object of height 11 cm is placed 44 cm in front of a converging lens with a focal length of 24 cm.

(a) Where's the image?
(b) Is it real or virtual?
(c) Is it upright or inverted?
(d) What's the height of the image?

Solution.

(a) With s_o = 44 cm and f = 24 cm, the lens equation gives us

$$\frac{1}{s_o} + \frac{1}{s_i} = \frac{1}{f} \Rightarrow \frac{1}{44\text{ cm}} + \frac{1}{s_i} = \frac{1}{24\text{ cm}} \Rightarrow \frac{1}{s_i} = 0.0189\text{ cm}^{-1} \Rightarrow s_i = 53\text{ cm}$$

So, the image is located 53 cm from the lens, on the opposite side from the object.

(b) Because s_i is positive, the image is real.

(c) Real images are inverted.

(d) $$\frac{h_i}{h_o} = \frac{-s_i}{s_o} \Rightarrow h_i = \frac{-s_i h_o}{s_o} \Rightarrow h_i = \frac{(-53\text{ cm})(11\text{ cm})}{44\text{ cm}} = -13\text{ cm}$$

The negative h_i reaffirms that we have an inverted image.

Example 20 An object of height 11 cm is placed 48 cm in front of a diverging lens with a focal length of –24.5 cm.

(a) Where's the image?
(b) Is it real or virtual?
(c) Is it upright or inverted?
(d) What's the height of the image?

Solution.

(a) With s_o = 48 cm and f = –24.5 cm, the lens equation gives us

$$\frac{1}{s_o}+\frac{1}{s_i}=\frac{1}{f} \Rightarrow \frac{1}{48\text{ cm}}+\frac{1}{s_i}=\frac{1}{-24.5\text{ cm}} \Rightarrow \frac{1}{s_i}=-0.0616\text{ cm}^{-1} \Rightarrow s_i=-16\text{ cm}$$

The image is 16 cm from the lens, on the same side as the object.

(b) Because s_i is negative, the image is virtual.

(c) Virtual images are upright.

(d) $$\frac{h_i}{h_o}=\frac{-s_i}{s_o} \Rightarrow h_i=\frac{-s_i h_o}{s_o} \Rightarrow h_i=\frac{-(-16\text{ cm})(11\text{ cm})}{48\text{ cm}}=3.7\text{ cm}$$

The positive h_i reaffirms that we have an upright image.

What Produces What?
Only converging lenses can produce both real images (if $s_o > f$) and virtual images (if $s_o < f$). Diverging lenses can produce only virtual images.

Chapter 9 Review Questions

Solutions can be found in Chapter 11.

Section I: Multiple Choice

1. A beam of light in air is incident upon the smooth surface of a piece of flint glass, as shown:

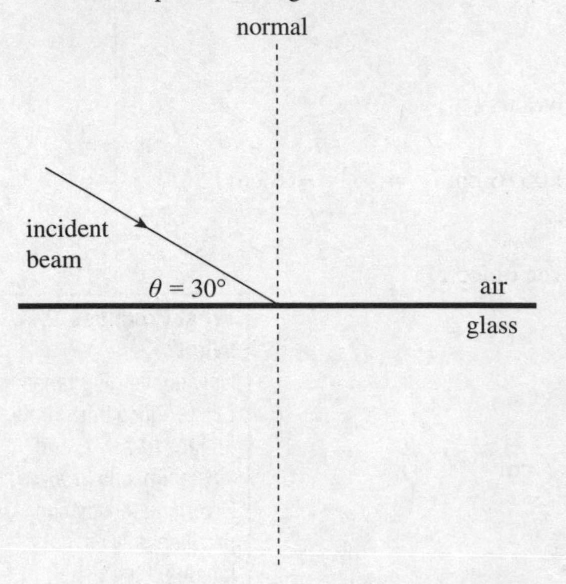

As the incident angle is increased toward $\theta = 90°$, what observation is made of the refracted ray? (All angle references are relative to the surface as shown for both rays.)

(A) The refracted ray angle increases as the incident angle increases, but the value of the refracted angle is always smaller than the incident angle.
(B) The refracted ray angle increases as the incident angle increases, but the value of the refracted angle is always larger than the incident angle.
(C) The refracted ray angle increases as the incident angle increases until at some angle total internal reflection begins to occur.
(D) The refracted ray angle decreases as the incident angle increases, but the value of the refracted angle is always smaller than the incident angle.

2. A convex lens constructed of glass makes a real image of an object when it is in air. When the object is located d_o in front of the lens, the image appears in air at a distance d_i behind the lens. What occurs if the object is still at d_o, but the object and the lens are submerged in water with an index of refraction between that of air and the glass of the lens?

(A) The image is still at d_i and is still real.
(B) The image is at a position closer to the lens than d_i and is real.
(C) The image is at a position farther from the lens than d_i and is real.
(D) The image becomes virtual.

3. A beam of light traveling in Medium 1 strikes the interface to another transparent medium with a lower index of refraction, Medium 2. If the intensity of light is measured to be less in Medium 2 than in Medium 1, what can be concluded?

(A) The decrease in intensity was caused by the change in the speed of light in the different media.
(B) Total internal reflection occurred at the interface.
(C) The angle the light travels relative to the normal in Medium 2 will be the same as the angle the light had traveled relative to the normal in Medium 1.
(D) Some part of the light was reflected or absorbed at the interface.

4. If a clear liquid has a refractive index of 1.45 and a transparent solid has an index of 2.90, then, for total internal reflection to occur at the interface between these two media, which of the following must be true?

	incident beam originates in	at an angle of incidence greater than
(A)	The solid	30°
(B)	The liquid	30°
(C)	The liquid	60°
(D)	Total internal reflection cannot occur.	

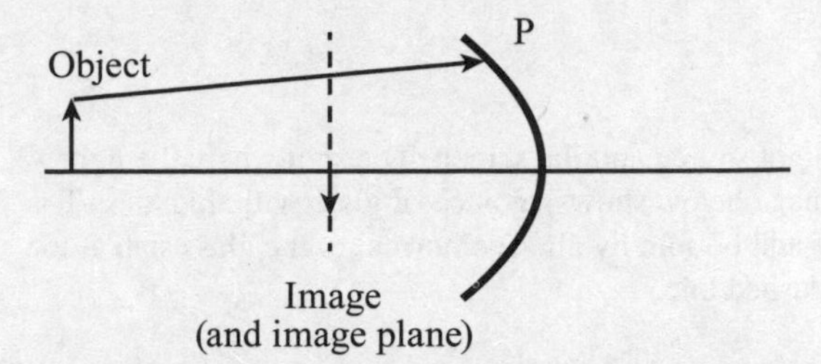

5. The above picture shows a converging mirror and an object, as well as the image, which is formed from standard ray tracing. Also shown, a dotted vertical line indicates the image place for this configuration. Also shown is a ray that travels from the top of the object to the point P on the curved mirror. The point P lies above the line which is parallel to the optic axis from the tip of the image. Where will the reflected ray from point P intersect the image plane?

(A) On the optic axis
(B) Below the optic axis but above the tip of the image
(C) At the tip of the image
(D) Farther from the optic axis than the tip of the image

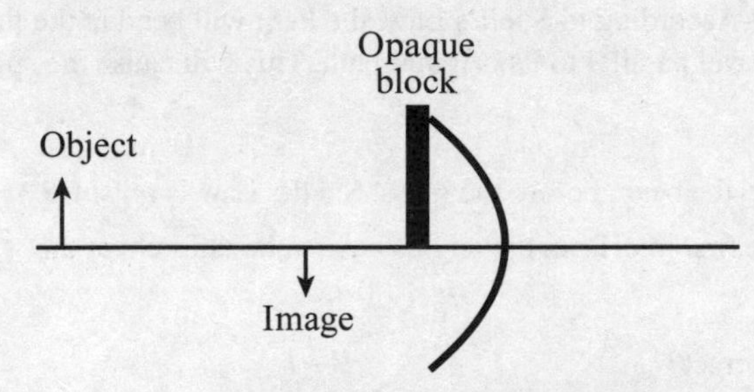

6. The above picture shows a converging mirror and an object, as well as the image, which is formed from standard ray tracing. After the image is formed, an opaque block, also shown in the picture, is inserted to block the top half of the mirror. What changes, if any, will be observed in the image?

(A) The image will remain complete but will be diminished in brightness.
(B) Part of the image will be absent, but the rest will be as bright as it was before the insertion of the block.
(C) Part of the image will be absent and the part which remains will be diminished in brightness.
(D) No change will be observed.

7. You are tasked with creating a real image using a diverging lens as your imaging system. Which of the following criteria is true about both the image and the object?

(A) A real image can be created only if the object is farther away from the lens than the focal length.
(B) A real image can be created only if the object is closer to the lens than the focal length.
(C) A real image can be created regardless of whether the object is farther away from the lens or closer to the lens than the focal length.
(D) A real image cannot be created using only a diverging lens.

Section II: Free Response

1. A beam of light shines on a screen with only air between the light source and the screen. The spot where the light strikes the screen is marked and called point P. Then, as the image below shows, a piece of glass with thickness T is placed in the path of the beam. The glass is surrounded on top and bottom by air. The normal to the glass slab at the point where the incident beam strikes the glass is drawn as a dashed line.

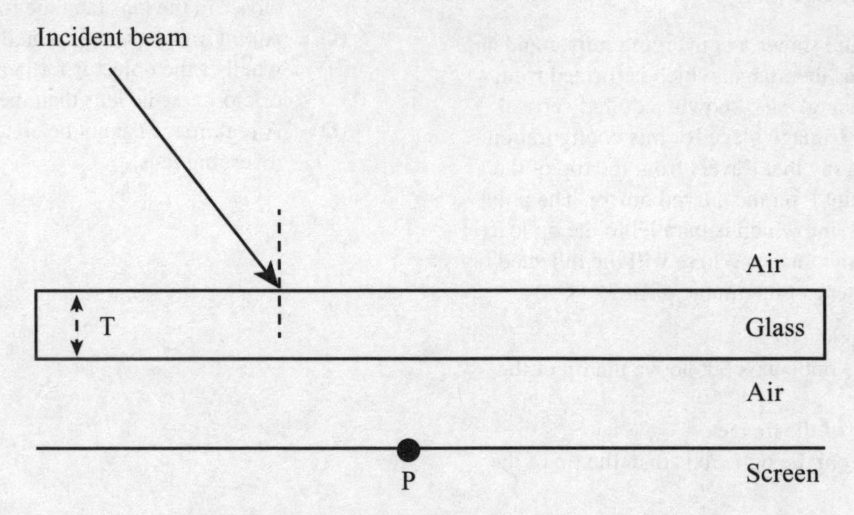

(A) Two students are discussing what observations will be made in this new arrangement.

Student 1: The spot on the screen will still be at point P. According to Snell's Law, the light will bend at the first interface and then unbend at the second interface and travel parallel to its original path. This will cause the spot on the screen to be at the same location.

Student 2: The beam after the glass cannot be parallel to the beam before the glass. Snell's Law is $n_{in} \sin(\theta_{in}) = n_{out} \sin(\theta_{out})$, and because θ_{out} is different from θ_{in} at the first interface, the angles have to be different at the second interface as well.

i. Which parts, if any, of Student 1's reasoning are correct?

ii. Which parts, if any, of Student 2's reasoning are correct?

iii. Which parts, if any, of Student 1's reasoning are incorrect?

iv. Which parts, if any, of Student 2's reasoning are incorrect?

(B) Use ray tracing to draw the path of the beam from where it is incident on the glass to where it strikes the screen. Use appropriate labels to indicate which angles are the same and which are different.

(C) In a paragraph-length explanation without relying solely on equations, explain how your answer in part (b) would be different if the area between the screen and the glass were filled with water (whose index of refraction is between that of air and that of glass).

2. An experiment is set up with an object, a single converging lens, and a screen to make an image appear on the screen. The lens has a focal length of 3 cm. The arrangement is configured so that the image which appears on the screen is the same size as the object. The lens is placed at a position labeled 10 cm on the optic axis.

   (A) Draw a diagram of the experimental setup along the optic axis below. Justify your answer.

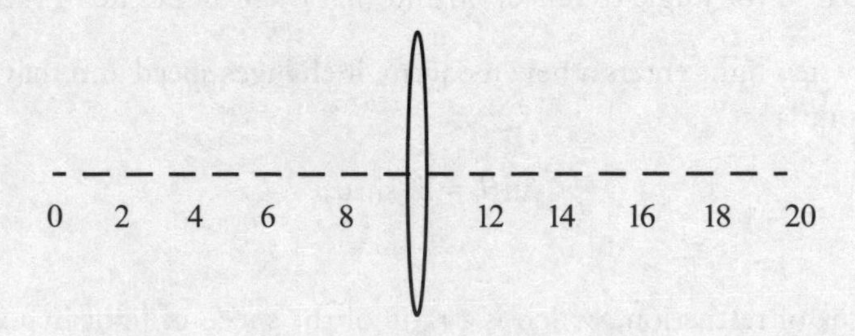

   (B) If the converging lens were replaced with a convex mirror of the same focal length, explain how (if at all) your experimental setup would have to change to produce an image of the same height as the object.

   (C) In a clear, coherent paragraph-length response, explain what a virtual image is. Explain what steps would have to be taken to experimentally verify the existence of a virtual image.

Summary

- Geometric optics always measures angles from the normal line. Topics include the fact that the angle of incidence is equal to the angle of reflection and that these angles lie in the same plane.
- Snell's Law states when light enters a new medium, it changes speed and may change direction. This is stated in the formula

$$n_1 \sin\theta_1 = n_2 \sin\theta_2$$

where n is the index of refraction, which is a ratio of the speed of light in a vacuum to the speed of light in the substance ($n = \frac{c}{v}$). The index of refraction is always greater than 1 and has no units. When going from a higher index of refraction to a lower index of refraction, light may experience total internal reflection if the angle of incidence is larger than the critical angle ($\sin\theta_c = \frac{n_2}{n_1}$).

- For both thin lenses and curved mirrors, you can use the mirror equation: $\frac{1}{s_o} + \frac{1}{s_i} = \frac{1}{f}$.

Note that f is positive for a convex lens or a concave mirror, f is negative for a concave lens or a convex mirror, and s_o is always positive.

- For a lens, if s_i is negative, the lens image is virtual and located on the same side of the lens as the object. If s_i is positive, the lens image is real and located on the opposite side of the lens from the object.
- For a mirror, if s_i is negative, the image is virtual and located on the opposite side of the mirror from the object. If s_i is positive, the image is real and located on the same side of the mirror as the object.
- The magnification of a lens is given by $M = \frac{h_i}{h_o} = \frac{-s_i}{s_o}$.

- A **diverging optical device** could be either a convex mirror or a concave lens. The thing these two have in common is that no matter where the object is located, they form images that are virtual and upright. The sign of h_i will always be positive and the sign of s_i will always be negative. The sign of the magnification will be positive and its value will be less than one. The image will always be located between the focal length and the optical device.

- A **converging optical device** could be either a concave mirror or a convex lens. One thing these two have in common is that when s_o is *outside* the focal length, they form images that are real and inverted. The sign of h_i will be negative and the sign of s_i will be positive. The sign of the magnification will be negative and its absolute value can be any number greater than zero. Another thing they have in common is that when s_o is *inside* the focal length, they form images that are virtual and upright. The sign of h_i will be positive and the sign of s_i will be negative. The sign of the magnification will be positive and its value will be greater than one.

Chapter 10
Quantum, Atomic, and Nuclear Physics

This chapter explores CED Unit 7.5: Properties of Waves and Particles. The chapter on Optics looked at two descriptions of light, one as a wave and one as a ray. Here we explore a third description, "light as a particle."

INTRODUCTION

The subject matter of the previous chapters was developed in the 17th, 18th, and 19th centuries, but as we delve into the physics of the very small, we enter the 20th century.

PHOTONS AND THE PHOTOELECTRIC EFFECT

CED Unit 7.6
Photoelectric Effect

Max Planck first proposed the idea of light being emitted as individual packets of constant energy, called **quanta**. This is where the name quantum mechanics comes from. The particle nature of light was pioneered by Einstein in 1905, when he showed light transferred energy like a particle, and Arthur Compton in 1923, when he showed light has momentum and can undergo elastic collisions. A quantum of electromagnetic energy is known as a **photon**. Light behaves like a stream of photons, and this is illustrated by the **photoelectric effect**.

When a piece of metal is illuminated by electromagnetic radiation (specifically visible light, ultraviolet light, or X-rays), the energy absorbed by electrons near the surface of the metal can liberate them from their bound state, and these electrons can fly off. The released electrons are known as **photoelectrons**. In this case, the classical, wave-only theory of light would predict three results:

(1) There would be a significant time delay between the moment of illumination and the ejection of photoelectrons. This is because the metal would have to heat up to the point that the thermal energy was enough to overcome the binding energy.

(2) Increasing the intensity of the light would cause the electrons to leave the metal surface with greater kinetic energy. This is because the energy carried by a wave is related to its intensity.

(3) Photoelectrons would be emitted regardless of the frequency of the incident energy, as long as the intensity was high enough.

Surprisingly, none of these predictions was observed. Photoelectrons were ejected within just a few billionths of a second after illumination, disproving prediction (1). Secondly, increasing the intensity of the light did not cause photoelectrons to leave the metal surface with greater kinetic energy. Although more electrons were ejected as the intensity was increased, there was a maximum photoelectron kinetic energy; prediction (2) was false. And, for each metal, there was a certain **threshold frequency**, f_0: if light of frequency lower than f_0 were used to illuminate the metal surface, *no* photoelectrons were ejected, regardless of how intense the incident radiation was; prediction (3) was also false. Clearly, something was wrong with the wave-only theory of light.

Einstein explained these observations by borrowing from Planck's idea that light came in individual quanta of energy, which were later called photons by Gilbert Lewis. The energy of a photon is proportional to the frequency of the wave,

$$E = hf$$

Equation Sheet

where h is **Planck's constant** (about $6.63 \cdot 10^{-34}$ J·s). A certain amount of energy had to be imparted to an electron on the metal surface in order to liberate it; this was known as the metal's **work function**, or ϕ. If an electron absorbed a photon whose energy E was greater than ϕ, it would leave the metal with a maximum kinetic energy equal to $E - \phi$. This process could occur *very* quickly, which accounts for the rapidity with which photoelectrons are produced after illumination. This explains why prediction (1) was incorrect.

In this view, increasing the intensity (and therefore the energy) just means increasing the number of photons and results in the ejection of more photoelectrons—but since the energy of each incident photon is fixed by the equation $E = hf$, the value of K_{max} will still be $E - \phi$. This can be expressed as

$$K_{max} = hf - \phi$$

Equation Sheet

This accounts for the observation that disproved prediction (2).

Finally, if the incoming photon's energy was less than the work function (or, $E = hf < \phi$), the photon energy would not be enough to liberate electrons. Blasting the metal surface with more photons (that is, increasing the intensity of the incident beam) would also do nothing; none of the photons would have enough energy to eject electrons. This accounts for the observation of a threshold frequency, which we now know is ϕ/h. This can be expressed as

$$f_o = \phi/h$$

and is why prediction (3) was incorrect. Before we get to some examples, it's worthwhile to introduce a new unit of energy. The SI unit for energy is the joule, but it's too large to be convenient in the domains we're studying now. We'll use a much smaller unit, the **electronvolt** (abbreviated **eV**). The eV is equal to the energy gained (or lost) by an electron accelerated through a potential difference of one volt. Using the equation $\Delta U_E = qV$, we find that

$$1 \text{ eV} = (1 \text{ e})(1 \text{ V}) = (1.6 \times 10^{-19} \text{ C})(1 \text{ V}) = 1.6 \times 10^{-19} \text{ J}$$

In terms of electronvolts, the value of Planck's constant is 4.14×10^{-15} eV·s.

Example 1 The work function, ϕ, for aluminum is 4.08 eV.

(a) What is the threshold frequency required to produce photoelectrons from aluminum?

(b) Classify the electromagnetic radiation that can produce photoelectrons.

(c) If light of frequency $f = 4.00 \times 10^{15}$ Hz is used to illuminate a piece of aluminum,

(i) what is K_{max}, the maximum kinetic energy of ejected photoelectrons?

(ii) what's the maximum speed of the photoelectrons? (Electron mass = 9.11×10^{-31} kg)

(d) If the light described in part (b) were increased by a factor of 2 in intensity, what would happen to the value of K_{max}?

Solution.

(a) We know from the statement of the question that, in order for a photon to successfully liberate an electron from the surface of the aluminum, its energy cannot be less than 4.08 eV. Therefore, the minimum frequency of the incident light—the threshold frequency—must be

$$f_0 = \frac{\phi}{h} = \frac{4.08 \text{ eV}}{4.14 \times 10^{-15} \text{ eV} \cdot \text{s}} = 9.86 \times 10^{14} \text{ Hz}$$

(b) Based on the electromagnetic spectrum given in the previous chapter, the electromagnetic radiation used to produce photoelectrons from aluminum must be at least in the ultraviolet region of the EM spectrum.

(c) (i) The maximum kinetic energy of photoelectrons is found from the equation

$$\begin{aligned} K_{max} &= hf - \phi \\ &= (4.14 \times 10^{-15} \text{ eV} \cdot \text{s})(4.00 \times 10^{15} \text{ Hz}) - (4.08 \text{ eV}) \\ &= 12.5 \text{ eV} \end{aligned}$$

(ii) Using the above result and $K = \frac{1}{2}mv^2$, we can find v_{max}:

$$\begin{aligned} v_{max} &= \sqrt{\frac{2}{m_e}K_{max}} \\ &= \sqrt{\frac{2}{9.11 \times 10^{-31} \text{ kg}}\left(12.5 \text{ eV} \cdot \frac{1.6 \times 10^{-19} \text{ J}}{1 \text{ eV}}\right)} \\ &= 2.1 \times 10^6 \text{ m/s} \end{aligned}$$

(d) Nothing will happen. Increasing the intensity of the illuminating radiation will cause more photons to impinge on the metal surface, thereby ejecting more photoelectrons, but their maximum kinetic energy will remain the same. The only way to increase K_{max} would be to increase the frequency of the incident energy.

THE BOHR MODEL OF THE ATOM

In the years immediately following Rutherford's announcement of his nuclear model of the atom, a young physicist, Niels Bohr, would add an important piece to the atomic puzzle. Rutherford told us where the positive charge of the atom was located; Bohr would tell us about the electrons.

Chemistry
Modern Physics begins to explore atoms, which bridges into chemistry. Fortunately, you will not need to know much about chemistry for the AP Physics 2 Exam—just a basic understanding of some topics.

For 50 years, it had been known that atoms in a gas discharge tube emit and absorb light only at specific wavelengths. The light from a glowing gas, passed through a prism to disperse the beam into its component wavelengths, produces patterns of sharp lines called **atomic spectra**. The visible wavelengths that appear in the emission spectrum of hydrogen had been summarized by the *Balmer formula*

$$\frac{1}{\lambda_n} = R\left(\frac{1}{2^2} - \frac{1}{n^2}\right)$$

where R is the *Rydberg constant* (about 1.1×10^7 m^{-1}). The formula worked—that is, it fit the observational data—but it had no theoretical basis. So, the question was, *why* do atoms emit (or absorb) radiation only at certain discrete wavelengths?

Bohr's model of the atom explains the spectroscopists' observations. Using the simplest atom, hydrogen (with only one electron), Bohr postulated that the electron would orbit the nucleus only at certain discrete radii. When the electron is in one of these special orbits, it does not radiate away energy (as the classical theory would predict). However, if the electron absorbs a certain amount of energy, it is **excited** to a higher orbit, one with a greater radius. After a short time in this excited state, it returns to a lower orbit, emitting a photon in the process. Since each allowed orbit—or **energy level**—has a specific radius (and corresponding energy), the photons emitted in each jump have only specific wavelengths.

When an excited electron drops from energy level $n = j$ to a lower one, $n = i$, the transition causes a photon of energy to be emitted, and the energy of the photon is the difference between the two energy levels. When an electron absorbs a photon and transitions from a lower energy level, $n = i$, to a higher energy level, $n = j$, the energy of the absorbed photon is the difference between the two energy levels.

$$E_{\text{emitted or absorbed photon}} = |\Delta E| = E_j - E_i$$

The wavelength of this photon is

$$\lambda = \frac{c}{f} = \frac{c}{|E_{\text{photon}}|/h} = \frac{hc}{|E_j - E_i|}$$

Speed of Light
Photons travel at the speed of light.

Example 2

The first five energy levels of an atom are shown in the diagram below:

-3 eV ____________ $n = 5$

-4 eV ____________ $n = 4$

-7 eV ____________ $n = 3$

-15 eV ____________ $n = 2$

-62 eV ____________ $n = 1$ ground state

(a) If the atom begins in the $n = 3$ level, what photon energies could be emitted as it returns to its ground state?

(b) What could happen if this atom, while in an undetermined energy state, were bombarded with a photon of energy 10 eV?

Solution.

(a) If the atom is in the $n = 3$ level, it could return to ground state by a transition from $3 \rightarrow 1$, or from $3 \rightarrow 2$ and then $2 \rightarrow 1$. The energy emitted in each of these transitions is simply the difference between the energies of the corresponding levels:

$$E_{3\rightarrow1} = E_3 - E_1 = (-7 \text{ eV}) - (-62 \text{ eV}) = 55 \text{ eV}$$

$$E_{3\rightarrow2} = E_3 - E_2 = (-7 \text{ eV}) - (-15 \text{ eV}) = 8 \text{ eV}$$

$$E_{2\rightarrow1} = E_2 - E_1 = (-15 \text{ eV}) - (-62 \text{ eV}) = 47 \text{ eV}$$

(b) Since no two energy levels in this atom are separated by 10 eV, the atom could not absorb a 10 eV photon, and as a result, nothing would happen. This atom would be transparent to light of energy 10 eV.

Example 3 The ground state of hydrogen is -13.6 eV. The first excited state is -3.4 eV. The second excited state is -1.5 eV. The third excited state is -0.85 eV.

(a) How much energy must a ground state electron in a hydrogen atom absorb to be excited to the $n = 4$ energy level?

(b) With the electron in the $n = 4$ level, what wavelengths are possible for the photon emitted when the electron drops to a lower energy level? In what regions of the EM spectrum do these photons lie?

Solution.

(a) The ground-state energy level ($n = 1$) for hydrogen is –13.6 eV, and $E_4 = -0.85$ eV.

Therefore, in order for an electron to make the transition from E_1 to E_4, it must absorb energy in the amount $E_4 - E_1 = (-0.85 \text{ eV}) - (-13.6 \text{ eV}) = 12.8$ eV.

(b) An electron in the $n = 4$ energy level can make several different transitions: it can drop to $n = 3$, $n = 2$, or all the way down to the ground state, $n = 1$.

The following diagram shows the electron dropping from $n = 4$ to $n = 3$:

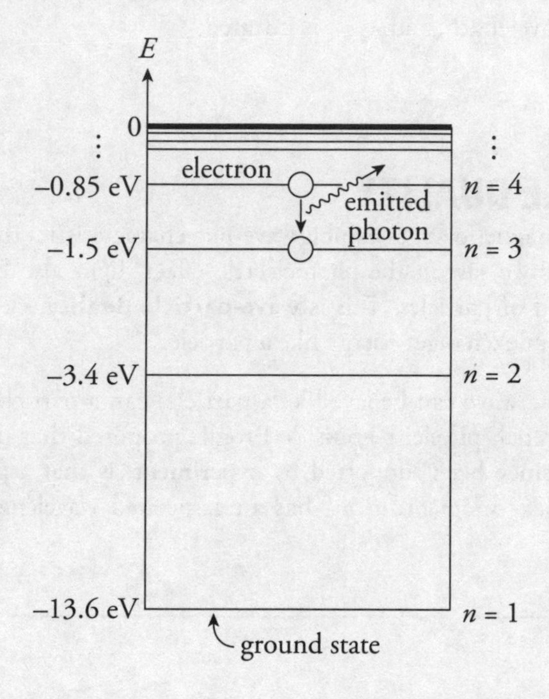

There are three possible values for the energy of the emitted photon, $E_{4\to3}$, $E_{4\to2}$, or $E_{4\to1}$:

$$E_{4\to3} = E_4 - E_3 = (-0.85 \text{ eV}) - (-1.5 \text{ eV}) = 0.65 \text{ eV}$$
$$E_{4\to2} = E_4 - E_2 = (-0.85 \text{ eV}) - (-3.4 \text{ eV}) = 2.55 \text{ eV}$$
$$E_{4\to1} = E_4 - E_1 = (-0.85 \text{ eV}) - (-13.6 \text{ eV}) = 12.8 \text{ eV}$$

From the equation $E = hf = hc/\lambda$, we get $\lambda = hc/E$, so

$$\lambda_{4\to3} = \frac{hc}{E_{4\to3}} = \frac{(4.14\times10^{-15}\text{ eV}\cdot\text{s})(3.00\times10^{8}\text{ m/s})}{0.65\text{ eV}} = 1{,}910\text{ nm}$$

$$\lambda_{4\to2} = \frac{hc}{E_{4\to2}} = \frac{(4.14\times10^{-15}\text{ eV}\cdot\text{s})(3.00\times10^{8}\text{ m/s})}{2.55\text{ eV}} = 487\text{ nm}$$

$$\lambda_{4\to1} = \frac{hc}{E_{4\to1}} = \frac{(4.14\times10^{-15}\text{ eV}\cdot\text{s})(3.00\times10^{8}\text{ m/s})}{12.8\text{ eV}} = 97\text{ nm}$$

Note that $\lambda_{4\to2}$ is in the visible spectrum; this wavelength corresponds to the color blue-green; $\lambda_{4\to1}$ is an ultraviolet wavelength, and $\lambda_{4\to3}$ is infrared.

WAVE-PARTICLE DUALITY

Light and other electromagnetic waves exhibit wave-like characteristics through interference and diffraction. However, as we saw in the photoelectric effect, light also behaves as if its energy were granular, composed of particles. This is **wave-particle duality**: electromagnetic radiation propagates like a wave but exchanges energy like a particle.

Since an electromagnetic wave can behave like a particle, can a particle of matter behave like a wave? In 1924, the French physicist Louis de Broglie proposed that the answer is "yes." His conjecture, which has since been supported by experiment, is that a particle of mass *m* and speed *v*—and thus, linear momentum *p*—has an associated wavelength, which is called its **de Broglie wavelength**:

Equation Sheet

$$\lambda = \frac{h}{p}$$

Particles in motion can display wave characteristics and behave as if they have a wavelength.

Since the value of *h* is so small, ordinary macroscopic objects do not display wave-like behavior. For example, a baseball (mass = 0.15 kg) thrown at a speed of 40 m/s has a de Broglie wavelength of

$$\lambda = \frac{h}{p} = \frac{h}{mv} = \frac{6.63\times10^{-34}\text{ J}\cdot\text{s}}{(0.15\text{ kg})(40\text{ m/s})} = 1.1\times10^{-34}\text{ m}$$

This is much too small to measure. However, with subatomic particles, the wave nature is clearly evident. The 1937 Nobel prize in physics was awarded for experiments by C. J. Davisson and G. P. Thomson that revealed that a stream of electrons exhibited diffraction patterns when scattered by crystals—a behavior that's characteristic of waves.

Example 4 Electrons in a diffraction experiment are accelerated through a potential difference of 200 V. What is the de Broglie wavelength of these electrons?

Solution. By definition, the kinetic energy of these electrons is 200 eV. Since the relationship between linear momentum and kinetic energy is $p = \sqrt{2mK}$,

$$\lambda = \frac{h}{p} = \frac{h}{\sqrt{2mK}} = \frac{6.63 \times 10^{-34}\ \text{J} \cdot \text{s}}{\sqrt{2(9.11 \times 10^{-31}\ \text{kg})\ 200\ \text{eV} \cdot \dfrac{1.6 \times 10^{-19}\ \text{J}}{1\ \text{eV}}}} = 8.7 \times 10^{-11}\ \text{m} = 0.087\ \text{nm}$$

This wavelength is characteristic of X-rays.

THE WAVE FUNCTION

CED Unit 7.7
Wave Functions and Probability

There are several major differences between classical physics and modern physics. One of the most important is that the mathematics that underpins modern physics (which is too complicated for treatment here) no longer specifies a definite position at a definite time for a particle such as an electron. Instead, the mathematics only gives the *probability* that a particle will be measured to be at a particular position when the position is measured. That probability is related to a new physical parameter called the *wave function*, Ψ.

In one interpretation of that mathematics, the act of measuring the position changes the wave function, Ψ, so that it has a single position with a probability of 100%, and that location corresponds to the result of the measurement (this is referred to as "collapsing the wave-function"). However, before the position is measured (or after a measurement has taken place and some time has passed, so the position is no longer certain), there are a range of probable locations that can be represented as a graph of Ψ vs. *x*. To interpret such a graph, the position with the largest absolute value is the most probable location for the result of a measurement. Any position where the graph has a height of 0, it is certain that the position will not be observed.

Interpreting wave-function graphs
Wave-function graphs will often have values that are positive as well as negative. When interpreting the graph, only the absolute value is important in ranking relative likelihoods of experimental outcomes (technically the likelihood is proportional to Ψ^2, but the absolute value will give correct interpretations and is easier to read from a graph).

RELATIVITY

Another important idea in modern physics is the Theory of Relativity. Relativity has only two postulates:

1. The results of physical experiments will be the same in any non-accelerating reference frames.
2. The speed of light is constant.

For the AP Exam, you are not required to know any of the mathematical framework surrounding relativity, but only have a qualitative understanding of the consequences of these postulates.

The major consequence of relativity is that time is not absolute, but that space and time are linked together. To see why this must be the case, imagine a car driving down the highway at 65 mph. In non-relativistic physics, the light from the headlight would be moving at the speed of light plus 65 mph. However, the second postulate of relativity states that this cannot be true, and the light from a headlight of a moving car must also be exactly the speed of light.

As the name indicates, relativity describes the consequences of performing experiments in different reference frames and measuring the same results.

When a clock placed on a fast-moving airplane is compared to a clock at rest on the ground, the clock in the airplane shows that less time has passed than the time recorded by the clock on the ground. This phenomenon is known as **time dilation**, and it has been demonstrated experimentally using synchronized atomic clocks.

The speed of light is constant according to the Theory of Relativity. However, time dilation shows there are disagreements about the amount of time that passes in different reference frames. To be consistent with time dilation, there must also be disagreement about distances. This is known as **length contraction**. A scientist moving along with the object he measures will observe the object to have a different length than a scientist in a different reference frame.

NUCLEAR PHYSICS

The nucleus of the atom is composed of particles called **protons** and **neutrons**, which are collectively called **nucleons**. The number of protons in a given nucleus is called the atom's **atomic number**, and is denoted Z, and the number of neutrons (the **neutron number**) is denoted N. The total number of nucleons, $Z + N$, is called the **mass number** (or **nucleon number**), and is denoted A. The number of protons in the nucleus of an atom defines the element. For example, the element chlorine (abbreviated Cl) is characterized by the fact that the nucleus of every chlorine atom contains 17 protons, so the atomic number of chlorine is 17; but, different chlorine atoms may contain different numbers of neutrons. In fact, about three-fourths of all naturally occurring chlorine atoms have 18 neutrons in their nuclei (mass number = 35), and most of the remaining one-fourth contain 20 neutrons (mass number = 37).

Nuclei that contain the same numbers of protons but different numbers of neutrons are called **isotopes**. The notation for a **nuclide**—the term for a nucleus with specific numbers of protons and neutrons—is to write Z and A before the chemical symbol of the element:

$$^{A}_{Z}\text{X}$$

The isotopes of chlorine mentioned above would be written as follows:

$$^{35}_{17}\text{Cl} \quad \text{and} \quad ^{37}_{17}\text{Cl}$$

Example 5 How many protons and neutrons are contained in the nuclide $^{63}_{29}\text{Cu}$?

Solution. The subscript (the atomic number, Z) gives the number of protons, which is 29. The superscript (the mass number, A) gives the total number of nucleons. Since $A = 63 = Z + N$, we find that $N = 63 - 29 = 34$.

Example 6 The element neon (abbreviated Ne, atomic number 10) has several isotopes. The most abundant isotope contains 10 neutrons, and two others contain 11 and 12. Write symbols for these three nuclides.

Solution. The mass numbers of these isotopes are 10 + 10 = 20, 10 + 11 = 21, and 10 + 12 = 22. So, we'd write them as follows:

$$^{20}_{10}\text{Ne}, \quad ^{21}_{10}\text{Ne}, \quad \text{and} \quad ^{22}_{10}\text{Ne}$$

Another common notation—which we also use—is to write the mass number after the name of the element. These three isotopes of neon would be written as neon-20, neon-21, and neon-22.

The Nuclear Force

CED Unit 7.1
Systems and Fundamental Forces

Why wouldn't any nucleus that has more than one proton be unstable? After all, protons are positively charged and would therefore experience a repulsive Coulomb force from each other. Why don't these nuclei explode? And what holds neutrons—which have no electric charge—in the nucleus? These issues are resolved by the presence of another fundamental force, the **strong nuclear force**, which binds neutrons and protons together to form nuclei. Although the strength of the Coulomb force can be expressed by a simple mathematical formula (it's inversely proportional to the square of their separation), the nuclear force is much more complicated; no simple formula can be written for the strength of the nuclear force.

Binding Energy

CED Unit 7.3
Energy in Modern Physics

CED Unit 7.4
Mass-Energy Equivalence

The masses of the proton and neutron are listed below.

$$\text{proton: } m_p = 1.6726 \times 10^{-27} \text{ kg}$$

$$\text{neutron: } m_n = 1.6749 \times 10^{-27} \text{ kg}$$

Because these masses are so tiny, a much smaller mass unit is used. With the most abundant isotope of carbon (carbon-12) as a reference, the **atomic mass unit** (abbreviated **amu** or simply **u**) is defined as 1/12 the mass of a ^{12}C atom. The conversion between kg and u is 1 u = 1.6605×10^{-27} kg. In terms of atomic mass units,

$$m_p = 1.00728 \text{ u}$$

$$m_n = 1.00867 \text{ u}$$

Now consider the **deuteron**, the nucleus of **deuterium**, an isotope of hydrogen that contains 1 proton and 1 neutron. The mass of a deuteron is 2.01356 u, which is a little *less* than the sum of the individual masses of the proton and neutron. The difference between the mass of any bound nucleus and the sum of the masses of its constituent nucleons is called the **mass defect**, Δm. In the case of the deuteron (symbolized **d**), the mass defect is

$$\begin{aligned}\Delta m &= (m_p + m_n) - m_d \\ &= (1.00728 \text{ u} + 1.00867 \text{ u}) - (2.01356 \text{ u}) \\ &= 0.00239 \text{ u}\end{aligned}$$

What happened to this missing mass? It was converted to energy when the deuteron was formed. In 1905, Einstein gave us the famous equation, which tells us how much energy mass contains:

Equation Sheet

$$E = mc^2$$

Because of this, the mass-difference gives us the amount of energy needed to break the deuteron into a separate proton and neutron. Since this tells us how strongly the nucleus is bound, it is called the **binding energy** of the nucleus.

$$E_B = (\Delta m)c^2$$

Using $E = mc^2$, the energy equivalent of 1 atomic mass unit is

$$\begin{aligned} E &= (1.6605 \times 10^{-27}\ \text{kg})(2.9979 \times 10^{8}\ \text{m/s})^2 \\ &= 1.4924 \times 10^{-10}\ \text{J} \\ &= 1.4924 \times 10^{-10}\ \text{J} \cdot \frac{1\ \text{eV}}{1.6022 \times 10^{-19}\ \text{J}} \\ &= 9.31 \times 10^{8}\ \text{eV} \\ &= 931\ \text{MeV} \end{aligned}$$

In terms of electronvolts, then, the binding energy of the deuteron is

$$E_B\ (\text{deuteron}) = 0.00239\ \text{u} \times \frac{931\ \text{MeV}}{1\ \text{u}} = 2.23\ \text{MeV}$$

Since the deuteron contains 2 nucleons, the **binding-energy-per-nucleon** is

$$\frac{2.23\ \text{MeV}}{2\ \text{nucleons}} = 1.12\ \text{MeV/nucleon}$$

This is the lowest value of all nuclides. The highest, 8.8 MeV/nucleon, is for an isotope of nickel, ^{62}Ni. Typically, when nuclei smaller than nickel are fused to form a single nucleus, the binding energy per nucleon increases, which tells us that energy is released in the process. On the other hand, when nuclei *larger* than nickel are *split*, binding energy per nucleon again increases, releasing energy. This is the basis of nuclear fission.

Example 7 What is the maximum wavelength of EM radiation that could be used to photodisintegrate a deuteron?

Solution. The binding energy of the deuteron is 2.23 MeV, so a photon would need to have at least this much energy to break the deuteron into a proton and neutron. Since $E = hf$ and $f = c/\lambda$,

$$E = \frac{hc}{\lambda} \longrightarrow \lambda_{\text{max}} = \frac{hc}{E_{\text{min}}} = \frac{(4.14 \times 10^{-15}\ \text{eV} \cdot \text{s})(3.00 \times 10^{8}\ \text{m/s})}{2.23 \times 10^{6}\ \text{eV}} = 5.57 \times 10^{-13}\ \text{m}$$

Example 8 The atomic mass of $^{27}_{13}\text{Al}$ is 26.9815 u. What is its nuclear binding energy per nucleon? (Mass of electron = 0.0005486 u)

Solution. The nuclear mass of ${}^{27}_{13}\text{Al}$ is equal to its atomic mass minus the mass of its electrons. Since an aluminum atom has 13 protons, it must also have 13 electrons. So,

$$\begin{aligned}\text{nuclear mass of } {}^{27}_{13}\text{Al} &= (\text{atomic mass of } {}^{27}_{13}\text{Al}) - 13m_e \\ &= 26.9815 \text{ u} - 13(0.0005486 \text{ u}) \\ &= 26.9744 \text{ u}\end{aligned}$$

Now, the nucleus contains 13 protons and 27 – 13 = 14 neutrons, so the total mass of the individual nucleons is

$$\begin{aligned}M &= 13m_p + 14m_n \\ &= 13(1.00728 \text{ u}) + 14(1.00867 \text{ u}) \\ &= 27.2160 \text{ u}\end{aligned}$$

and the mass defect of the aluminum nucleus is

$$\Delta m = M - m = 27.2160 \text{ u} - 26.9744 \text{ u} = 0.2416 \text{ u}$$

Converting this mass to energy, we can see that

$$E_B = 0.2416 \text{ u} \times \frac{931 \text{ MeV}}{1 \text{ u}} = 225 \text{ MeV}$$

so the binding energy per nucleon is

$$\frac{225 \text{ MeV}}{27} = 8.3 \text{ MeV/nucleon}$$

CED Unit 7.2
Radioactive Decay

NUCLEAR REACTIONS

Natural radioactive decay provides one example of a nuclear reaction. Other examples of nuclear reactions include the bombardment of target nuclei with subatomic particles to artificially induce radioactivity, such as the emission of a particle or the splitting of the nucleus (this is **nuclear fission**), and the **nuclear fusion** of small nuclei at extremely high temperatures. In all cases of nuclear reactions that we'll study, nucleon number and charge must be conserved. In order to balance nuclear reactions, we write ${}^{1}_{1}\text{p}$ or ${}^{1}_{1}\text{H}$ for a proton and ${}^{1}_{0}\text{n}$ for a neutron. Gamma-ray photons can also be produced in nuclear reactions; they have no charge or nucleon number and are represented as ${}^{0}_{0}\gamma$.

Matter Cannot be Created or Destroyed
But it *can* undergo other forms—which does not break this rule. The quickest way to determine whether something undergoes alpha, beta, or gamma decay is to find the amount of nucleons that have escaped and then to determine which kind of decay happened.

Alpha Decay

When a nucleus undergoes alpha decay, it emits an alpha particle, which consists of two protons and two neutrons and is the same as the nucleus of a helium-4 atom. An alpha particle can be represented as

$$\alpha,\ {}^{4}_{2}\alpha,\ \text{or}\ {}^{4}_{2}\text{He}$$

Very large nuclei can shed nucleons quickly by emitting one or more alpha particles; for example, radon-222 (${}^{222}_{86}\text{Rn}$) is radioactive and undergoes alpha decay.

$${}^{222}_{86}\text{Rn} \rightarrow {}^{218}_{84}\text{Po} + {}^{4}_{2}\alpha$$

This reaction illustrates two important features of a nuclear reaction.

(1) Mass number is conserved.
(2) Charge is conserved.

The decaying nuclide is known as the **parent**, and the resulting nuclide is known as the **daughter**. (Here, radon-222 is the parent nuclide and polonium-218 is the daughter.) Alpha decay decreases the mass number by 4 and the atomic number by 2. Therefore, alpha decay looks like the following:

$${}^{A}_{Z}\text{X} \rightarrow {}^{A-4}_{Z-2}\text{X}' + {}^{4}_{2}\alpha$$

Beta Decay

There are three subcategories of **beta** (β) decay, called β^-, β^+, and **electroncapture** (**EC**).

β^- Decay When the neutron-to-proton ratio is too large, the nucleus undergoes β^- decay, which is the most common form of beta decay. β^- decay occurs when a neutron transforms into a proton and releases an electron. The expelled electron is called a **beta particle**. The transformation of a neutron into a proton and an electron (and another particle, the **electron-antineutrino**, $\bar{\nu}_e$) is caused by the action of the **weak nuclear force**, another of nature's fundamental forces. A common example of a nuclide that undergoes β^- decay is carbon-14, which is used to date archaeological artifacts.

$${}^{14}_{6}\text{C} \rightarrow {}^{14}_{7}\text{N} + e^- + \bar{\nu}_e$$

The reaction is balanced, since 14 = 14 + 0 and 6 = 7 + (–1).

β^+ Decay When the neutron-to-proton ratio is too small, the nucleus will undergo β^+ decay. In this form of beta decay, a proton is transformed into a neutron and a **positron,** ${}^{0}_{+1}e$ (the electron's **antiparticle**), plus another particle, the **electron-neutrino,** ν_e, which are then both ejected from the nucleus. An example of a positron emitter is fluorine-17.

$${}^{17}_{9}\text{F} \rightarrow {}^{17}_{8}\text{O} + e^{+} + \nu_e$$

Electron Capture Another way in which a nucleus can increase its neutron-to-proton ratio is to capture an orbiting electron and then cause the transformation of a proton into a neutron. Beryllium-7 undergoes this process.

$${}^{7}_{4}\text{Be} + e^{-} \rightarrow {}^{7}_{3}\text{Li} + \nu_e$$

Gamma Decay

In each of the decay processes defined above, the daughter was a different element than the parent. Radon becomes polonium as a result of α decay, carbon becomes nitrogen as a result of β^- decay, fluorine becomes oxygen from β^+ decay, and beryllium becomes lithium from electron capture. By contrast, gamma decay does not alter the identity of the nucleus; it just allows the nucleus to relax and shed energy. Imagine that potassium-42 undergoes β^- decay to form calcium-42.

$${}^{42}_{19}\text{K} \rightarrow {}^{42}_{20}\text{Ca}^{*} + e^{-} + \bar{\nu}_e$$

The asterisk indicates that the daughter calcium nucleus is left in a high-energy, excited state. For this excited nucleus to drop to its ground state, it must emit energy in the form of a photon, a **gamma ray**, symbolized by γ.

$${}^{42}_{20}\text{Ca}^{*} \rightarrow {}^{42}_{20}\text{Ca} + \gamma$$

Let's sum up the three types of radiation:

Type of Radiation		What Happens?	Charge
Alpha		Particle—a helium nucleus, containing 2 protons and 2 neutrons, is released from the nucleus.	Positive 2+
Beta	Beta-minus	Particle—a highly energetic, massless electron is released from the nucleus after a neutron divides into a proton and an electron.	Negative 1–
	Beta-plus	Particle—a highly energetic, massless positron is released from the nucleus after a proton divides into a neutron and a positron.	Positive 1+
Gamma		Wave—energy is emitted in the form of a photon.	No Charge

Example 9 A mercury-198 nucleus is bombarded by a neutron, which causes a nuclear reaction:

$$^{1}_{0}\text{n} + {}^{198}_{80}\text{Hg} \rightarrow {}^{197}_{79}\text{Au} + {}^{?}_{?}\text{X}$$

What's the unknown product particle, X?

Solution. In order to balance the superscripts, we must have $1 + 198 = 197 + A$, so $A = 2$, and the subscripts are balanced if $0 + 80 = 79 + Z$, so $Z = 1$:

$$^{1}_{0}\text{n} + {}^{198}_{80}\text{Hg} \rightarrow {}^{197}_{79}\text{Au} + {}^{2}_{1}\text{X}$$

Therefore, X must be a deuteron, $^{2}_{1}\text{H}$ (or just d).

DISINTEGRATION ENERGY

Nuclear reactions not only produce new nuclei and other subatomic product particles, but they also involve the absorption or emission of energy. Nuclear reactions must conserve total energy, so changes in mass are accompanied by changes in energy according to Einstein's equation

$$\Delta E = (\Delta m)c^2$$

A general nuclear reaction is written

$$\text{A} + \text{B} \rightarrow \text{C} + \text{D} + Q$$

where Q denotes the **disintegration energy**. If Q is positive, the reaction is **exothermic** (or **exoergic**) and the reaction can occur spontaneously; if Q is negative, the reaction is **endothermic** (or **endoergic**) and the reaction cannot occur spontaneously. The energy Q is calculated as follows:

$$Q = [(m_A + m_B) - (m_C + m_D)]c^2 = (\Delta m)c^2$$

For spontaneous reactions—ones that liberate energy—most of the energy is revealed as kinetic energy of the least massive product nuclei.

Example 10 The process that powers the Sun—and upon which all life on Earth is dependent—is the fusion reaction:

$$4\,{}^{1}_{1}\text{H} \rightarrow {}^{4}_{2}\alpha + 2e^{+} + 2\nu_e + \gamma$$

(a) Show that this reaction releases energy.

(b) How much energy is released per proton?

Use the fact that $m_\alpha = 4.0015$ u and ignore the mass of the electron-neutrino, ν_e.

Solution.

(a) We need to find the mass difference between the reactants and products:

$$\begin{aligned}\Delta m &= 4m_{\text{p}} - (m_{\alpha} + 2m_{\text{e}}) \\ &= 4(1.00728\text{ u}) - [4.0015\text{ u} + 2(0.0005486\text{ u})] \\ &= 0.02652\text{ u}\end{aligned}$$

Since Δm is positive, the reaction is exothermic: energy is released.

(b) Converting the mass difference to energy gives

$$Q = 0.02652\text{ u} \times \frac{931\text{ MeV}}{1\text{ u}} = 24.7\text{ MeV}$$

Since four protons went into the reaction, the energy liberated per proton is

$$\frac{24.7\text{ MeV}}{4\text{ p}} = 6.2\text{ MeV/proton}$$

Example 11 Can the following nuclear reaction occur spontaneously?

$${}^{4}_{2}\alpha + {}^{14}_{7}\text{N} \rightarrow {}^{17}_{8}\text{O} + {}^{1}_{1}\text{H}$$

(The mass of the nitrogen nucleus is 13.9992 u, and the mass of the oxygen nucleus is 16.9947 u.)

Solution. We first figure out the mass equivalent of the disintegration energy:

$$\begin{aligned}\Delta m &= (m_{\alpha} + m_{\text{N}}) - (m_{\text{O}} + m_{\text{p}}) \\ &= (4.0015\text{ u} + 13.9992\text{ u}) - (16.9947\text{ u} + 1.00728\text{ u}) \\ &= -0.00128\text{ u}\end{aligned}$$

Since Δm is negative, this reaction is nonspontaneous; energy must be *supplied* in order for this reaction to proceed. But how much?

$$|Q| = 0.00128\text{ u} \times \frac{931\text{ MeV}}{1\text{ u}} = 1.19\text{ MeV}$$

Want Even More Physics Review?
Check out *High School Physics Unlocked*, our skill-building guide to mastering complex physics concepts.

Chapter 10 Review Questions

Solutions can be found in Chapter 11.

Section I: Multiple Choice

1. According to the theory put forth by Louis de Broglie, all matter has wave-like properties such as interference, but these properties are seen only at a microscopic scale. Why are these properties not typically observed at a macroscopic scale?
   (A) The wavelength of matter is typically too large to observe this interference.
   (B) Interference of matter occurs only when these waves interact with other objects comparable to their wavelength and those things are microscopic.
   (C) There are no energy level transitions available to allow for this interference to be observed.
   (D) At the macroscopic scale, the interference is always destructive, so it cannot be observed.

2. An experiment is conducted on the photoelectric effect. A metal whose work function is 6.0 eV is struck with a beam of light with power 1.0 mW of a frequency 7.2×10^{15} Hz. A photoelectron is ejected from the surface of the metal and is found to require a stopping potential of 24 eV. Which of the following changes could be made so that a photoelectron was not ejected from the surface of the metal? (Select two answers.)
   (A) Alter the work function of the metal.
   (B) Alter the power of the light.
   (C) Alter the frequency of the light.
   (D) Alter the stopping potential.

3. An atom with one electron has an ionization energy of 25.0 eV. An electron in this atom makes a transition from an excited energy level, where $E = -16.0$ eV, to the ground state. What is the wavelength of the emitted photon from this transition?
   (A) 138 nm
   (B) 112 nm
   (C) 77.5 nm
   (D) 49.6 nm

4. The single electron in an atom has an energy of –40 eV when it's in the ground state, and the first excited state for the electron is at –10 eV. What will happen to this electron if the atom is struck by a stream of photons, each of energy 15 eV?
   (A) The electron will absorb the energy of one photon and become excited halfway to the first excited state, then quickly return to the ground state, emitting a 15 eV photon in the process.
   (B) The electron will absorb the energy of one photon and become excited halfway to the first excited state, then quickly absorb the energy of another photon to reach the first excited state.
   (C) The electron will absorb two photons and be excited to the first excited state.
   (D) Nothing will happen.

5. The products of several radioactive decays are being studied. Each particle starts with the same speed and enters into a region with a uniform magnetic field directed perpendicular to the initial velocity of the particles. Which observation could be made?
   (A) A neutron and an electron are deflected in the same direction, but with the electron turning with a larger radius.
   (B) An alpha particle and an electron are deflected in the same direction, but with the alpha particle turning with a larger radius.
   (C) An alpha particle and a neutron are both undeflected.
   (D) An alpha particle is deflected and a neutron is undeflected.

6. A partial energy-level diagram for an atom is shown below. What photon energies could this atom emit if it begins in the $n = 3$ state?

-3 eV ________ $n = 4$

-5 eV ________ $n = 3$

-8 eV ________ $n = 2$

-12 eV ________ $n = 1$ ground state

(A) 5 eV only
(B) 3 eV or 7 eV only
(C) 2 eV, 3 eV, or 7 eV
(D) 3 eV, 4 eV, or 7 eV

7. Which of the following transitions between energy levels results in emission of the shortest wavelength photon?

(A) A large energy transition to a higher energy level
(B) A small energy transition to a higher energy level
(C) A large energy transition to a lower energy level
(D) A small energy transition to a lower energy level

8. The de Broglie hypothesis, that $\lambda = h/p$, was experimentally confirmed by Davisson, Germer, and Thompson. Which of the following describes this hypothesis?

(A) A photon carries a momentum that depends on the wavelength of that photon.
(B) Momentum will be conserved during quantum mechanical processes.
(C) Photons may behave like particles under certain circumstances.
(D) Electrons will undergo diffraction in certain circumstances.

9. In nuclear reactions, all of the following are true about mass and mass defect EXCEPT

(A) the mass defect is directly proportional to the binding energy
(B) the mass of the unbound nucleons will be less than their mass when combined
(C) the energy associated with the mass defect is significantly greater than the energy levels for electrons in the atom
(D) the mass defect is significantly less than the rest mass of the nucleons

10. What's the missing particle in the following nuclear reaction?

$${}^{2}_{1}\text{H} + {}^{63}_{29}\text{Cu} \longrightarrow {}^{64}_{30}\text{Zn} + (?)$$

(A) Proton
(B) Neutron
(C) Electron
(D) Positron

11. A particular isotope of platinum ${}^{175}_{78}\text{Pt}$ has a half-life of just over 2.5 seconds. It can decay either via alpha decay or beta(+) decay. What are the daughter nuclei from each of these processes?

(A) Alpha decay results in ${}^{171}_{76}\text{Os}$ and beta(+) decay results in ${}^{175}_{77}\text{Ir}$.
(B) Alpha decay results in ${}^{171}_{76}\text{Os}$ and beta(+) decay results in ${}^{175}_{79}\text{Au}$.
(C) Alpha decay results in ${}^{179}_{80}\text{Hg}$ and beta(+) decay results in ${}^{175}_{77}\text{Ir}$.
(D) Alpha decay results in ${}^{179}_{80}\text{Hg}$ and beta(+) decay results in ${}^{175}_{79}\text{Au}$.

Section II: Free Response

1. An experiment is carried out with a series of different colored LED lights (which glow by emitting photoelectrons), which are connected one at a time to a variable voltage supply. When the voltage supply is set to 0 V, none of the LED bulbs light. As the voltage is increased, each LED turns on at a different voltage setting. As the voltage is further increased, the brightness of the LED increases.

   Data is collected on the color wavelength (and corresponding frequency) for each LED and the smallest value of the voltage where the light is illuminated.

Color	Wavelength (nm)	Frequency (10^{14} Hz)	Lighting Voltage (V)
Red	650	4.62	0.51
Yellow	580	5.17	0.74
Green	532	5.64	0.93
Blue	400	7.50	1.71

(A) Explain why an increase of $\Delta V = 0.1$ V in the voltage after the light is illuminated causes the brightness to increase, but the bulb remains dark for any of the LEDs with the same voltage increase $\Delta V = 0.1$ V from 0 V to 0.1 V.

(B) Plot the data on the axes below. Then calculate the slope and the y-intercept of the plot.

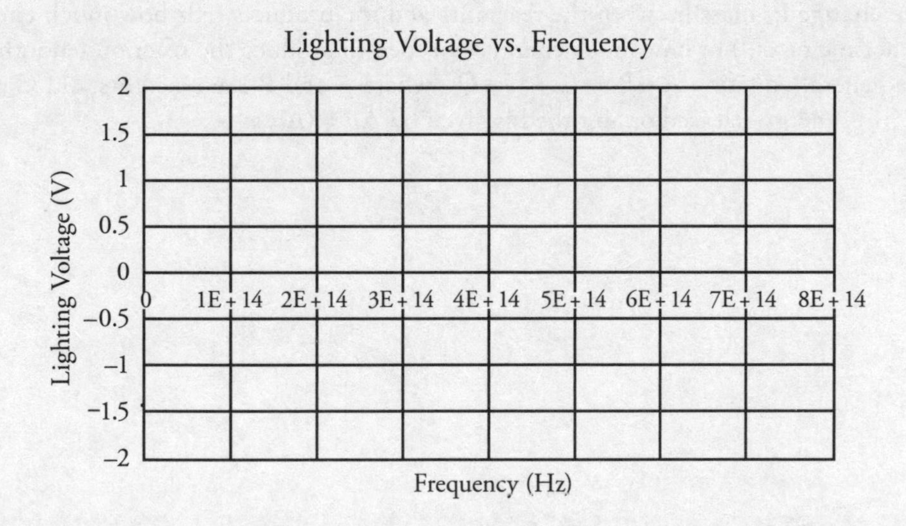

(C) Answer the following:

i. Explain how, if possible, this experiment could be repeated to get a plot with the same slope but a different intercept. If this is not possible, explain why.

ii. Explain how, if possible, this experiment could be repeated to get a plot with a different slope but the same intercept. If this is not possible, explain why.

Summary

- The energy available to liberate electrons near the surface of a metal (the photoelectric effect) is proportional to the frequency of the incident photon. This idea is expressed in $E = hf$, where h is Planck's constant ($h = 6.63 \times 10^{-34}$ J·s).

- The work function (ϕ) indicates the amount of energy needed to liberate the electron. There is a minimum frequency needed to liberate the electrons: $f_0 = \frac{\phi}{h}$. The kinetic energy of the emitted electron is $K_{max} = hf - \phi$.

- Particles in motion have wavelike properties: $\lambda = \frac{h}{p}$, where p is the momentum of the particle. Combining this with $E = hf$ and $c = f\lambda$ yields $E = pc$.

- The standard notation for an element is ${}^{A}_{Z}X$, where A is the mass number, Z is the number of protons in the nucleus, and $A = Z + N$, where N is the number of neutrons in the nucleus.

- A nuclear reaction produces new nuclei, other subatomic particles, and the absorption or emission of energy. The change in mass between the reactants and the products tells how much energy is released (exothermic or $+Q$) or how much energy is needed to produce the reaction (endothermic or $-Q$) in the general equation $\text{A} + \text{B} \rightarrow \text{C} + \text{D} + Q$, where A and B are reactants and C and D are products. This energy released or absorbed is given by $\Delta E = (\Delta m)c^2$.

Chapter 11
Solutions to Chapter Review Questions

CHAPTER 3 REVIEW QUESTIONS

Section I: Multiple Choice

1. **D** Since Point X is 5 m below the surface of the water, the pressure due to the water at X, P_X, is $\rho g h_X = \rho g(5 \text{ m})$, where ρ is the density of water. Because Point Y is 4 m below the surface of the water, the pressure due to the water at Y, P_Y, is $\rho g(4 \text{ m})$. Therefore,

$$\frac{P_X}{P_Y} = \frac{\rho g(5 \text{ m})}{\rho g(4 \text{ m})} = \frac{5}{4} \quad \Rightarrow \quad 4P_X = 5P_Y$$

2. **C** Because the top of the box is at a depth of $D - z$ below the surface of the liquid, the pressure on the top of the box is

$$P = P_{atm} + \rho g h = P_{atm} + \rho g(D - z)$$

The area of the top of the box is $A = xy$, so the force on the top of the box is

$$F = PA = [P_{atm} + \rho g(D - z)]xy$$

3. **B** The scale reading is the difference between the weight (which is constant) and the buoyant force. The buoyant force is calculated from $F_{buoy} = \rho_{fluid} V_{sub} g$ and will increase as the volume submerged is increased. As the cube is lowered, the scale reading will drop at a constant rate, since V is directly proportional to h for a cube.

4. **B** The independent variable and dependent variable are not changed because those are determined by the experimental setup, so you can eliminate (C) and (D). Changing the cube to a sphere makes V no longer directly proportional to h, so the graphs will not be the same, so you can eliminate (A).

5. **B** The buoyant force on the Styrofoam block is $F_{buoy} = \rho_L V g$, and the weight of the block is $F_g = m_S g = \rho_S V g$. Because $\rho_L > \rho_S$, the net force on the block is upward and has magnitude

$$F_{net} = F_{buoy} - F_g = (\rho_L - \rho_S)Vg$$

Therefore, by Newton's Second Law, you have

$$a = \frac{F_{net}}{m} = \frac{(\rho_L - \rho_S)Vg}{\rho_S V} = \left(\frac{\rho_L}{\rho_S} - 1\right) g$$

6. **D** The upward force on the ball is the buoyant force. There is no change in the water density, so you can eliminate (A). The change in tension results from a change in the buoyant force and not because of flow speed, so you can eliminate (B). The downward forces are gravity and tension. Gravity will remain constant, so you can eliminate (C).

7. **A** If the object weighs 100 N less when completely submerged in water, the buoyant force must be 100 N; therefore,

$$F_{buoy} = \rho_{water} V_{sub} g = \rho_{water} V g = 100 \text{ N} \Rightarrow V = \frac{100 \text{ N}}{\rho_{water} g} = \frac{100 \text{ N}}{(1{,}000 \frac{\text{kg}}{\text{m}^3})(10 \frac{\text{N}}{\text{kg}})} = 10^{-2} \text{ m}^3$$

Now that you know the volume of the object, you can figure out its weight:

$$F_g = mg = \rho_{object} V g = (2{,}000 \frac{\text{kg}}{\text{m}^3})(10^{-2} \text{ m}^3)(10 \frac{\text{N}}{\text{kg}}) = 200 \text{ N}$$

8. **A** The cross-sectional diameter at Y is 3 times the cross-sectional diameter at X, so the cross-sectional area at Y is $3^2 = 9$ times that at X. The Continuity Equation states that the flow speed, v, is inversely proportional to the cross-sectional area, A. So, if A is 9 times greater at Point Y than it is at X, then the flow speed at Y is 1/9 the flow speed at X; that is, $v_Y = (1/9)v_X = (1/9)(6 \text{ m/s}) = 2/3 \text{ m/s}$.

9. **D** Each side of the rectangle at the bottom of the conduit is 1/4 the length of the corresponding side at the top. Therefore, the cross-sectional area at the bottom is $(1/4)^2 = 1/16$ the cross-sectional area at the top. The Continuity Equation states that the flow speed, v, is inversely proportional to the cross-sectional area, A. So, if A at the bottom is 1/16 the value of A at the top, then the flow speed at the bottom is 16 times the flow speed at the top.

10. **D** Apply Bernoulli's Equation to a point at the pump (Point 1) and at the nozzle (the exit point, Point 2). Choose the level of Point 1 as the horizontal reference level; this makes $y_1 = 0$ and $y_2 = 1$ m. Now, because the cross-sectional diameter decreases by a factor of 10 between Points 1 and 2, the cross-sectional area decreases by a factor of $10^2 = 100$, so flow speed must increase by a factor of 100; that is, $v_2 = 100v_1 = 100(0.4 \text{ m/s}) = 40 \text{ m/s}$. Because Point 2 is exposed to the air, the pressure there is P_{atm}. Bernoulli's Equation becomes

$$P_1 + \frac{1}{2}\rho v_1^2 = P_{atm} + \rho g y_2 + \frac{1}{2}\rho v_2^2$$

Therefore,

$$\begin{aligned} P_1 - P_{atm} &= \rho g y_2 + \frac{1}{2}\rho v_2^2 - \frac{1}{2}\rho v_1^2 \\ &= (1{,}000 \frac{\text{kg}}{\text{m}^3})(10 \text{ m/s}^2)(1 \text{ m}) + \frac{1}{2}(1{,}000 \frac{\text{kg}}{\text{m}^3})(40 \text{ m/s})^2 - \frac{1}{2}(1{,}000 \frac{\text{kg}}{\text{m}^3})(0.4 \text{ m/s})^2 \\ &\approx (1{,}000 \frac{\text{kg}}{\text{m}^3})(10 \text{m/s}^2)(1 \text{ m}) + \frac{1}{2}(1{,}000 \frac{\text{kg}}{\text{m}^3})(40 \text{ m/s})^2 \\ &= (10{,}000 \text{ Pa}) + (800{,}000 \text{ Pa}) \\ &= 810{,}000 \text{ Pa} \\ &= 810 \text{ kPa} \end{aligned}$$

Section II: Free Response

1. (A) The pressure at the top surface of the block is $P_{top} = P_{atm} + \rho_L gh$. Since the area of the top of the block is $A = xy$, the force on the top of the block has magnitude

$$F_{top} = P_{top}A = (P_{atm} + \rho_L gh)xy$$

The pressure at the bottom of the block is $P_{bottom} = P_{atm} + \rho_L g(h + z)$. Since the area of the bottom face of the block is also $A = xy$, the force on the bottom surface of the block has magnitude

$$F_{bottom} = P_{bottom}A = [P_{atm} + \rho_L g(h + z)]xy$$

These forces are sketched below:

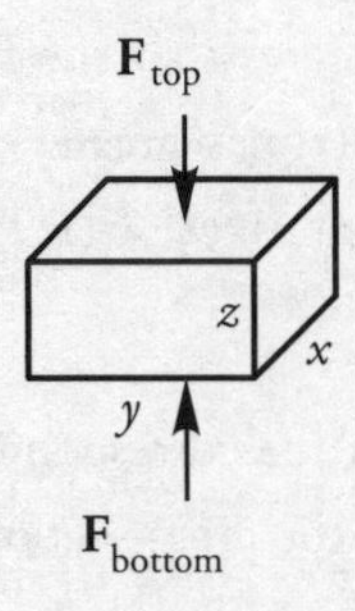

(B) Each of the other four faces of the block (left and right, front and back) is at an average depth of $h + \frac{1}{2}z$, so the average pressure on each of these four sides is

$$\bar{P}_{sides} = P_{atm} + \rho_L g(h + \frac{1}{2}z)$$

The left and right faces each have area $A = xz$, so the magnitude of the average force on this pair of faces is

$$\bar{F}_{\text{left and right}} = \bar{P}_{sides}A = [P_{atm} + \rho_L g(h + \frac{1}{2}z)]xz$$

The front and back faces each have area $A = yz$, so the magnitude of the average force on this pair of faces is

$$\bar{F}_{\text{front and back}} = \bar{P}_{sides}A = [P_{atm} + \rho_L g(h + \frac{1}{2}z)]yz$$

These forces are sketched below:

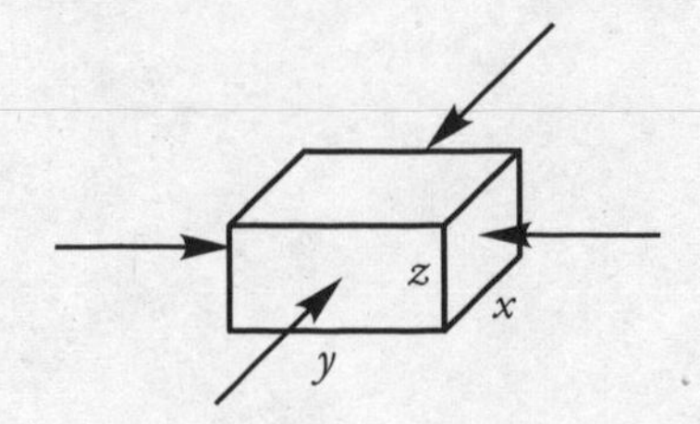

(C) The four forces sketched in part (b) add up to zero, so the total force on the block due to the pressure is the sum of F_{top} and F_{bottom}; because $F_{bottom} > F_{top}$, this total force points upward and its magnitude is

$$F_{bottom} - F_{top} = [P_{atm} + \rho_L g(h+z)]xy - (P_{atm} + \rho_L gh)xy = \rho_L gxyz$$

(D) By Archimedes' Principle, the buoyant force on the block is upward and has magnitude

$$F_{buoy} = \rho_L V_{sub} g = \rho_L Vg = \rho_L xyzg$$

This is the same as the result you found in part (c).

(E) The weight of the block is

$$F_g = mg = \rho_B Vg = \rho_B xyzg$$

If F_T is the tension in the string, then the total upward force on the block, $F_T + F_{buoy}$, must balance the downward force, F_g; that is, $F_T + F_{buoy} = F_g$, so

$$F_T = F_g - F_{buoy} = \rho_B xyzg - \rho_L xyzg = xyzg(\rho_B - \rho_L)$$

2. (A) Applying Bernoulli's Equation to a point on the surface of the water in the tank (Point 1) and a point at the hole (Point 2), the assumption that $v_1 \approx 0$ leads to the result

$$P_1 + \rho gh_1 + 0 = P_2 + \rho gh_2 + \frac{1}{2}\rho v_2^2$$

The external pressure P_1 and P_2 are both P_{atm} so they cancel. Then all the ρ terms cancel. Identifying the difference in the hole heights as h results in

$$v_2 = \sqrt{2gh}$$

(B) The initial velocity of the water, as it emerges from the hole, is horizontal. Since there's no initial vertical velocity, the time t required to drop the distance $y = D - h$ to the ground is found as follows:

$$y = \frac{1}{2}gt^2 \Rightarrow D - h = \frac{1}{2}gt^2 \Rightarrow t = \sqrt{\frac{2(D-h)}{g}}$$

Therefore, the horizontal distance the water travels is

$$x = v_x t = \sqrt{2gh} \cdot \sqrt{\frac{2(D-h)}{g}} = 2\sqrt{h(D-h)}$$

(C) The second hole would be at a depth of $h/2$ below the surface of the water, so the horizontal distance it travels—from the edge of the tank to the point where it hits the ground—is given by the same formula you found in part (B) except it will have $h/2$ in place of h; that is,

$$x_2 = 2\sqrt{\frac{1}{2}h(D - \frac{1}{2}h)}$$

If both streams land at the same point, then the value of x from part (B) is the same as x_2:

$$
\begin{aligned}
2\sqrt{h(D-h)} &= 2\sqrt{\frac{1}{2}h(D-\frac{1}{2}h)} \\
h(D-h) &= \frac{1}{2}h(D-\frac{1}{2}h) \\
D-h &= \frac{1}{2}(D-\frac{1}{2}h) \\
4D-4h &= 2D-h \\
-3h &= -2D \\
h &= \frac{2}{3}D
\end{aligned}
$$

(D) Once again, apply Bernoulli's Equation to a point on the surface of the water in the tank (Point 1) and to a point at the hole (Point 2). Choose the ground level as the horizontal reference level; then $y_1 = D$ and $y_2 = D - h$. If v_1 is the flow speed of Point 1—that is, the speed with which the water level in the tank drops—and v_2 is the efflux speed from the hole, then, by the Continuity Equation, $A_1v_1 = A_2v_2$, where A_1 and A_2 are the cross-sectional areas at Points 1 and 2, respectively. Therefore, $v_1 = (A_2/A_1)v_2$. Bernoulli's Equation then becomes

$$P_1 + \rho gD + \frac{1}{2}\rho v_1^2 = P_2 + \rho g(D-h) + \frac{1}{2}\rho v_2^2$$

Since $P_1 = P_2 = P_{atm}$, these terms cancel out; and substituting $v_1 = (A_2/A_1)v_2$, you have

$$
\begin{aligned}
\rho gD + \frac{1}{2}\rho\left(\frac{A_2}{A_1}v_2\right)^2 &= \rho g(D-h) + \frac{1}{2}\rho v_2^2 \\
\frac{1}{2}\rho v_2^2\left[\left(\frac{A_2}{A_1}\right)^2 - 1\right] &= -\rho gh \\
v_2 &= \sqrt{\frac{2gh}{1-\left(\frac{A_2}{A_1}\right)^2}}
\end{aligned}
$$

Now, since $A_1 = \pi R^2$ and $A_2 = \pi r^2$, this final equation can be written as

$$v_2 = \sqrt{\frac{2gh}{1-\left(\frac{r}{R}\right)^4}}$$

[Note that if $r << R$, then $(r/R)^4 \approx 0$, and the equation above reduces to $v_2 = \sqrt{2gh}$, as in part (A).]

3. (A) Point X is at a depth of h_1 below Point 1, where the pressure is P_1. Therefore, the hydrostatic pressure at X is $P_X = P_1 + \rho_F gh_1$.

(B) Point Y is at a depth of $h_2 + d$ below Point 2, where the pressure is P_2. The column of static fluid above Point Y is comprised of two parts. For the depth h_2, the fluid has density ρ_F. For the depth d, the fluid has density ρ_V. Therefore, the hydrostatic pressure at Y is $P_Y = P_2 + \rho_F gh_2 + \rho_V gd$.

(C) First, notice that Points 1 and 2 are at the same horizontal level; therefore, the heights y_1 and y_2 are the same, and the terms $\rho_F g y_1$ and $\rho_F g y_2$ will cancel out of the equation. Bernoulli's Equation then becomes

$$P_1 + \frac{1}{2}\rho_F v_1^2 = P_2 + \frac{1}{2}\rho_F v_2^2$$

By the Continuity Equation, you have $A_1 v_1 = A_2 v_2$, so $v_1 = (A_2/A_1)v_2$. Therefore,

$$\begin{aligned} P_1 - P_2 &= \frac{1}{2}\rho_F v_2^2 - \frac{1}{2}\rho_F v_1^2 \\ &= \frac{1}{2}\rho_F v_2^2 - \frac{1}{2}\rho_F \left(\frac{A_2}{A_1}v_2\right)^2 \\ &= \frac{1}{2}\rho_F v_2^2 \left[1 - \left(\frac{A_2}{A_1}\right)^2\right] \end{aligned}$$

(D) In parts (A) and (B) above, you found that $P_X = P_1 + \rho_F g h_1$ and $P_Y = P_2 + \rho_F g h_2 + \rho_v g d$. Since $P_X = P_Y$, you have

$$P_1 + \rho_F g h_1 = P_2 + \rho_F g h_2 + \rho_V g d$$

so

$$\begin{aligned} P_1 - P_2 &= \rho_F g(h_2 - h_1) + \rho_V g d \\ &= \rho_F g(-d) + \rho_V g d \\ &= (\rho_V - \rho_F) g d \end{aligned}$$

(E) In parts (C) and (D), you found two expressions for $P_1 - P_2$. Setting them equal to each other gives

$$\frac{1}{2}\rho_F v_2^2 \left[1 - \left(\frac{A_2}{A_1}\right)\right]^2 = (\rho_V - \rho_F) g d$$

$$v_2^2 = \frac{\rho_V - \rho_F}{\rho_F} \cdot \frac{2gd}{1 - \left(\frac{A_2}{A_1}\right)^2}$$

$$v_2 = \sqrt{\frac{2gd\left(\frac{\rho_V}{\rho_F} - 1\right)}{1 - \left(\frac{A_2}{A_1}\right)^2}}$$

The flow rate in the pipe is

$$f = A_2 v_2 = A_2 \sqrt{\frac{2gd\left(\frac{\rho_V}{\rho_F} - 1\right)}{1 - \left(\frac{A_2}{A_1}\right)^2}}$$

Since

$$f = A_2 \sqrt{\frac{2g\left(\frac{\rho_V}{\rho_F} - 1\right)}{1 - \left(\frac{A_2}{A_1}\right)^2}} \cdot \sqrt{d}$$

you see that f is proportional to $\sqrt{d}$, as desired.

CHAPTER 4 REVIEW QUESTIONS

Section I: Multiple Choice

1. **B** Because the average kinetic energy of a molecule of gas is directly proportional to the temperature of the sample, the fact that the gases are at the same temperature—since they're in the same container at thermal equilibrium—tells you that the molecules have the same average kinetic energy. The ratio of their kinetic energies is therefore equal to 1.

2. **A** *P–V* diagrams have pressure on the vertical axis and volume on the horizontal axis. You know volume is changing, so the graph will have to show a change in the horizontal axis, so you can eliminate (B). Then, from the Ideal Gas Law, $PV = nRT$, you see that pressure is unchanged if the volume and temperature are both doubled. The vertical value of the *P–V* diagram remains constant, resulting in a horizontal line on the graph.

3. **A** The work done on the gas during a thermodynamic process is equal to the area of the region in the *P–V* diagram above the *V*-axis and below the path the system takes from its initial state to its final state. Since the area below path 1 is the greatest, the work done on the gas during the transformation along path 1 is the greatest.

4. **C** During an isothermal change, ΔU is always zero.

5. **B** There is no way to measure a pressure of 0 Pa, so you can eliminate (A) and (C) as answers. The Ideal Gas Law states $PV = nRT$, so a graph of P versus n will be linear when V and T are both held constant, so you can eliminate (D).

6. **C** In order for the temperature to remain constant, the entire container needs to be kept in thermal equilibrium with the surroundings. This allows heat to flow into and out of the system and keeps the temperature of the container constant.

7. **A** By convention, work done *on* the gas sample is designated as positive, so in the First Law of Thermodynamics, $\Delta U = Q + W$, you must write $W = +320$ J. Therefore, $Q = \Delta U - W = 560 \text{ J} - 320 \text{ J} = +240$ J. Positive Q denotes heat *in*.

8. **C** No work is done during the step from state *a* to state *b* because the volume doesn't change. Therefore, the work done from *a* to *c* is equal to the work done from *b* to *c*. Since the pressure remains constant (this step is isobaric), find that

   $W = -P\Delta V = -(3.0 \times 10^5 \text{ Pa})[(10 - 25) \times 10^{-3} \text{ m}^3] = 4{,}500 \text{ J}$

9. **C** Choice (A) is wrong because *no heat is exchanged between the gas and its surroundings* is the definition of *adiabatic*, not *isothermal*. Choice (B) cannot be correct since the step described in the question is isothermal; by definition, the temperature does not change. This also eliminates (D) and supports (C). If the sample could be brought back to its initial state *and* have a 100% conversion of heat to work, *that* would violate the Second Law of Thermodynamics, which states that heat cannot be completely converted to work with no other change taking place. In this case, there are changes taking place: the pressure decreases and the volume increases.

10. **C** The Second Law of Thermodynamics indicates that energy will flow from a hot object to a cool object, making the final temperature of the hot object cooler and the cold object warmer.

Section II: Free Response

1. (A) First, calculate ΔU_{acb}. Using path *acb*, the question tells you that $Q = +70$ J and $W = -30$ J (W is negative here because it is the *system* that does the work). The First Law, $\Delta U = Q + W$, tells you that $\Delta U_{acb} = +40$ J. Because $\Delta U_{a\to b}$ does not depend on the path taken from *a* to *b*, you must have $\Delta U_{ab} = +40$ J, and $\Delta U_{ba} = -\Delta U_{ab} = -40$ J. Thus, $-40 \text{ J} = Q_{ba} + W_{ba}$, where Q_{ba} and W_{ba} are the values along the curved path from *b* to *a*. Since $Q_{ba} = -60$ J, it follows that $W_{ba} = +20$ J. Therefore, the surroundings do 20 J of work on the system.

   (B) Again, using the fact that $\Delta U_{a\to b}$ does not depend on the path taken from *a* to *b*, you know that $\Delta U_{adb} = +40$ J, as computed above. Writing $\Delta U_{adb} = QW_{adb} + W_{adb}$, if $W_{adb} = -10$ J, it follows that $Q_{adb} = +50$ J. That is, the system absorbs 50 J of heat.

   (C) For the process *db*, there is no change in volume, so $W_{db} = 0$. Therefore, $\Delta U_{db} = Q_{db} + W_{db} = Q_{db}$. Now, since $U_{ab} = +40$ J, the fact that $U_a = 0$ J implies that $U_b = 40$ J, so $\Delta U_{db} = U_b - U_d = 40 \text{ J} - 30 \text{ J} = 10$ J. Thus, $Q_{db} = 10$ J. Now let's consider the process *ad*. Since $W_{adb} = W_{ad} + W_{db} = W_{ad} + 0 = W_{ad}$, the fact that $W_{adb} = -10$ J [computed in part (b)] tells you that $W_{ad} = -10$ J. Because $\Delta U_{ad} = U_d - U_a = 30$ J, it follows from $\Delta U_{ad} = Q_{ad} + W_{ad}$ that $Q_{ad} = \Delta U_{ad} - W_{ad} = 30 \text{ J} - (-10 \text{ J}) = 40$ J.

(D) The process *adbca* is cyclic, so ΔU is zero. Because this cyclic process is traversed *counterclockwise* in the P–V diagram, you know that W is *positive*. Then, since $\Delta U = Q + W$, it follows that Q must be negative.

2. (A) (i) Use the Ideal Gas Law:

$$T_a = \frac{P_a V_a}{nR} = \frac{(2.4 \times 10^5 \text{ Pa})(12 \times 10^{-3} \text{ m}^3)}{(0.4 \text{ mol})(8.31 \text{ J/mol}\cdot\text{K})} = 870 \text{ K}$$

(ii) Since state b is on the isotherm with state a, the temperature of state b must also be 870 K.

(iii) Use the Ideal Gas Law:

$$T_c = \frac{P_c V_c}{nR} = \frac{(0.6 \times 10^5 \text{ Pa})(12 \times 10^{-3} \text{ m}^3)}{(0.4 \text{ mol})(8.31 \text{ J/mol}\cdot\text{K})} = 220 \text{ K}$$

(B) There is heat flow into the system during the step *ab*. As the volume expands, the system does work on the surroundings (uses up some of its internal energy). As a result, the system would drop in temperature if no heat flowed into the gas to make up for this use of energy. During an isothermal process, $\Delta T = 0$ K, so the change in internal energy of the gas, $\Delta U = \frac{3}{2} Nk_b \Delta T$, is zero. From the First Law of Thermodynamics, $\Delta U = W + Q$, any energy that is used by expanding (work done by the system on the surroundings) must be replaced by heat flowing into the system.

(C) Using the equation given, find that

$$W_{ab} = -nRT \cdot \ln \frac{V_b}{V_a} = -(0.4 \text{ mol})(8.31 \text{ J/mol} \cdot \text{K})(870 \text{ K}) \cdot \ln \frac{48 \times 10^{-3} \text{ m}^3}{12 \times 10^{-3} \text{ m}^3}$$
$$= -4{,}000 \text{ J}$$

(D) The total work done over the cycle is equal to the sum of the values of the work done over each step:

$$\begin{aligned} W_{\text{cycle}} &= W_{ab} + W_{bc} + W_{ca} \\ &= W_{ab} + W_{bc} \\ &= (-4{,}000 \text{ J}) + (2{,}200 \text{ J}) \\ &= -1{,}800 \text{ J} \end{aligned}$$

CHAPTER 5 REVIEW QUESTIONS

Section I: Multiple Choice

1. **D** Electrostatic force obeys an inverse-square law: $F_E \propto 1/r^2$. Therefore, a plot should be made of F versus the inverse square of r to get a line.

2. **C** The strength of the electric force is given by kq^2/r^2, and the strength of the gravitational force is Gm^2/r^2. Since both of these quantities have r^2 in the denominator, you simply need to compare the numerical values of kq^2 and Gm^2. There's no contest. Since

   $kq^2 = (9 \times 10^9 \text{ N·m}^2/\text{C}^2)(1 \text{ C})^2 = 9 \times 10^9 \text{ N·m}^2$

   and

   $Gm^2 = (6.7 \times 10^{-11} \text{ N·m}^2/\text{kg}^2)(1 \text{ kg})^2 = 6.7 \times 10^{-11} \text{ N·m}^2$

   you can see that $kq^2 >> Gm^2$, so F_E is much stronger than F_G.

3. **C** If the net electric force on the center charge is zero, the electrical repulsion by the $+2q$ charge must balance the electrical repulsion by the $+3q$ charge:

   $$\frac{1}{4\pi\varepsilon_0}\frac{(2q)(q)}{x^2} = \frac{1}{4\pi\varepsilon_0}\frac{(3q)(q)}{y^2} \Rightarrow \frac{2}{x^2} = \frac{3}{y^2} \Rightarrow \frac{y^2}{x^2} = \frac{3}{2} \Rightarrow \frac{y}{x} = \sqrt{\frac{3}{2}}$$

4. **D** Since P is equidistant from the two charges, and the magnitudes of the charges are identical, the strength of the electric field at P due to $+Q$ is the same as the strength of the electric field at P due to $-Q$. The electric field vector at P due to $+Q$ points away from $+Q$, and the electric field vector at P due to $-Q$ points toward $-Q$. Since these vectors point in the same direction, the net electric field at P is (E to the right) + (E to the right) = ($2E$ to the right).

5. **C** The acceleration of the small sphere is

   $$a = \frac{F_E}{m} = \frac{1}{4\pi\varepsilon_0}\frac{Qq}{mr^2}$$

   As r increases (that is, as the small sphere is pushed away), a decreases. However, since a is always positive, the small sphere's speed, v, is always increasing.

6. **B** Since $\mathbf{F}_E$ (on q) = $q\mathbf{E}$, it must be true that $\mathbf{F}_E$ (on $-2q$) = $-2q\mathbf{E} = -2\mathbf{F}_E$.

7. **C** All excess electric charge on a conductor resides on the outer surface.

Section II: Free Response

1. (A) (i) The force from the charge at A will be repulsive and must act at a 45° angle because the force acts along the bisector of the square. The balancing forces from the charges q will be at right angles to one another, so each must balance a component of the force from Q alone. Since $\sin(45°) = \cos(45°)$, the forces from each q must be equal, and therefore the charges must also be equal.

(ii) In order for the force on C to be 0, the repulsive force from charge A must be balanced by an attractive force from each q, so the charges at positions B and D must have a negative sign.

A free body diagram of the charge at point C is drawn below.

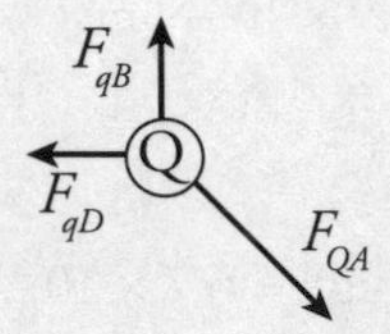

The distance between opposite vertices of a square of side length s is $\sqrt{2}\,s$. You can look at either the x- or y-components because each direction yields identical equations. From the x direction, $F_{QA}\sin(45°) = F_{qD}$. Therefore, $F_{qD} = k\dfrac{Q^2}{(\sqrt{2}\,s)^2} \times \dfrac{\sqrt{2}}{2}$.

$$F_{qD} = k\frac{\sqrt{2}Q^2}{4s^2}$$

Now that you know the force from the charge at D, use Coulomb's Law to determine q.

$$F_{qD} = k\frac{Qq}{s^2} = k\frac{\sqrt{2}Q^2}{4s^2}$$

$$Qq = \frac{\sqrt{2}Q^2}{4}$$

$$q = \frac{\sqrt{2}Q}{4}$$

(B) Within the square, the field from the charge Q at position A will point into the fourth quadrant (the x-component will be positive and the y-component will be negative). With the charges positive, within the square the field from D will point into the first quadrant (the x-component will be positive and the y-component will be positive) and from B will point into the third quadrant (the x-component will be negative and the y-component will be negative). To cancel the fields, the x-coordinate would have to be closer to point B than to point D, and the y-coordinate would have to be closer to D than to B. Some point that satisfies these two requirements will have a field of 0 N/C.

(C) At the center of the square, the net electric field is the vector sum of the contributions from the four charges. The field from the charges at A and C will always balance one another as long as those charges have the same sign, regardless of that sign. First, the charges are equidistant from the center of the square, so r will be equal regardless of the sign or magnitude of the charges. If they are both positive, the field from A will be in the fourth quadrant and the field from C will be in the second quadrant. If both charges are negative, then the field from C will be in the fourth quadrant and the field from A will be in the second quadrant. Therefore, regardless of the magnitude or sign of the charges, the contributions from the charges Q will cancel one another. An identical argument can be made for the charges q showing that their fields will also cancel with one another. Therefore, the net field at the center of the square must be 0 N/C regardless of the signs or magnitudes of Q or q.

2. (A) The electric force on charge 1 is $F_1 = k\dfrac{Q \cdot 2Q}{(a+2a)^2} = k\dfrac{2Q^2}{9a^2}$. The force between the charges is repulsive, so the charge at 1 is pushed upward.

(B) $\overrightarrow{E_{net}} = \overrightarrow{E_1} + \overrightarrow{E_2} = k\dfrac{Q}{a^2}$ down $+ k\dfrac{2Q}{(2a)^2}$ up. Making upward positive results in $E_{net} = k\dfrac{Q}{2a^2} - k\dfrac{Q}{a^2} = -k\dfrac{Q}{a^2}$.

The net field is $k\dfrac{Q}{2a}$ in the downward direction.

(C) No. The only point on the x-axis where the individual electric field vectors due to each of the two charges point in exactly opposite directions is the origin (0, 0). But at that point, the two vectors are not equal and thus do not cancel.

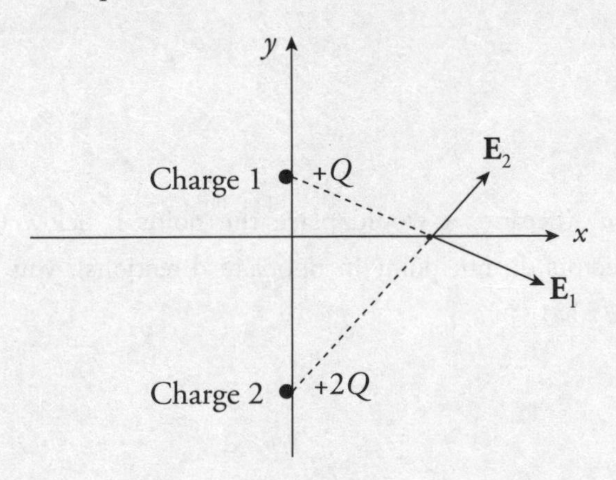

Therefore, at no point on the x-axis could the total electric field be zero.

(D) Yes. There will be a Point P on the y-axis between the two charges,

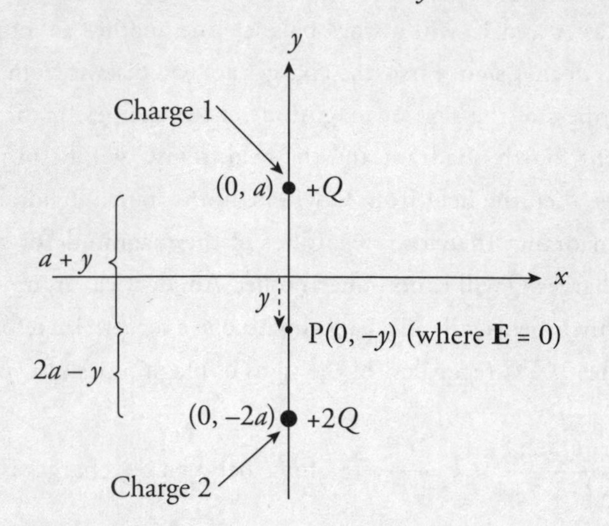

where the electric fields due to the individual charges will cancel each other out.

$$E_1 = E_2$$

$$\frac{1}{4\pi\varepsilon_0}\frac{Q}{(a+y)^2} = \frac{1}{4\pi\varepsilon_0}\frac{2Q}{(2a-y)^2}$$

$$\frac{1}{(a+y)^2} = \frac{2}{(2a-y)^2}$$

$$(2a-y)^2 = 2(a+y)^2$$

$$0 = y^2 + 8ay - 2a^2$$

$$y = \frac{-8a \pm \sqrt{(8a)^2 - 4(-2a^2)}}{2}$$

$$= (-4 \pm 3\sqrt{2})a$$

Disregarding the value $y = (-4-3\sqrt{2})a$ (because it would place the point P below Charge 2 on the y-axis, where the electric field vectors do not point in opposite directions), you find that $\mathbf{E} = 0$ at the point $P = (0, -y) = (0, (4-3\sqrt{2})a)$.

CHAPTER 6 REVIEW QUESTIONS

Section I: Multiple Choice

1. **B** For a point particle, which a uniform sphere will behave like, the relationship between the electric potential and the distance is $V = \frac{1}{4\pi\varepsilon_0}\frac{q}{r}$, so with V on the vertical axis and $\frac{1}{r}$ on the horizontal axis, the graph produced is a straight line whose slope is $\frac{1}{4\pi\varepsilon_0}q$.

2. **D** Equipotential curves for a parallel-plate capacitor are horizontal lines. Both (A) and (B) are at the same potential, so their potential difference is 0 V. As one moves closer to the $+Q$ plate, the potential will increase, so (D) is correct.

3. **B** Use the definition $\Delta V = -W_E/q$. If an electric field accelerates a negative charge doing positive work on it, then $W_E > 0$. If $q < 0$, then $-W_E/q$ is positive. Therefore, ΔV is positive, which implies that V increases.

4. **D** By definition, $V_{A\rightarrow B} = \Delta U_E/q$, so $V_B - V_A = \Delta U_E/q$

5. **B** Because **E** is uniform, the potential varies linearly with distance from either plate ($\Delta V = Ed$). Since Points 2 and 4 are at the same distance from the plates, they lie on the same equipotential. (The equipotentials in this case are planes parallel to the capacitor plates.)

6. **B** The charge Q must be positive because the potential decreases as the distance from the charge increases. A negative charge will be attracted to a positive charge. This means that the electric field of Q would do positive work if q were brought closer, and it would do negative work if q were moved farther away.

Section II: Free Response

1. (A) Greatest <u>*A* and *B*</u> <u>*A*</u> <u>*A*, *B*, and *C*</u> <u>All four</u> Least

The work required to move a charge through an existing electric field can be related to the voltage of that field at the initial and final positions of the charge. All voltages are defined to be 0 when the charges are very far away from one another, so V_i is always 0 V. Charge A is placed first, and there is no electric field that charge A is moved through, so the charge distribution has no energy. At location B, in the presence of charge A, there is an electric field, and you can calculate the voltage at point B as $V_B = \frac{+kQ}{s}$. Moving the charge to point B adds $W = \frac{kQ^2}{s}$ to the charge distribution. At location C, in the presence of charges at A and B, calculate the voltage as the sum of the contributions from A and B as $V_C = \frac{kQ}{\sqrt{2}s} + \frac{kQ}{s}$. Charge C is negative, so putting it at position C removed $W = \frac{kQ^2}{s} + \frac{kQ^2}{\sqrt{2}s}$, making the energy of the distribution $-\frac{kQ^2}{\sqrt{2}s} = -\frac{\sqrt{2}kQ^2}{2s}$. Finally, the voltage at position D from the charges at A, B, and C, respectively, is $V_D = \frac{kQ}{s} + \frac{kQ}{\sqrt{2}s} - \frac{kQ}{s} = \frac{kQ}{\sqrt{2}s}$. Charge D is negative, so putting it at position D removed $W = \frac{kQ^2}{\sqrt{2}s}$ from the distribution. The energy in the distribution of all four charges is $U_E = -\frac{2kQ^2}{\sqrt{2}s} = -\frac{2\sqrt{2}kQ^2}{2s}$.

(B) At the exact center of the square, this distance from each charge is d, where from the Pythagorean Theorem, $d^2 = \left(\frac{s}{2}\right)^2 + \left(\frac{s}{2}\right)^2$, so $d = \frac{\sqrt{2}s}{2}$. The voltage at the center is then the sum of the voltages from each charge: $V_{center} = V_A + V_B + V_C + V_D = \frac{+kQ}{d} + \frac{+kQ}{d} + \frac{-kQ}{d} + \frac{-kQ}{d} = 0\text{ V}$.

(C) The equipotential will be described as always the same distance from the pair at AD as well as the same distance from the pair at BD. Thus, the equipotential line is a horizontal line through the square.

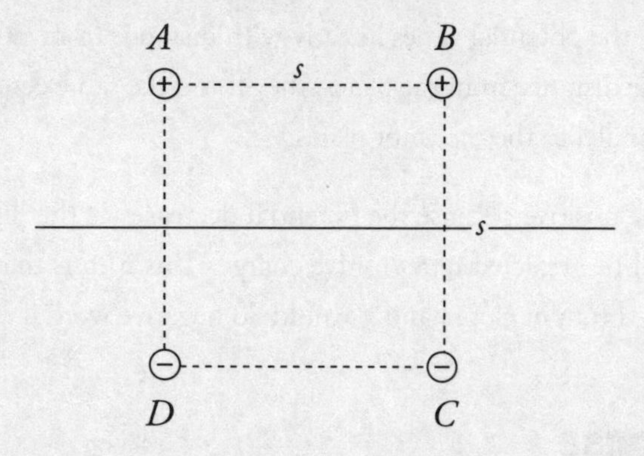

(D) The electric field along the center line will be uniform because any diminishment from the influence of the top plate is counteracted by an increased influence from the lower plate. However, at a position not between the plates, this is no longer true, and so for the bottom dots, the lower plate is more influential. This causes the electric field at these locations to tend to point toward the negatively charged plate.

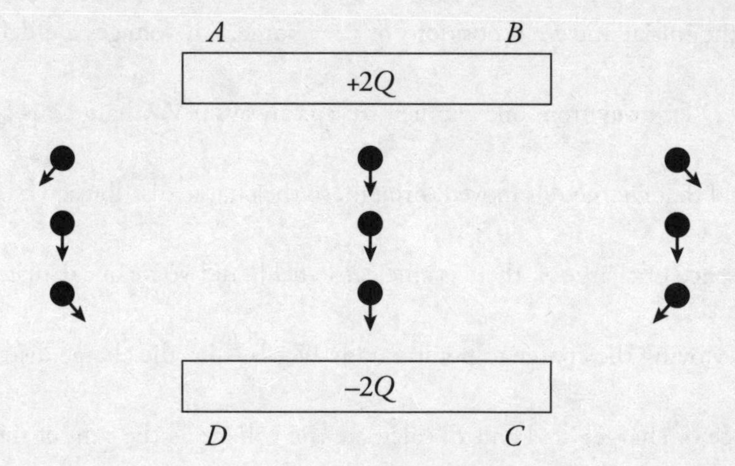

2.

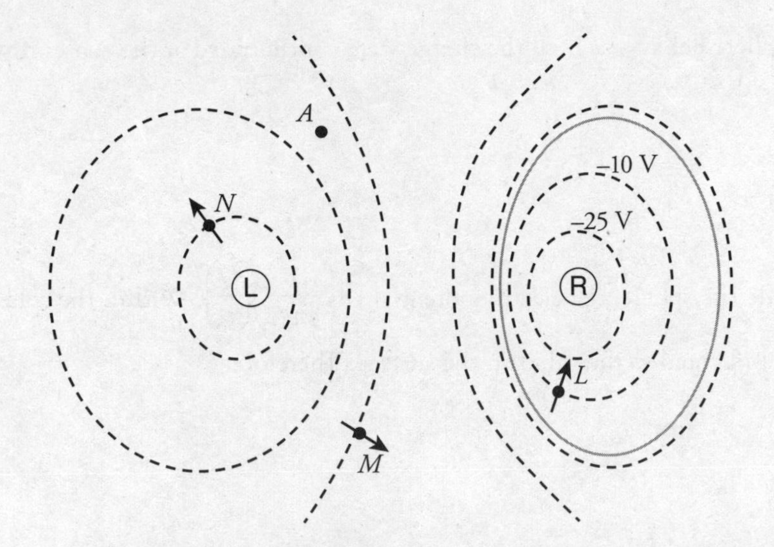

(A) Because the potential difference between adjacent lines is $\Delta V = 15$ V and the potential is going from –25 V to –10 V on the two lines closest to R, the 0 V isoline must be in between the line labeled –10 V and the next line in the image. The light gray solid line is the line where the potential is 0 V.

(B) 42.5 V

(C) (i) The potential is increasing as the isolines get farther away from R, so R must be negative and L must be positive.

(ii) R carries the greater magnitude of charge, as the isolines are closer together in the neighborhood of R, which indicates a stronger magnitude of electric field.

(D) (i) See image above.

(ii) $L > M > N$. At L, the isolines are closest together, and at N, they are farthest apart.

(E) Sphere L carries positive charge, and potential is larger (relative to positions near negative charges) at positions near positive charges. Making Sphere L negatively charged would therefore make the potential at point N lower than it is when Sphere L is positive.

3. (A) Outside the sphere, the sphere behaves as if all the charge were concentrated at the center. Inside the sphere, the electrostatic field is zero:

$$E(r) = \begin{cases} 0 & (r < a) \\ \dfrac{1}{4\pi\varepsilon_0}\dfrac{Q}{r^2} & (r > a) \end{cases}$$

(B) On the surface and outside the sphere, the electric potential is $\left(\dfrac{1}{4\pi\varepsilon_0}\right)\left(\dfrac{Q}{r}\right)$. Within the sphere, V is constant (because $E = 0$) and equal to the value on the surface. Therefore,

$$V(r) = \begin{cases} \dfrac{1}{4\pi\varepsilon_0}\dfrac{Q}{a} & (r \le a) \\ \dfrac{1}{4\pi\varepsilon_0}\dfrac{Q}{r} & (r > a) \end{cases}$$

(C) See diagrams:

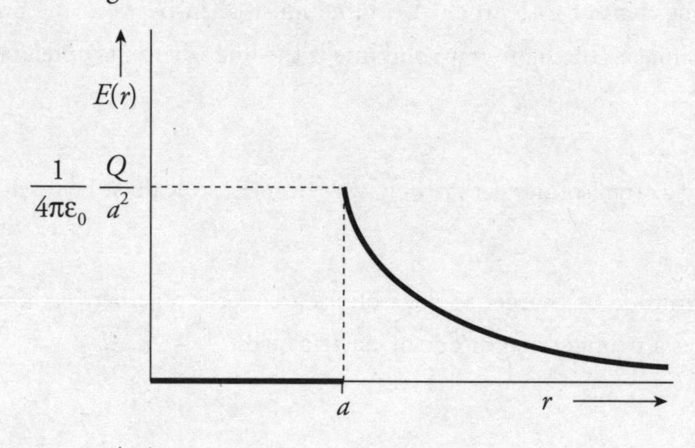

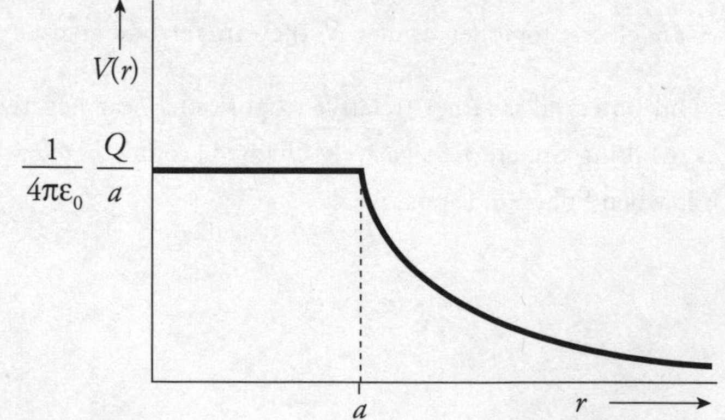

CHAPTER 7 REVIEW QUESTIONS

Section I: Multiple Choice

1. **A** Let ρ_S denote the resistivity of silver and let A_S denote the cross-sectional area of the silver wire. Then

$$R_B = \frac{\rho_B L}{A_B} = \frac{(5\rho_S)L}{4^2 A_S} = \frac{5}{16}\frac{\rho_S L}{A_S} = \frac{5}{16} R_S.$$

2. **C** The equation $I = V/R$ implies that increasing V by a factor of 2 will cause I to increase by a factor of 2.

3. **D** In order for one resistor to have current flowing through it immediately when the circuit is connected, it must be in series with the capacitor as all of the current will flow to the capacitor immediately upon constructing the circuit. In order for there to be current in the other resistor after a long time, the second resistor should be in parallel to the branch with the capacitor and the first resistor.

4. **C** The resistivity ρ is found from the resistance $R = \rho L/A$. To find the resistance, Ohm's Law $V = IR$ is needed. In addition to V and L, you need to find both I and A. The ammeter will yield I, and the diameter of the wire will allow you to determine A.

5. **C** Because the voltage has been applied for a long time, the capacitor will be fully charged, so it behaves like an open switch. This puts the 4 Ω resistor and the top 3 Ω resistor in series with one another, making an equivalent resistor of 7 Ω. That combination is in parallel with the other 3 Ω resistor, so

$$\frac{1}{R_{eq}} = \frac{1}{7\Omega} + \frac{1}{3\Omega} \rightarrow \frac{1}{R_{eq}} = \frac{0.48}{\Omega} \rightarrow R_{eq} = 2.1\Omega.$$

6. **C** If each of the identical bulbs has resistance R, then the current through each bulb is ε/R. This is unchanged if the middle branch is taken out of the parallel circuit. (What *will* change is the total amount of current supported by the battery.)

7. **B** The three parallel resistors are equivalent to a single 2 Ω resistor, because $\frac{1}{8\,\Omega} + \frac{1}{4\,\Omega} + \frac{1}{8\,\Omega} = \frac{1}{2\,\Omega}$. This 2 Ω resistance is in series with the given 2 Ω resistor, so their equivalent resistance is 2 Ω + 2 Ω = 4 Ω. Therefore, three times as much current will flow through this equivalent 4 Ω resistance in the top branch as through the parallel 12 Ω resistor in the bottom branch, which implies that the current through the bottom branch is 3 A, and the current through the top branch is 9 A. The voltage drop across the 12 Ω resistor is therefore $V = IR$ = (3 A)(12 Ω) = 36 V.

8. **B** Ohm's Law states that $V = IR$, but with an internal resistance in the battery, the total resistance in the circuit is $R_{resistor} + R_{internal}$. Therefore, $V_{total} = I(R_{resistor} + R_{internal})$. The voltage drop across the resistor is $IR_{resistor}$, so when you isolate the voltage drop across the resistor, you get a line with a negative intercept of $IR_{internal}$.

9. **D** The equation $P = I^2R$ gives

$P = (0.5 \text{ A})^2(100\ \Omega) = 25 \text{ W} = 25 \text{ J/s}$

Therefore, in 20 s, the energy dissipated as heat is

$E = Pt = (25 \text{ J/s})(20 \text{ s}) = 500 \text{ J}$

10. **D**

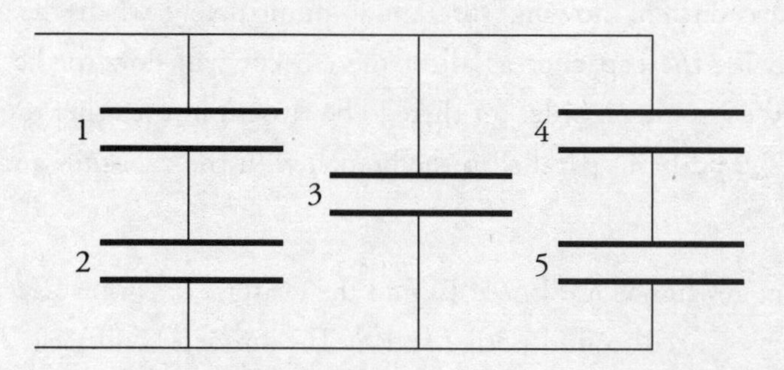

Capacitors 1 and 2 are in series, so their equivalent capacitance is $C_{1\text{-}2} = C/2$. (This is obtained from the equation $1/C_{1\text{-}2} = 1/C_1 + 1/C_2 = 1/C + 1/C = 2/C$.) Capacitors 4 and 5 are also in series, so their equivalent capacitance is $C_{4\text{-}5} = C/2$. The capacitances $C_{1\text{-}2}$, C_3, and $C_{4\text{-}5}$ are in parallel, so the overall equivalent capacitance is $(C/2) + C + (C/2) = 2C$.

Section II: Free Response

1. Begin by labeling each resistor and making a chart of the resistance, voltage, current, and power for each resistor.

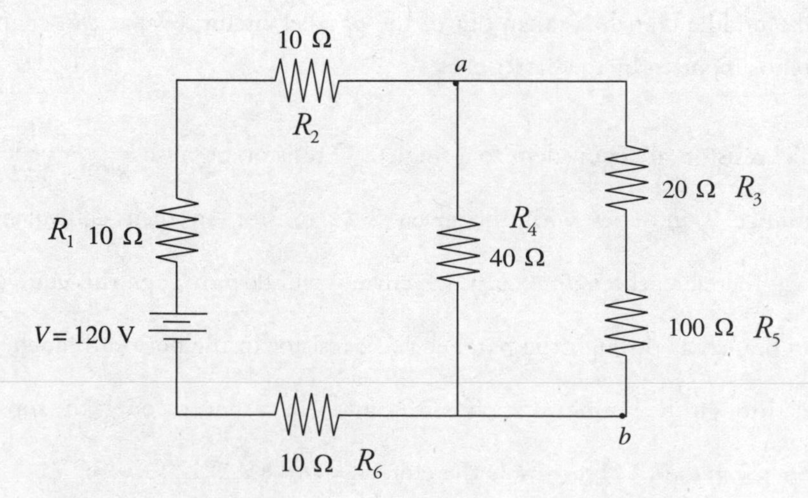

	Resistance (Ω)	**Voltage (V)**	**Current (A)**	**Power (W)**
R_1	10			
R_2	10			
R_3	20			
R_4	40			
R_5	100			
R_6	10			
Entire Circuit		120		

The two parallel branches, the one containing the 40 Ω resistor and the other a total of 120 Ω, are equivalent to a single 30 Ω resistance. This 30 Ω resistance is in series with the three 10 Ω resistors, giving an overall circuit resistance of 10 Ω + 10 Ω + 30 Ω + 10 Ω = 60 Ω.

	Resistance (Ω)	**Voltage (V)**	**Current (A)**	**Power (W)**
R_1	10			
R_2	10			
R_3	20			
R_4	40			
R_5	100			
R_6	10			
Entire Circuit	**60**	120	120/60 = 2	$120^2/60 = 240$

R_1, R_2 and R_6 are all in series with the battery, so they all have the same current as the current supported by the battery.

	Resistance (Ω)	**Voltage (V)**	**Current (A)**	**Power (W)**
R_1	10	2 * 10 = 20	**2**	$2^2 * 10 = 40$
R_2	10	2 * 10 = 20	**2**	$2^2 * 10 = 40$
R_3	20			
R_4	40			
R_5	100			
R_6	10	2 * 10 = 20	**2**	$2^2 * 10 = 40$
Entire Circuit	60	120	2	240

There is a loop $V - V_1 - V_2 - V_4 - V_6 = 0$, so $120 - 20 - 20 - V_4 - 20 = 0$ and $V_4 = 60$ V.

	Resistance (Ω)	Voltage (V)	Current (A)	Power (W)
R_1	10	20	2	40
R_2	10	20	2	40
R_3	20			
R_4	40	**60**	60/40 = 1.5	$60^2/40 = 90$
R_5	100			
R_6	10	20	2	40
Entire Circuit	60	120	2	240

Finally, you know that 2 A of current flows into the junction point labeled *a* and 1.5 A flows into R_4, which leaves 2 A – 1.5 A = 0.5 A to travel down the other branch.

	Resistance (Ω)	Voltage (V)	Current (A)	Power (W)
R_1	10	20	2	40
R_2	10	20	2	40
R_3	20	0.5 * 20 = 10	**0.5**	$0.5^2 * 20 = 5$
R_4	40	60	1.5	90
R_5	100	0.5 * 100 = 50	**0.5**	$0.5^2 * 100 = 25$
R_6	10	20	2	40
Entire Circuit	60	120	2	240

Check the power sum: 40 + 40 + 5 + 90 + 25 + 40 = 240 is correct. Use the completed chart to answer the questions.

(A) The battery delivers 240 W of power.

(B) There is 1.5 A of current through the 40 Ω resistor.

(C) (i) The potential difference between points *a* and *b* is the voltage across R_4, so it is 60 V.

(ii) The higher potential is near the long pole of the battery in the diagram, so *a* is 60 V higher in potential than *b*.

(D) Using $R = \rho L/A$ gives $100 = 0.45(0.04)/\pi r^2$ so $r = \sqrt{\dfrac{0.45(0.04)}{100\pi}} = 0.0076\ \text{m}$.

(E) In order for there to be no current flow supported by the battery after the capacitor is fully charged, the capacitor would have to be in series with the battery. Then, once the capacitor is full, it would have the same voltage drop as the battery supplies and the current in the circuit would go to 0 A. This happens if you replace any of the 10 Ω resistors with a capacitor.

2. Begin by making a chart of the resistance, voltage, current, and power for each resistor.

	Resistance (Ω)	Voltage (V)	Current (A)	Power (W)
R_1	10			
R_2	60			
R_3	20			
R_4	40			
R_5	60	$0.5 * 60 = 30$	0.5	$0.5^2 * 60 = 15$
Entire Circuit				

To calculate the equivalent resistance of the entire circuit, there are three parallel branches each with a resistance of 60 Ω. Those are equivalent to a single 20 Ω resistor. That is in series with a 10 Ω resistor.

	Resistance (Ω)	Voltage (V)	Current (A)	Power (W)
R_1	10			
R_2	60			
R_3	20			
R_4	40			
R_5	60	30	0.5	15
Entire Circuit	**30**			

As the standard approach did not result in a row with two entries in a column, you must examine the circuit to see whether the Loop Rule or Junction Rule will be more useful as the next step. You can see that the three parallel branches all have the same resistance, and therefore they must all have the same current of 0.5 A flowing through them. This is a result of the Loop Rule stating that parallel branches must have equal voltages.

	Resistance (Ω)	Voltage (V)	Current (A)	Power (W)
R_1	10			
R_2	60	$0.5 * 60 = 30$	**0.5**	$0.5^2 * 60 = 15$
R_3	20	$0.5 * 20 = 10$	**0.5**	$0.5^2 * 20 = 5$
R_4	40	$0.5 * 40 = 20$	**0.5**	$0.5^2 * 40 = 10$
R_5	60	30	0.5	15
Entire Circuit	30			

As you just used the Loop Rule, you can now use the Junction Rule to see that $I_1 = I_2 + I_3 + I_5 = 1.5$ A. Also, since R_1 is in series with the battery, you can complete the chart.

	Resistance (Ω)	Voltage (V)	Current (A)	Power (W)
R_1	10	1.5 * 10 = 15	**1.5**	1.5^2 * 10 = 22.5
R_2	60	30	0.5	15
R_3	20	10	0.5	5
R_4	40	20	0.5	10
R_5	60	30	0.5	15
Entire Circuit	30	1.5 * 30 = 45	**1.5**	1.5^2 * 30 = 67.5

The answers to parts (A), (B), and (C) can now simply be read off the chart.

(D) (i) When the capacitor is uncharged, the leftmost 60 Ω resistor (along with the 20 Ω and 40 Ω series combination) will be shorted out and will have a current of 0 A flowing through them.

(ii) The current will be greater than 0.5 A. The full capacitor acts like a resistor with infinite resistance, which essentially removes a parallel branch from the circuit. Removing a parallel branch lowers the current throughout the circuit, so it lowers the voltage drop across the 10 Ω resistor and increases the voltage drop across the remaining parallel branches. Thus, it increases the current through the remaining 60 Ω resistor.

3. (A) First, find the equivalent resistance for the whole circuit. First, the total resistance of the two parallel resistors can be found by using $\frac{1}{R_T} = \frac{1}{R_2} + \frac{1}{R_3} = \frac{R_2 + R_3}{R_2R_3}$, which means $R_T = \frac{R_2R_3}{R_2 + R_3}$. Then this would be in series with the first resistor, so the equivalent resistance of the whole circuit, R_{eq}, would be $R_{eq} = R_1 + R_T = R_1 + \frac{R_2R_3}{R_2 + R_3} = \frac{R_1R_2 + R_1R_3 + R_2R_3}{R_2 + R_3}$.

Next, you know from Ohm's Law that $V = IR$, so you can say $I = V/R = \frac{V}{\frac{R_1R_2 + R_1R_3 + R_2R_3}{R_2 + R_3}} = \frac{V(R_2 + R_3)}{R_1R_2 + R_1R_3 + R_2R_3}$. This will be the total current of the circuit. Because there are no branches before R_1, this will also be the current flowing through that resistor. So the voltage drop of that resistor, V_1, can be found by again using Ohm's Law for that location rather than the whole resistor. This gives $V_1 = I_1R_1 = \frac{V(R_2 + R_3)}{R_1R_2 + R_1R_3 + R_2R_3}R_1$.

(B) The diagram should look like this:

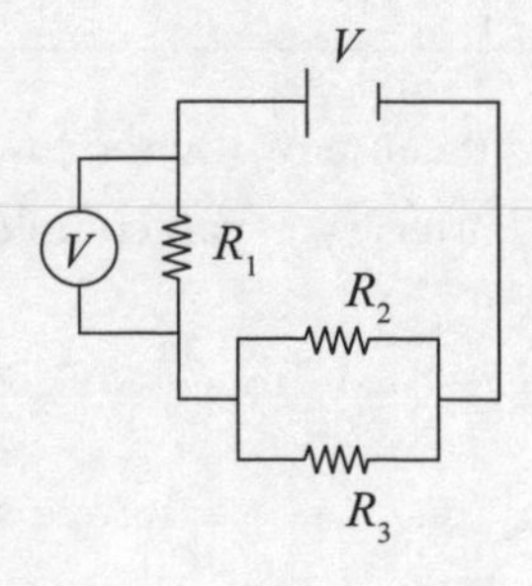

In order to measure the voltage loss of a particular resistor, the voltmeter must be arranged in parallel with that resistor. This is because parallel elements always have equal voltage drops. Thus, whatever voltage drop the device measures will be the same as the voltage drop in the resistor being measured.

(C) The diagram should look like this:

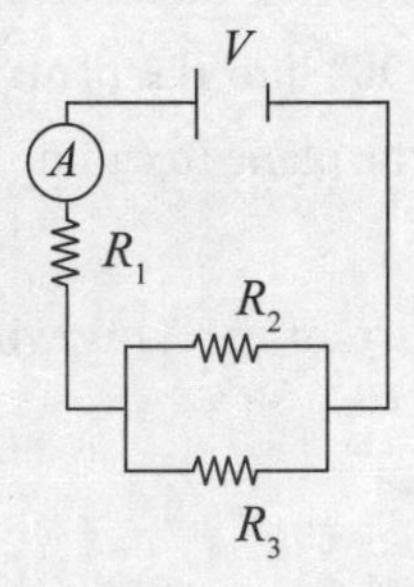

In order to measure the current passing through a particular resistor, the ammeter must be arranged in series with that resistor. This is because elements in series always have equal currents. Thus, whatever current the device measures will be the same as the current flowing through the resistor being measured.

(D) For the voltmeter, infinite resistance would be ideal. As in all measurements, you don't want to disturb the system in any way as you take the measurement. Otherwise, your readings would not be accurate for the original system. Thus, in order to leave the circuit as it was, you need to maintain the flow of current that exists before you started your measurement. Having an infinite resistance would ensure no current flows into the voltmeter, leaving it all on its original path.

For the ammeter, 0 resistance would be ideal. Again, you want to have minimal disturbance on the system. If the ammeter did have resistance, then it would be a source of voltage loss that did not previously exist. In essence, it would be an additional resistor that would need to be considered, forcing you to recalculate everything from the ground up.

CHAPTER 8 REVIEW QUESTIONS

Section I: Multiple Choice

1. **C** Current-carrying wires produce fields that are circles around the wire. The direction of the field is found from the right-hand rule. Choice (A) is false because perpendicular wires will produce perpendicular fields, which will not be able to cancel out. It is not necessary to make solenoids, so (B) is also false. To get the fields to cancel out at a location between the wires, they must carry current in the same direction, so (C) is correct.

2. **B** All of the conditions for the second particle have the same magnitude of force as the original. In order to have the new experiment result in the same magnitude of the force, $F_B = |q||\mathbf{v}||\mathbf{B}| \sin(\theta)$ must be unchanged. Changing the signs of q, **v**, or **B** does not change the strength of the force. Rotating **v** and **B** by the same amount leaves θ unchanged, and rotating **B** by 180° results in the same value for $\sin(\theta)$. Thus, the invalid conditions must have a different direction of the magnetic force. Using the right-hand rule, rotating your thumb and fingers by 90° into the plane of the paper causes the magnetic force to rotate from the original direction of out of the plane (recalling q is negative) to finally point to the left.

3. **D** The magnetic field strength for a current-carrying wire is given by $\mathbf{B} = \frac{\mu_0}{2\pi}\frac{I}{r}$. A graph of **B** versus I will give a line with a slope of $\frac{\mu_0}{2\pi}\frac{1}{r}$.

4. **C** Relative to a current-carrying wire, the measured value of the magnetic field is directly proportional to the current and inversely proportional to the distance the measurement is taken. Therefore, a large current at a small distance will result in the largest measured magnetic field value.

5. **D** The strength of the magnetic field at a distance r from a long, straight wire carrying a current I is given by the equation $B = (\mu_0/2\pi)(I/r)$. Therefore, the strength of the magnetic field at Point P due to either wire is $B = (\mu_0/2\pi)(I/\frac{1}{2}d)$. By the right-hand rule, the direction of the magnetic field at P due to the top wire is into the plane of the page and the direction of the magnetic field at P due to the bottom wire is out of the plane of the page. Since the two magnetic field vectors at P have the same magnitude but opposite directions, the net magnetic field at Point P is zero.

6. **C** Use the right-hand rule for wires. If you point your thumb to the right and wrap your fingers along the wire, you will note that the magnetic field goes into the page below the wire and comes out of the page above the wire. This allows you to eliminate (A) and (B). Because $B = \frac{\mu_o I}{2\pi r}$, the closer you are to the wire, the stronger the magnetic field. Choice (C) is closer, so it is the correct answer.

7. **C** Magnetic fields point from north to south. Therefore, the magnetic field between the two magnets is toward the right of the page. Use the right-hand rule. Because the B field is to the right and the charges through the wire flow to the bottom of the page, the force must be out of the page.

8. **C** Since **v** is upward and **B** is into the page, the direction of **v** × **B** is to the left. Therefore, free electrons in the wire will be pushed to the right, leaving an excess of positive charge at the left. Therefore, the potential at point a will be higher than that at point b, by $\varepsilon = BLv$.

9. **A** The magnitude of the emf induced between the ends of the rod is $\varepsilon = B\ell v = (0.5\text{ T})(0.2\text{ m})(3\text{ m/s}) = 0.3\text{ V}$. Since the resistance is 10 Ω, the current induced will be $I = \varepsilon/R = (0.3\text{ V})/(10\ \Omega) = 0.03\text{ A}$. To determine the direction of the current, note that since positive charges in the rod are moving to the left and the magnetic field points into the plane of the page, the right-hand rule states that the magnetic force, $q\boldsymbol{v} \times \boldsymbol{B}$, points downward. Since the resulting force on the positive charges in the rod is downward, so is the direction of the induced current.

10. **C** First, you can eliminate (A) and (B). By definition, magnetic field lines emerge from the north pole and enter at the south pole. The magnetic field from the bar magnet will always point toward a viewer who is looking down at the loop from above. As the north pole gets closer to the loop, the field at the loop grows in strength, but as the south pole recedes from the loop, the magnetic field strength shrinks. Because of this, the flux changes from growing to shrinking as the magnet passes through the loop and the induced current must also change directions. Therefore, (A) and (B) are wrong. To determine whether (C) or (D) is correct, look at the first half of the motion. As the north pole gets closer to the loop, the magnetic flux increases. To oppose an increasing flux, the direction of the magnetic field generated by induced current must be downward. Looking from above, a clockwise induced current generates a downward magnetic field. Therefore, (C) is correct.

11. **C** An induced emf requires that the magnetic flux through a loop of wire change over time. Changing the area of the loop of wire will always cause the flux to change. Choice (B) can cause an emf, but will not necessarily depending on the axis of rotation. Choice (D) is also incorrect because while a time varying field will cause an emf, it is not necessary to have such a field to produce an emf.

Section II: Free Response

1. (A) The acceleration of an ion of charge q is equal to F_E/m. The electric force is equal to qE, where $E = V/d$. Therefore, $a = qV/(dm)$.

   (B) Using $a = qV/(dm)$ and the equation $v^2 = v_0^2 + 2ad = 2ad$, you get

   $$v^2 = 2\frac{qV}{dm}d \quad \Rightarrow \quad v = \sqrt{\frac{2qV}{m}}$$

   As an alternate solution, notice that the change in the electrical potential energy of the ion from the source S to the entrance to the magnetic-field region is equal to qV; this is equal to the gain in the particle's kinetic energy.

   Therefore,

   $$qV = \frac{1}{2}mv^2 \quad \Rightarrow \quad v = \sqrt{\frac{2qV}{m}}$$

   (C) (i) and (ii) Use the right-hand rule. Since **v** points to the right and **B** is into the plane of the page, the direction of **v** × **B** is upward. Therefore, the magnetic force on a positively charged particle (cation) will be upward, and the magnetic force on a negatively charged particle (anion) will be downward. The magnetic force provides the centripetal force that causes the ion to travel in a circular path. Therefore, a cation would follow Path 1 and an anion would follow Path 2.

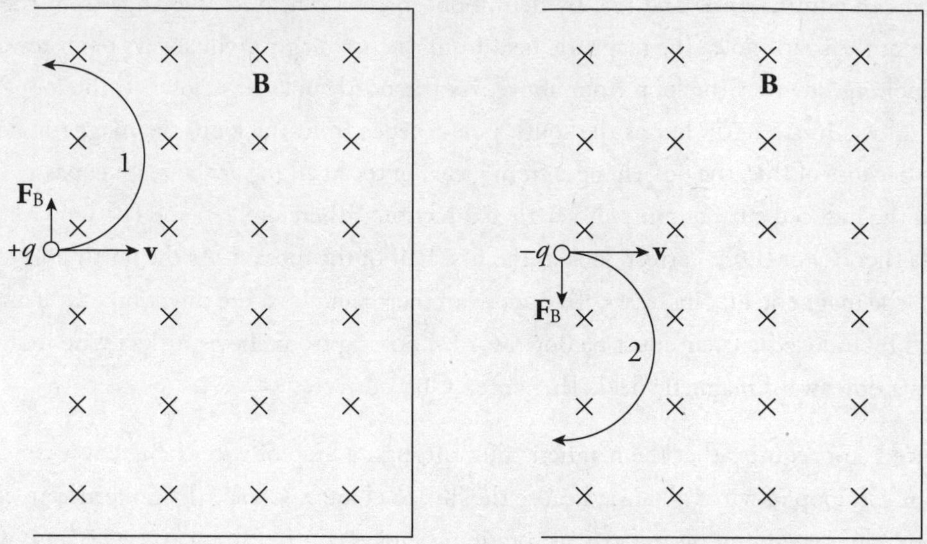

(D) Since the magnetic force on the ion provides the centripetal force,

$$qvB = \frac{mv^2}{r} \Rightarrow qvB = \frac{mv^2}{\frac{1}{2}y} \Rightarrow m = \frac{qBy}{2v}$$

Now, by the result of part (B),

$$m = \frac{qBy}{2\sqrt{\frac{2qV}{m}}} \Rightarrow m^2 = \frac{q^2B^2y^2}{\frac{8qV}{m}} \Rightarrow m^2 = \frac{mq^2B^2y^2}{8qV} \Rightarrow m = \frac{qB^2y^2}{8V}$$

(E) Since the magnetic force cannot change the speed of a charged particle, the time required for the ion to hit the photographic plate is equal to the distance traveled (the length of the semicircle) divided by the speed computed in part (b):

$$t = \frac{s}{v} = \frac{\pi \cdot \frac{1}{2}y}{\sqrt{\frac{2qV}{m}}} = \frac{1}{2}\pi y\sqrt{\frac{m}{2qV}}$$

(F) Since the magnetic force $\mathbf{F}_B$ is always perpendicular to a charged particle's velocity vector **v**, it can do no work on the particle. Thus, the answer is zero.

2. (A) The force is directed in the +y direction. The particle is charged positively and is moving in the +x direction, so the thumb points to the right. With your fingers facing inward toward the paper, your palm is toward the top of the paper.

(B) (i) The new speed is equal to the old speed because the magnetic force cannot do work, and therefore is unable to change the kinetic energy, and thus the speed, of the particle.

(ii) The new velocity is different from the old velocity as the direction has changed. The particle is traveling along a circular arc, and at any instant the velocity direction will be tangential to that arc. The new velocity will have the same magnitude as the initial velocity, but its x-component will be smaller as the y-component becomes larger. If the particle passes through 90° of the circular arc, the y-component will again begin to shrink as the x-component grows.

(C) Both A and B could be exit points from the magnetic field region.

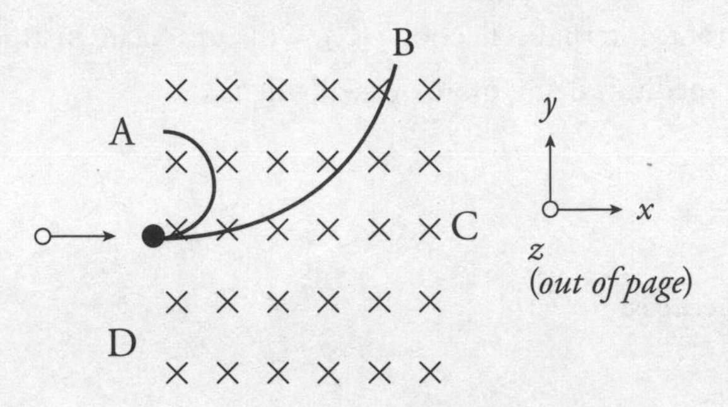

3. (A) $\varepsilon = B\ell v = (2\text{ T})(0.4\text{ m})(1\text{ m/s}) = 0.8\text{ V}$

(B) $V = IR$ becomes $I = \frac{V}{R} = \frac{0.8\text{ V}}{20\,\Omega} = 0.04\text{ A}$. The direction is given by Lenz's Law. Current flows in order to oppose the change in the magnetic flux. Because there is suddenly a new flux "in," current flows to produce an outward flux. This would be in the counterclockwise direction.

(C) As the edge of the loop first encounters the magnetic field region, the induced current is established at 0.04 A, counterclockwise (found in part (B)). This continues as long as the flux is changing, which occurs until the left edge of the loop enters the magnetic field region. During the time the entire loop is within the region of the magnetic field, the flux is constant, and the emf and current are both 0. Once the loop begins to leave the region of the magnetic field, the flux begins to change again, and an induced current is established at 0.04 A. As the loop leaves the region of the magnetic field, the flux is shrinking instead of growing, so the current must be clockwise as the loop exits the magnetic field region.

(D) The only thing that has changed is that the length of the wire at the leading edge has decreased. Thus, to maintain the same induced current, the speed at which the wire is pulled would have to increase. Moreover, as the length is halved, the speed would need to double in order for the induced current to remain constant.

CHAPTER 9 REVIEW QUESTIONS

Section I: Multiple Choice

1. **B** As light travels into an optically dense medium, it will refract in toward the normal and away from the surface. So the light in the glass will always be at a greater angle from the surface than the light in the air, so (B) is correct. Note that total internal reflection, (C), will not occur in this situation because it happens only with the initial medium being more optically dense.

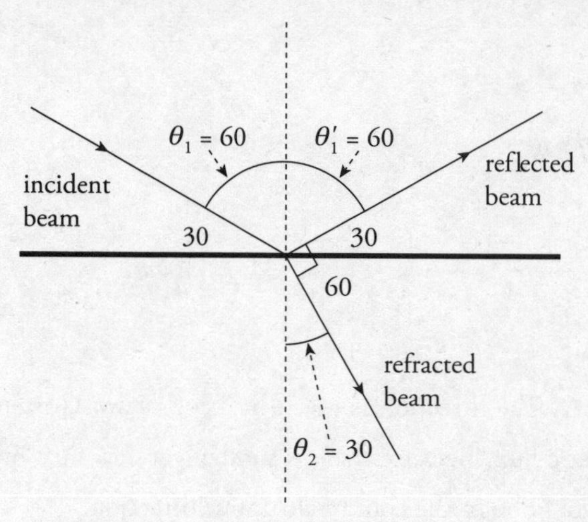

2. **C** Any convex lens made of a material with an index of refraction larger than the surroundings will create a real image, so (D) is incorrect. Because the glass-to-air difference is greater than the water-to-air difference, the glass-to-air transition results in a greater angle of refraction, so the light in air bends more when exiting the lens than the light in water. As a result, the point where the rays intersect will be farther from the lens than d_i.

3. **D** Choice (A) is false because a change in speed at an interface does not cause intensity to change. Choice (B) is false because had there been total internal reflection, the light would have remained in Medium 1. Choice (C) is false because the problem states that Medium 2 has a lower index of refraction, and from Snell's Law, $n_1 \sin(\theta_1) = n_2 \sin(\theta_2)$, if $n_2 < n_1$, then $\sin(\theta_2) > \sin(\theta_1)$ and the angles are not equal. Choice (D) is true because when light arrives at an interface, it must be absorbed, reflected, or transmitted. Therefore, any intensity not detected as transmitted into Medium 2 was either reflected back into Medium 1 or absorbed.

4. **A** The critical angle for total internal reflection is computed as follows:

$$n_1 \sin \theta_c = n_2 \sin 90^\circ$$

$$\sin \theta_c = \frac{n_2}{n_1} = \frac{1.45}{2.90} = \frac{1}{2} \quad \Rightarrow \quad \theta_c = 30^\circ$$

Total internal reflection can happen only if the incident beam originates in the medium with the higher index of refraction, and therefore the beam must originate in the solid. The refracted ray must strike the interface of the other medium at an angle of incidence greater than the critical angle.

5. **C** Every ray, regardless of whether it is a principal ray or not, which originates at the tip of an object and strikes a mirror, will converge at the tip of the image.

6. **A** All of the rays that originate at the tip of the image and travel along paths that remain above the optic axis will be blocked. However, all of the rays that originate at the tip of the image and travel along paths that are below the optic axis at the plane of the block will be unaltered. As a result, fewer rays will converge at the image location, resulting in a less bright image.

7. **D** Diverging lenses always create virtual images.

Section II: Free Response

1. (A) (i) Student 1 is correct that "According to Snell's Law, the light will bend at the first interface" and also "unbend at the second interface and travel parallel to its original path."

(ii) Student 2 is correct that "Snell's Law is n_{in} sin(θ_{in}) = n_{out} sin (θ_{out})" and that "θ_{out} is different from θ_{in} at the first interface."

(iii) Student 1 is incorrect that "The spot on the screen will still be at point P."

(iv) Student 2 is incorrect that "The beam after the glass cannot be parallel to the beam before the glass."

(B) Incident beam

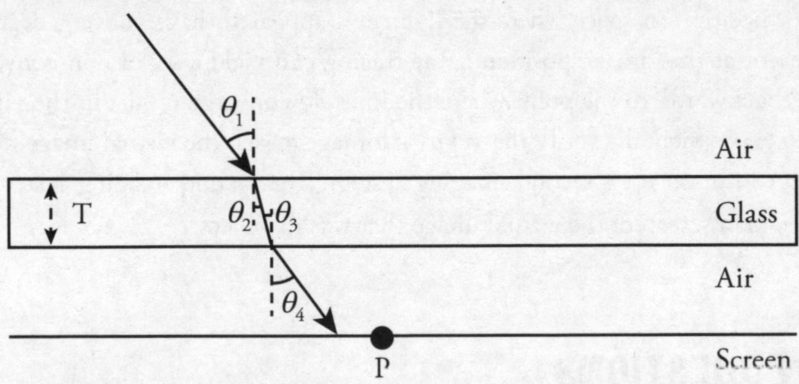

Here, $\theta_1 = \theta_4$ and $\theta_2 = \theta_3$.

(C) If the area between the glass and screen were filled with water, it would still be the case that $\theta_2 = \theta_3$ because there is no change to the glass. However, upon leaving the glass and entering the water, the amount that the light would bend away from the normal would be less than in the case of air because the difference in the indices of refraction is smaller with glass-to-water than glass-to-air. As a result, the beams in air and water would not be parallel, and the spot that would appear on the screen would be even farther away from point P than in the original experiment.

2. (A)

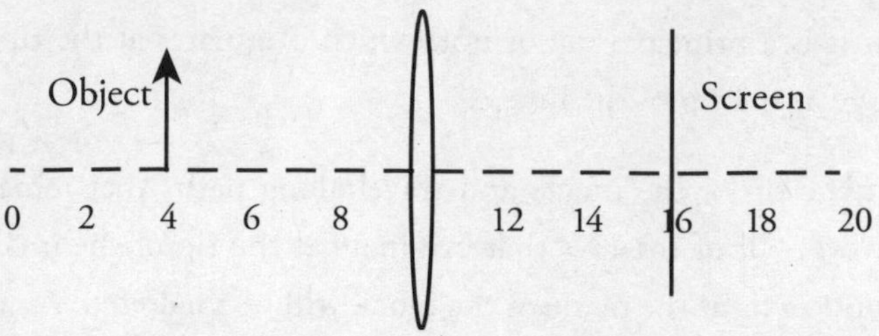

Object

Screen

0 2 4 6 8 12 14 16 18 20

In order for the object and image to be the same size, the magnitude of the magnification must be 1, so $|M| = \left|\left(\frac{s_i}{s_o}\right)\right| = 1$ and therefore $|s_o| = |s_i|$. The only way to satisfy the equation $\frac{1}{f} = \frac{1}{s_i} + \frac{1}{s_o}$ when the image distance is equal to the object distance is for both to be $2f$. The object should be placed $2f$ before the lens and the screen should be placed $2f$ after the lens. The image will be below the optical axis in this case, so the part of the screen where the image forms needs to also be placed below the optic axis. Note: switching the positions of the object and screen (so that the object is on the right and the screen is on the left) is also correct.

(B) A converging lens is similar to a convex mirror, but with the mirror the light always stays on one side of the mirror. Both the object and the screen should be placed at the 4 cm mark. The image will be below the optical axis in this case.

(C) A virtual image is a position in space where the light rays appear to have converged, although no light is actually present at the “image position.” Ray tracing can yield a set of non-converging rays. Tracing those rays “backward” to the point where the lines do converge results in the virtual image position. In order to experimentally verify that a virtual image exists, the virtual image location may be used as the object location for a second imaging system. The second imaging system may then produce a real image on a screen of the virtual image that was its object.

CHAPTER 10 REVIEW QUESTIONS

Section I: Multiple Choice

1. **B** Waves will interfere with apertures that are approximately the same size as the wave. The wavelengths of matter from the de Broglie theory are on the same order as the size of molecule spacing in solids, and therefore these effects are not visible at macroscopic scales.

2. **A, C** Photoelectrons will be emitted in the photoelectric effect any time that the energy of the incident light is greater than the work function of the metal. Altering the work function could cause the incident light energy to be below the work function. Altering the incident light frequency (since $E = hf$) could cause the energy of the incident light to be below the work function. Neither the stopping

potential nor the power of the incident light has any effect on whether photoelectrons will be emitted during a photoelectric effect experiment.

3. **A** The ground state of an atom is (–1) times its ionization energy, so the ground state would have to be at an energy level of –25 eV. To transition from –16 eV to –25 eV, a photon with an energy of 9 eV would have to be produced in order for energy to be conserved. To find the wavelength of the photon, $E = hf = h(c/\lambda) = hc/\lambda$, so $\lambda = hc/E$. From the equation sheet, hc = 1,240 nm • eV and $\lambda = \frac{1{,}240 \text{ nm} \cdot \text{eV}}{9 \text{ eV}} = 138 \text{ nm}$.

4. **D** The gap between the ground-state and the first excited state is

   –10 eV – (–40 eV) = 30 eV

   Therefore, the electron must absorb the energy of a 30 eV photon (at least) in order to move even to the first excited state. Since the incident photons have only 15 eV of energy, the electron will be unaffected.

5. **D** When a charged particle goes into a magnetic field, it will be deflected, but an uncharged particle will not be deflected. Particles with opposite signs will deflect in opposite directions. An alpha particle is charged and a neutron is not, so (D) is the correct answer.

6. **D** If the atom begins in the n = 3 state, it could lose energy by making any of the following transitions: 3 → 2, 3 → 1, or 3 → 2 → 1. The 3 → 2 transition would result in the emission of –5 eV – (–8 eV) = 3 eV; the 3 → 1 transition would emit a –5 eV – (–12 eV) = 7 eV photon; and the 2 → 1 transition would result in the emission of –8 eV – (–12 eV) = 4 eV. Therefore, if the atom is initially in the n = 3 state, it could emit photons of energy 3 eV, 4 eV, or 7 eV.

7. **C** To emit a photon, an energy level transition must go to a lower energy level (a transition to a higher level requires absorbing a photon), so (A) and (B) are incorrect. A long wavelength corresponds to a small energy photon, so (D) is also incorrect, and (C) is the correct answer.

8. **D** At the time that Louis de Broglie made his hypothesis, it was known that light, which often behaves like a wave, also has particle-like behavior. The de Broglie hypothesis is that while electrons (and other matter) most often demonstrate particle-like behaviors, they would also demonstrate wave-like behaviors. Diffraction is a phenomenon associated with wave interference, so de Broglie's hypothesis can be summarized as electrons will diffract in certain circumstances.

9. **B** The binding energy is given by $E_B = (\Delta m)c^2$, so (A) is true. The binding energy, which is the energy associated with the mass defect, is typically measured in MeV, while typical electronic transitions are measured in eV, so (C) is also true. Mass defect is also typically four orders of magnitude smaller than the mass of the involved nucleons, so (D) is true. Choice (B) is false because the binding energy is always positive, so that the mass of the bound nucleons plus the binding energy equals the mass of the constituent nucleons.

10. **B** In order to balance the mass number (the superscripts), you must have $2 + 63 = 64 + A$, so $A = 1$. In order to balance the charge (the subscripts), you need $1 + 29 = 30 + Z$, so $Z = 0$. A particle with a mass number of 1 and no charge is a neutron, ${}^{1}_{0}n$.

11. **A** The alpha decay process produces an alpha particle ${}^{4}_{2}He$ and a daughter nucleus. The daughter nucleus must have a mass number 4 below the original isotope and a charge number 2 below the original isotope, resulting in ${}^{171}_{76}Os$. The beta(+) decay process produces a positive beta particle ${}^{0}_{1}e$ and a daughter nucleus. The daughter nucleus must have a mass number equal to the original isotope and a charge number 1 below the original isotope, resulting in ${}^{175}_{77}Ir$.

Section II: Free Response

1. (A) The work function is the amount of energy that must be supplied before a photoelectron is released. For any amount of energy supplied by the power supply below the work function of the material that the LED light is made of, there will be no production of photoelectrons. Therefore, the work function of the material must be greater than 0.1 eV since there were no photoelectrons emitted at the voltage of 1 V.

(B)

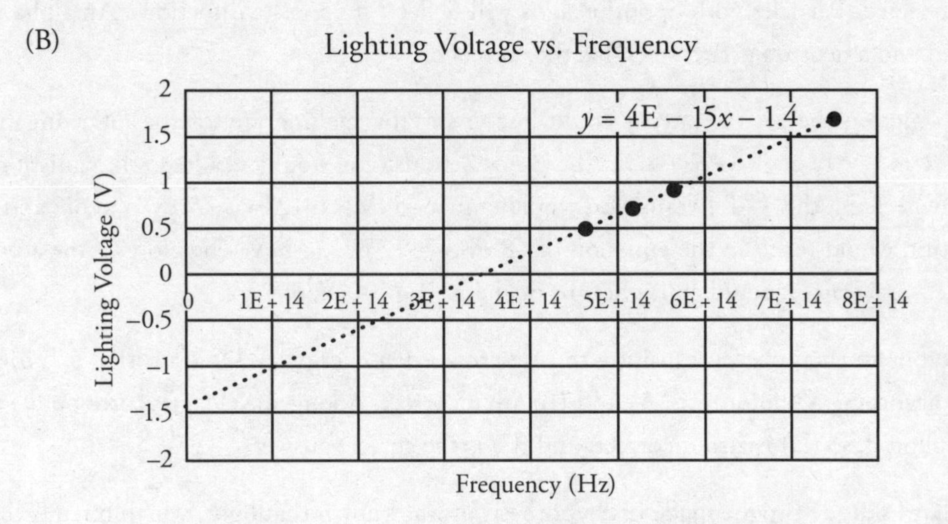

The slope of the graph is 4×10^{-15} V · s and the y-intercept is –1.4 V. The slope has the value of Planck's constant and the intercept is the work function of the material the LED lights are made of.

(C) (i) To change the intercept, the work function would have to be different, so a supply of LEDs made from a different material would have to be obtained.

(ii) It would not be possible to change the slope, as the slope determined from this experiment is Planck's constant.

Part VI
Practice Test 2

- Practice Test 2
- Practice Test 2: Diagnostic Answer Key and Explanations
- How to Score Practice Test 2

Practice Test 2

AP® Physics 2 Exam

SECTION I: Multiple-Choice Questions

DO NOT OPEN THIS BOOKLET UNTIL YOU ARE TOLD TO DO SO.

At a Glance

Total Time
90 minutes
Number of Questions
50
Percent of Total Grade
50%
Writing Instrument
Pen required

Instructions

Section I of this examination contains 50 multiple-choice questions. Fill in only the ovals for numbers 1 through 50 on your answer sheet.

CALCULATORS MAY BE USED IN BOTH SECTIONS OF THE EXAMINATION.

Indicate all of your answers to the multiple-choice questions on the answer sheet. No credit will be given for anything written in this exam booklet, but you may use the booklet for notes or scratch work. After you have decided which of the suggested answers is best, completely fill in the corresponding oval on the answer sheet. Give only one answer to each question. If you change an answer, be sure that the previous mark is erased completely. Here is a sample question and answer.

Sample Question

Chicago is a
(A) state
(B) city
(C) country
(D) continent

Sample Answer

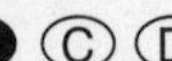

Use your time effectively, working as quickly as you can without losing accuracy. Do not spend too much time on any one question. Go on to other questions and come back to the ones you have not answered if you have time. It is not expected that everyone will know the answers to all the multiple-choice questions.

About Guessing

Many candidates wonder whether or not to guess the answers to questions about which they are not certain. Multiple-choice scores are based on the number of questions answered correctly. Points are not deducted for incorrect answers, and no points are awarded for unanswered questions. Because points are not deducted for incorrect answers, you are encouraged to answer all multiple-choice questions. On any questions you do not know the answer to, you should eliminate as many choices as you can, and then select the best answer among the remaining choices.

GO ON TO THE NEXT PAGE.

ADVANCED PLACEMENT PHYSICS 2 TABLE OF INFORMATION

CONSTANTS AND CONVERSION FACTORS

Proton mass, $m_p = 1.67 \times 10^{-27}$ kg	Electron charge magnitude, $e = 1.60 \times 10^{-19}$ C
Neutron mass, $m_n = 1.67 \times 10^{-27}$ kg	1 electron volt, $1 \text{ eV} = 1.60 \times 10^{-19}$ J
Electron mass, $m_e = 9.11 \times 10^{-31}$ kg	Speed of light, $c = 3.00 \times 10^{8}$ m/s
Avogadro's number, $N_A = 6.02 \times 10^{23}$ mol^{-1}	Universal gravitational constant, $G = 6.67 \times 10^{-11}$ m^3/kg$\cdot$s^2
Universal gas constant, $R = 8.31$ J/(mol$\cdot$K)	Acceleration due to gravity at Earth's surface, $g = 9.8$ m/s^2
Boltzmann's constant, $k_B = 1.38 \times 10^{23}$ J/K	

1 unified atomic mass unit,	$1 \text{ u} = 1.66 \times 10^{-27}$ kg $= 931$ MeV/c^2
Planck's constant,	$h = 6.63 \times 10^{-34}$ J$\cdot$s $= 4.14 \times 10^{-15}$ eV$\cdot$s
	$hc = 1.99 \times 10^{-25}$ J$\cdot$m $= 1.24 \times 10^{3}$ eV$\cdot$nm
Vacuum permittivity,	$\varepsilon_0 = 8.85 \times 10^{-12}$ C^2/N$\cdot$m^2
Coulomb's law constant,	$k = 1/4\pi\varepsilon_0 = 9.0 \times 10^{9}$ N$\cdot$m^2/C^2
Vacuum permeability,	$\mu_0 = 4\pi \times 10^{-7}$ (T$\cdot$m)/A
Magnetic constant,	$k' = \mu_0/4\pi = 1 \times 10^{-7}$ (T$\cdot$m)/A
1 atmosphere pressure,	$1 \text{ atm} = 1.0 \times 10^{5}$ N/m^2 $= 1.0 \times 10^{5}$ Pa

UNIT SYMBOLS								
	meter,	m	mole,	mol	watt,	W	farad,	F
	kilogram,	kg	hertz,	Hz	coulomb,	C	tesla,	T
	second,	s	newton,	N	volt,	V	degree Celsius,	°C
	ampere,	A	pascal,	Pa	ohm,	Ω	electron volt,	eV
	kelvin,	K	joule,	J	henry,	H		

PREFIXES

Factor	Prefix	Symbol
10^{12}	tera	T
10^{9}	giga	G
10^{6}	mega	M
10^{3}	kilo	k
10^{-2}	centi	c
10^{-3}	milli	m
10^{-6}	micro	μ
10^{-9}	nano	n
10^{-12}	pico	p

VALUES OF TRIGONOMETRIC FUNCTIONS FOR COMMON ANGLES

θ	0°	30°	37°	45°	53°	60°	90°
$\sin\theta$	0	1/2	3/5	$\sqrt{2}/2$	4/5	$\sqrt{3}/2$	1
$\cos\theta$	1	$\sqrt{3}/2$	4/5	$\sqrt{2}/2$	3/5	1/2	0
$\tan\theta$	0	$\sqrt{3}/3$	3/4	1	4/3	$\sqrt{3}$	∞

The following conventions are used in this exam.

I. The frame of reference of any problem is assumed to be inertial unless otherwise stated.
II. In all situations, positive work is defined as work done on a system.
III. The direction of current is conventional current: the direction in which positive charge would drift.
IV. Assume all batteries and meters are ideal unless otherwise stated.
V. Assume edge effects for the electric field of a parallel plate capacitor unless otherwise stated.
VI. For any isolated electrically charged object, the electric potential is defined as zero at infinite distance from the charged object.

GO ON TO THE NEXT PAGE.

ADVANCED PLACEMENT PHYSICS 2 EQUATIONS

MECHANICS

$v_x = v_{x0} + a_x t$

$x = x_0 + v_{x0} t + \frac{1}{2} a_x t^2$

$v_x^2 = v_{x0}^2 + 2a_x (x - x_0)$

$\vec{a} = \frac{\sum \vec{F}}{m} = \frac{\vec{F}_{net}}{m}$

$|\vec{F}_f| \le \mu |\vec{F}_n|$

$a_c = \frac{v^2}{r}$

$\vec{p} = m\vec{v}$

$\Delta\vec{p} = \vec{F}\,\Delta t$

$K = \frac{1}{2} m v^2$

$\Delta E = W = F_{\parallel} d = Fd\cos\theta$

$P = \frac{\Delta E}{\Delta t}$

$\theta = \theta_0 + \omega_0 t + \frac{1}{2}\alpha t^2$

$\omega = \omega_0 + \alpha t$

$x = A\cos(\omega t) = A\cos(2\pi f t)$

$x_{cm} = \frac{\sum m_i x_i}{\sum m_i}$

$\vec{\alpha} = \frac{\sum \vec{\tau}}{I} = \frac{\vec{\tau}_{net}}{I}$

$\tau = r_{\perp} F = rF\sin\theta$

$L = I\omega$

$\Delta L = \tau\,\Delta t$

$K = \frac{1}{2} I\omega^2$

$|\vec{F}_s| = k|\vec{x}|$

a = acceleration
A = amplitude
d = distance
E = energy
F = force
f = frequency
I = rotational inertia
K = kinetic energy
k = spring constant
L = angular momentum
ℓ = length
m = mass
P = power
p = momentum
r = radius or separation
T = period
t = time
U = potential energy
v = speed
W = work done on a system
x = position
y = height
α = angular acceleration
μ = coefficient of friction
θ = angle
τ = torque
ω = angular speed

$U_s = \frac{1}{2} k x^2$

$\Delta U_g = mg\,\Delta y$

$T = \frac{2\pi}{\omega} = \frac{1}{f}$

$T_s = 2\pi\sqrt{\frac{m}{k}}$

$T_p = 2\pi\sqrt{\frac{\ell}{g}}$

$|\vec{F}_g| = G\frac{m_1 m_2}{r^2}$

$\vec{g} = \frac{\vec{F}_g}{m}$

$U_G = -\frac{G m_1 m_2}{r}$

ELECTRICITY AND MAGNETISM

$|\vec{F}_E| = \frac{1}{4\pi\varepsilon_0}\frac{|q_1 q_2|}{r^2}$

$\vec{E} = \frac{\vec{F}_E}{q}$

$|\vec{E}| = \frac{1}{4\pi\varepsilon_0}\frac{|q|}{r^2}$

$\Delta U_E = q\Delta V$

$V = \frac{1}{4\pi\varepsilon_0}\frac{q}{r}$

$|\vec{E}| = \left|\frac{\Delta V}{\Delta r}\right|$

$\Delta V = \frac{Q}{C}$

$C = \kappa\varepsilon_0\frac{A}{d}$

$E = \frac{Q}{\varepsilon_0 A}$

$U_C = \frac{1}{2} Q\Delta V = \frac{1}{2} C(\Delta V)^2$

$I = \frac{\Delta Q}{\Delta t}$

$R = \frac{\rho\ell}{A}$

$P = I\,\Delta V$

$I = \frac{\Delta V}{R}$

$R_s = \sum_i R_i$

$\frac{1}{R_p} = \sum_i \frac{1}{R_i}$

$C_p = \sum_i C_i$

$\frac{1}{C_s} = \sum_i \frac{1}{C_i}$

$B = \frac{\mu_0}{2\pi}\frac{I}{r}$

A = area
B = magnetic field
C = capacitance
d = distance
E = electric field
$\mathcal{E}$ = emf
F = force
I = current
ℓ = length
P = power
Q = charge
q = point charge
R = resistance
r = separation
t = time
U = potential (stored) energy
V = electric potential
v = speed
κ = dielectric constant
ρ = resistivity
θ = angle
Φ = flux

$\vec{F}_M = q\vec{v} \times \vec{B}$

$|\vec{F}_M| = |q\vec{v}||\sin\theta||\vec{B}|$

$\vec{F}_M = I\vec{\ell} \times \vec{B}$

$|\vec{F}_M| = |I\vec{\ell}||\sin\theta||\vec{B}|$

$\Phi_B = \vec{B} \cdot \vec{A}$

$\Phi_B = |\vec{B}|\cos\theta|\vec{A}|$

$\mathcal{E} = -\frac{\Delta\Phi_B}{\Delta t}$

$\mathcal{E} = B\ell v$

GO ON TO THE NEXT PAGE.

ADVANCED PLACEMENT PHYSICS 2 EQUATIONS

FLUID MECHANICS AND THERMAL PHYSICS

$\rho = \frac{m}{V}$

$P = \frac{F}{A}$

$P = P_0 + \rho g h$

$F_b = \rho V g$

$A_1 v_1 = A_2 v_2$

$P_1 + \rho g y_1 + \frac{1}{2}\rho v_1^2 = P_2 + \rho g y_2 + \frac{1}{2}\rho v_2^2$

$\frac{Q}{\Delta t} = \frac{kA\,\Delta T}{L}$

$PV = nRT = Nk_BT$

$K = \frac{3}{2}k_BT$

$W = -P\Delta V$

$\Delta U = Q + W$

A = area
F = force
h = depth
k = thermal conductivity
K = kinetic energy
L = thickness
m = mass
n = number of moles
N = number of molecules
P = pressure
Q = energy transferred to a system by heating
T = temperature
t = time
U = internal energy
V = volume
v = speed
W = work done on a system
y = height
ρ = density

MODERN PHYSICS

$E = hf$

$K_{max} = hf - \phi$

$\lambda = \frac{h}{p}$

$E = mc^2$

E = energy
f = frequency
K = kinetic energy
m = mass
p = momentum
λ = wavelength
ϕ = work function

WAVES AND OPTICS

$\lambda = \frac{v}{f}$

$n = \frac{c}{v}$

$n_1 \sin\theta_1 = n_2 \sin\theta_2$

$\frac{1}{s_i} + \frac{1}{s_o} = \frac{1}{f}$

$|M| = \left|\frac{h_i}{h_o}\right| = \left|\frac{s_i}{s_o}\right|$

$\Delta L = m\lambda$

$d\sin\theta = m\lambda$

d = separation
f = frequency or focal length
h = height
L = distance
M = magnification
m = an integer
n = index of refraction
s = distance
v = speed
λ = wavelength
θ = angle

GEOMETRY AND TRIGONOMETRY

Rectangle
$A = bh$

Triangle
$A = \frac{1}{2}bh$

Circle
$A = \pi r^2$
$C = 2\pi r$

Rectangular solid
$V = \ell wh$

Cylinder
$V = \pi r^2 \ell$
$S = 2\pi r\ell + 2\pi r^2$

Sphere
$V = \frac{4}{3}\pi r^3$
$S = 4\pi r^2$

A = area
C = circumference
V = volume
S = surface area
b = base
h = height
ℓ = length
w = width
r = radius

Right triangle
$c^2 = a^2 + b^2$

$\sin\theta = \frac{a}{c}$

$\cos\theta = \frac{b}{c}$

$\tan\theta = \frac{a}{b}$

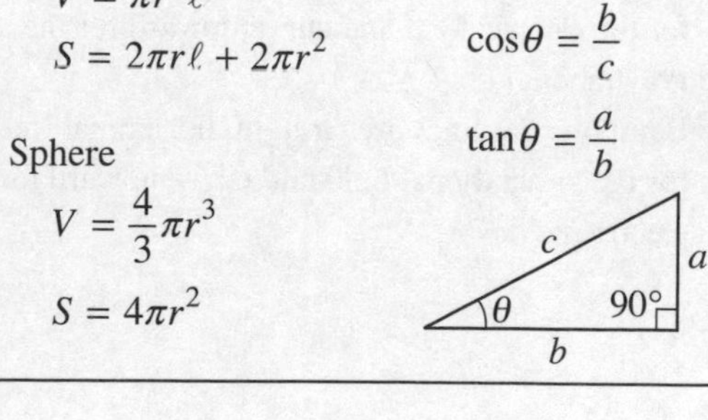

GO ON TO THE NEXT PAGE.

PHYSICS 2

SECTION I

Time—90 minutes

50 Questions

Note: To simplify calculations, you may use $g = 10$ m/s^2 in all problems.

Directions: Each of the questions or incomplete statements below is followed by four suggested answers or completions. Select the one that is best in each case and mark it on your sheet.

Questions 1 and 2 refer to the following situation.

A proton is traveling along a straight line at a constant speed through a uniform electric field near the surface of the Earth.

1. Which of the following choices correctly describes the direction of the electric field and the relative magnitudes of the electric and gravitational fields?

	Electric Field Direction	Field Strength
(A)	Down	$E_g = E_{el}$
(B)	Down	$E_g > E_{el}$
(C)	Up	$E_g = E_{el}$
(D)	Up	$E_g > E_{el}$

2. Which of the following describes the equipotential lines for the electric and gravitational fields that the proton experiences?

(A) Equipotential lines are straight, horizontal lines for both fields.

(B) Equipotential lines are straight, vertical lines for both fields.

(C) Equipotential lines are straight, horizontal lines for the electric field and curve upward for the gravitational field.

(D) Equipotential lines are straight, horizontal lines for the gravitational field and curve upward for the electric field.

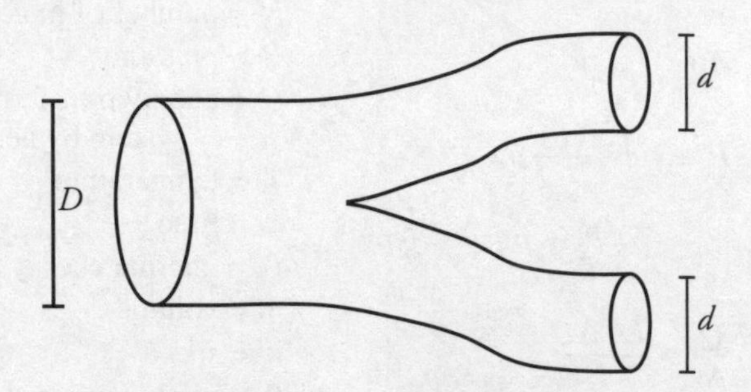

3. A pipe with a diameter of D splits into two smaller, identical pipes with diameter d. If the speed of the water in the small pipes is v, what is the speed of the water in the large pipe?

(A) $\frac{dv}{D}$

(B) $\frac{2dv}{D}$

(C) $\frac{d^2v}{D^2}$

(D) $\frac{2d^2v}{D^2}$

4. A fluid container is shaped as a rectangular prism. The areas of the three faces of the prism are different. Which face of the container should be placed on a horizontal table so that the fluid pressure against the face touching the table is at the lowest value?

(A) The face with the smallest area

(B) The face with the median area

(C) The face with the largest area

(D) All three orientations will result in the same fluid pressure.

GO ON TO THE NEXT PAGE.

5. An ohmmeter is used to measure resistance. Measurements are made of the cross-sectional area, A, and length, L, of each resistor. What should be plotted so that the slope of the plot will yield the resistivity, ρ, of the resistors?

   (A) Resistance on the y-axis versus ratio of L to A on the x-axis
   (B) Resistance on the y-axis versus ratio of A to L on the x-axis
   (C) Ratio of L to A on the y-axis versus resistance on the x-axis
   (D) Ratio of A to L on the y-axis versus resistance on the x-axis

6. An ideal gas is taken from an initial set of conditions with pressure P_i and volume V_i, to a final set of conditions, with pressure P_f and volume V_f, through several different processes. At the end of the process, the gas is at both a higher pressure and a larger volume than when it started. Which process requires the least amount of work on the system?

   (A) First, expansion at constant pressure P_i from V_i to V_f followed by increasing pressure from P_i to P_f at constant volume V_f
   (B) First, increasing pressure from P_i to P_f at constant V_i volume followed by expansion at constant pressure P_f from V_i to V_f
   (C) A series of small increases in volume alternating with small increases in pressure, resulting in a nearly straight line on a PV graph from the beginning to the end of the process
   (D) Any set of steps will require the same amount of work because all gases have the same change in pressure and volume.

Questions 7 and 8 refer to the following diagram.

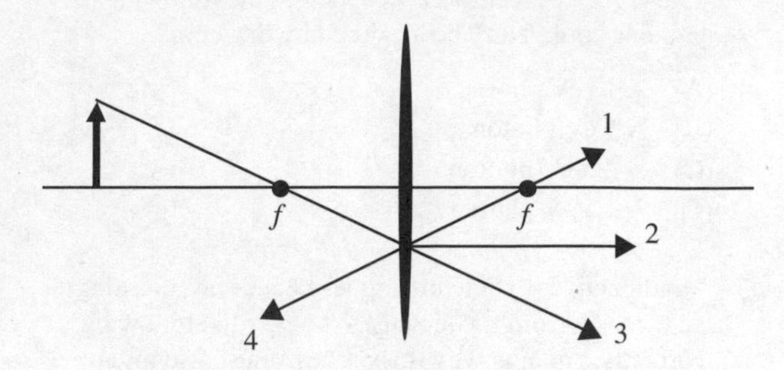

7. The image above shows a converging lens and an object represented as a bold vertical arrow. Which line correctly depicts the output path of the ray that is incident on the lens coming in through the focus?

   (A) 1
   (B) 2
   (C) 3
   (D) 4

8. The converging lens above has a focal length of 25 cm. The object is located at a distance of 65 cm from the lens. Where should a screen be placed so that the observer will see a focused image on the screen?

   (A) 65 cm to the right side of the lens
   (B) 40.6 cm to the right side of the lens
   (C) At the focus of the lens
   (D) The image in such an arrangement will be virtual and cannot be seen on a screen.

9. You are tasked with creating a real image using a concave mirror as your imaging system. Which of the following criteria is true about both the image and the object?

   (A) A real image can be created only if the object is farther away from the lens than the radius of curvature of the lens.
   (B) A real image can be created only with the object located between the center of the lens and the focal length of the lens.
   (C) A real image can be created with the object located anywhere farther from the lens than the focal point.
   (D) A real image cannot be created using only a concave lens.

GO ON TO THE NEXT PAGE.

10. An atom has its lowest four energy levels at –10 eV, –5 eV, –3.5 eV, and –2 eV. Which of the following photons could NOT be absorbed by the atom?

    (A) A 10 eV photon
    (B) A 5 eV photon
    (C) A 2.5 eV photon
    (D) A 1.5 eV photon

11. A radioactive particle undergoes beta decay, emitting an electron from its nucleus. Which of the following correctly explains why this process must also involve a neutron turning into a proton within the nucleus?

    (A) The total charge of the system before and after is not the same if the proton is not created.
    (B) The mass energy of the system is not conserved if the proton is not created.
    (C) The momentum of the system could not be conserved without the generation of a proton.
    (D) All nuclear decay processes involve the generation of two particles.

12. A thermodynamic process is conducted wherein an ideal gas is taken from State A to B to C to D to A. State A is at a pressure P and a volume 5V. State B is at pressure 4P and volume V. State C is at pressure 4P and volume 4V. State D is at pressure 2P and volume 10V. Which step in the process requires the largest change in internal energy of the system?

    (A) State A to State B
    (B) State B to State C
    (C) State C to State D
    (D) State D to State A

13. A circuit consists of a 50 V battery, a 100 Ω resistor and a 25 μF capacitor. Once the capacitor has become fully charged, how much energy is stored in the capacitor?

    (A) 0.0013 J
    (B) 0.031 J
    (C) 0.062 J
    (D) 0.125 J

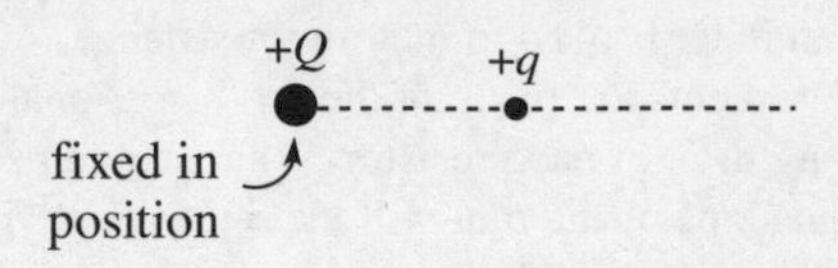

14. As shown above, the $+Q$ charge is fixed in position, and the $+q$ charge is brought close to $+Q$ and then released from rest. Which graph best shows the acceleration of the $+q$ charge as a function of its distance r from $+Q$?

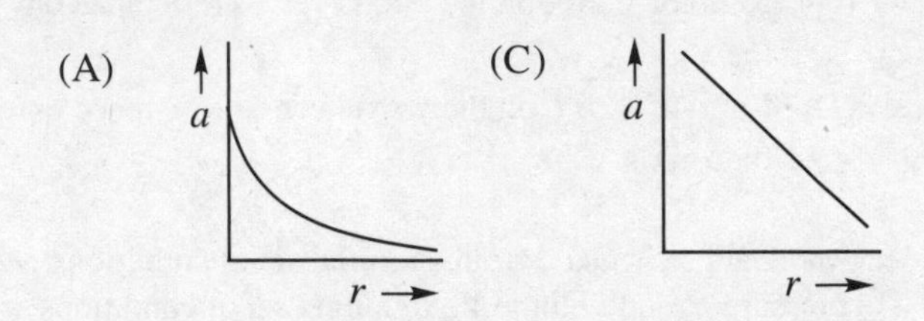

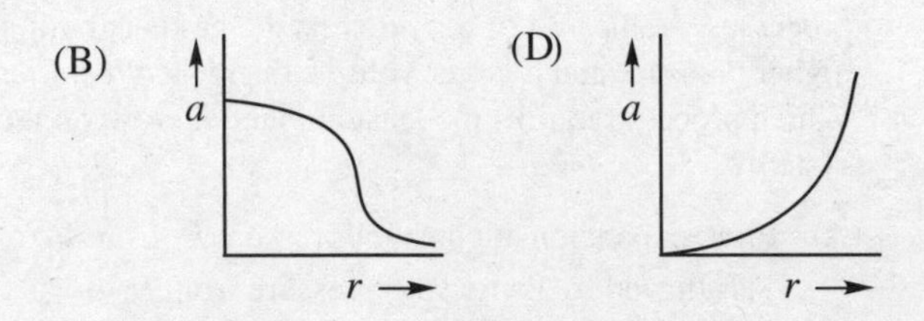

15. A charged particle with mass m is moving at a speed v at one particular instant in time. The particle is later found at a position with an electrical potential of ΔV higher than its initial position, and a gravitational potential equal to its original position. Which mathematical routine could be used to determine its speed in the final position?

    (A) Calculate the work from the electrical potential and use conservation of energy to find the final speed.
    (B) Calculate the work from the gravitational potential and use conservation of energy to find the final speed.
    (C) Calculate the impulse from the electric potential and use conservation of momentum to find the final speed.
    (D) Calculate the impulse from the gravitational potential and use conservation of momentum to find the final speed.

GO ON TO THE NEXT PAGE.

Questions 16, 17, and 18 refer to the following diagram.

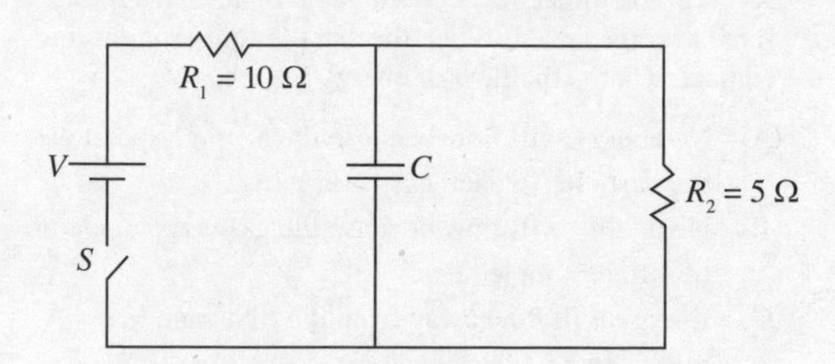

16. Voltmeters are placed across the Resistor R_1, the capacitor C, and the resistor R_2. The switch S has been closed a long time. What is the rank of the value readings on the voltmeters?

    (A) $V_{R_1} > (V_C = V_{R_2})$
    (B) $(V_{R_1} = V_C = V_{R_2})$
    (C) $(V_C = V_{R_2}) > V_{R_1}$
    (D) $V_{R_1} > V_C > V_{R_2}$

17. The circuit is reset and the capacitor is discharged. Then, the switch is closed again. At what time will the current through the resistor R_2 be greatest?

    (A) The current through the resistor will be constant.
    (B) The current through the resistor will be greatest before closing the switch.
    (C) The current through the resistor will be greatest immediately after closing the switch.
    (D) The current through the resistor will be greatest a long time after closing the switch.

18. The capacitance of the capacitor is known initially. The capacitor is now altered to have a larger capacitance. Which of the following observations will occur with the new capacitor in the circuit a long time after the switch is closed as compared to what was observed a long time after the switch was closed with the original capacitor?

    (A) The current flowing in R_1 will be greater.
    (B) The current flowing in R_2 will be greater.
    (C) The current flowing into the capacitor will be greater.
    (D) The charge stored in the capacitor will be greater.

19. An ideal gas is confined within a cube-shaped container. In addition to the length of the side of the container, which of the following sets of measurements will allow a student to determine the pressure of the gas in the container?

    I. The mass of gas in the container and the average speed of a gas molecule
    II. The impulse delivered to the gas by a wall in a measured time period
    III. The force of the gas against one of the walls

    (A) I only
    (B) III only
    (C) II or III
    (D) I, II, or III

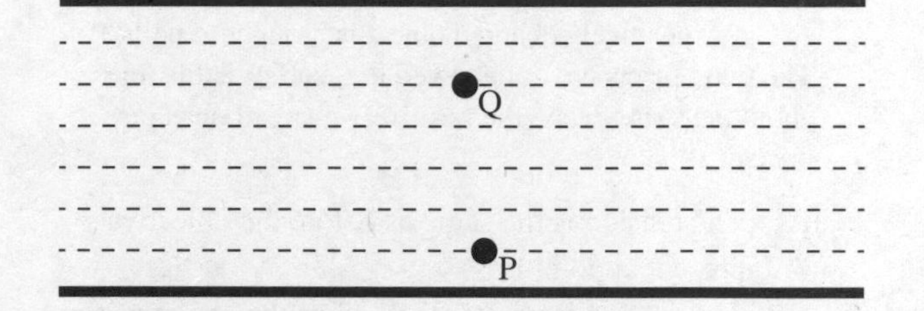

20. Points P and Q lie between the plates of a fully charged parallel-plate capacitor as shown above. The lower plate is negatively charged, and the upper plate is positively charged. How do the magnitudes of the electric fields at points P and Q compare?

    (A) The field is 0 N/C at both points.
    (B) The field is the same at both points, but not 0 N/C.
    (C) $E_P > E_Q$
    (D) $E_Q > E_P$

21. A wire carries a constant current to the right. A positively charged particle is a distance d above the wire, and it is moving in the same direction as the current. The particle will experience a magnetic force in which direction?

    (A) To the right
    (B) To the top of the page
    (C) To the left
    (D) To the bottom of the page

GO ON TO THE NEXT PAGE.

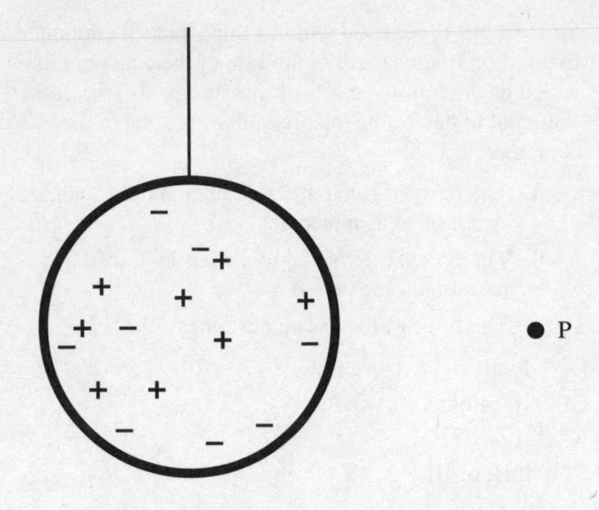

22. A neutral conducting sphere is hung from a thin insulating string. A positively charged object is brought to point P. The two objects are not allowed to touch. What is true about the string when the positively charged object is present at point P?

    (A) The tension is the same as before the object was present at point P.
    (B) The tension is greater than when the object was not at point P, and the string stretches to the left of its original orientation.
    (C) The tension is greater than when the object was not at point P, and the string stretches to the right of its original orientation.
    (D) The tension is less than when the object was not at point P, and the string stretches to the left of its original orientation.

23. An experiment is performed on a fixed volume of an ideal gas. The pressure, in pascals, of the gas is plotted on the vertical axis, and the temperature of the gas, in degrees Kelvin, is plotted on the horizontal axis. During a second performance of the experiment at a greater volume, the pressure-temperature gas is expected to

    (A) have a greater slope and the same intercept
    (B) have a smaller slope and the same intercept
    (C) have a greater slope and a greater intercept
    (D) have a smaller slope and a greater intercept

24. Two gas samples contain different gases. The first gas sample contains more massive molecules than the second. The molecules in both gas samples have the same average speed. When the samples are brought into contact, what is the flow of energy?

    (A) No energy will flow because the average speed of the particles in each gas is the same.
    (B) No energy will flow because the gases are made up of different molecules.
    (C) Energy will flow away from the first sample because its molecules are more massive.
    (D) Energy will flow away from the first sample because its molecules have more kinetic energy.

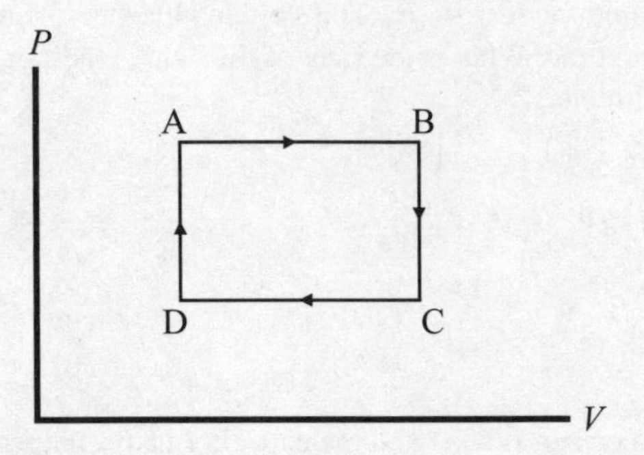

25. Which statement correctly characterizes the work done by the gas during the ABCDA cycle shown in the above *P*–*V* diagram?

    (A) There is no work done by the gas because the system both starts and concludes in state A.
    (B) There is no work done because the work done during the transition from A→B cancels out the work done in transition from C→D.
    (C) The work done by the gas is positive because the work done during the transition from A→B is greater than the work done in transition from C→D.
    (D) The work done by the gas is positive because the work done during the transition from B→C is greater than the work done in transition from D→A.

GO ON TO THE NEXT PAGE.

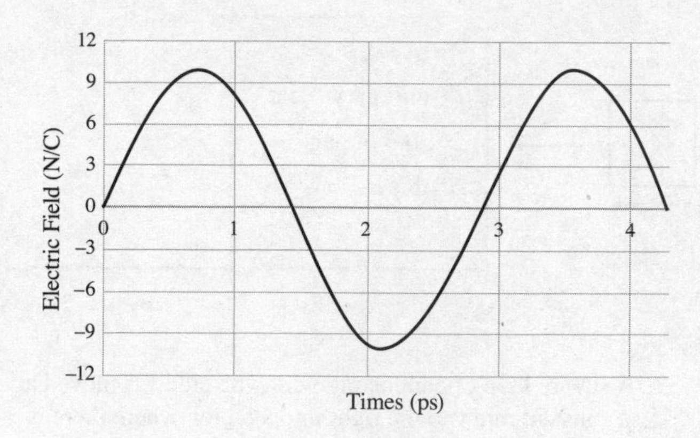

26. What is the equation for the electric field given by the above plot? The electric field crosses the axis at times 0 ps, 1.43 ps, 2.85 ps, and 4.28 ps.

(A) $E = \left(\frac{0\text{ N}}{\text{C}}\right)\sin((2.85\text{ THz})t)$

(B) $E = \left(\frac{10\text{ N}}{\text{C}}\right)\sin((2.85\text{ THz})t)$

(C) $E = \left(\frac{10\text{ N}}{\text{C}}\right)\sin((2.20\text{ THz})t)$

(D) $E = \left(\frac{10\text{ N}}{\text{C}}\right)\cos((0.45\text{ THz})t)$

27. A spherical balloon filled with helium is floating in air. If the balloon is inflated until its radius is doubled, how will the buoyant force on the balloon be affected?

(A) It will decrease by a factor of 4.
(B) It will increase by a factor of 4.
(C) It will increase by a factor of 8.
(D) It will not be affected.

28. Data is collected in an experiment performed on an ideal gas. In the experiment, temperature (in K) is the independent variable and volume (in m^3) is the dependent variable. If the data is graphed, which of the following is true about the slope and *y*-intercept of the graph?

(A) The slope will be directly proportional to the pressure of the gas and the intercept will be 0 m^3.
(B) The slope will be inversely proportional to the pressure of the gas and the intercept will be 0 m^3.
(C) The slope will be directly proportional to the pressure of the gas and the intercept will not be 0 m^3.
(D) The slope will be inversely proportional to the pressure of the gas and the intercept will not be 0 m^3.

29. Which of the following relationships, when plotted, will yield a curve which is inverse to the first power?

I. The electric potential versus distance from a positive point particle
II. The volume versus pressure for an ideal gas
III. The magnetic field from a current-carrying wire versus distance from the wire

(A) I only
(B) I and III
(C) II only
(D) I, II, or III

30. Two massive charged objects are fixed in space with a separation of *d* meters. The quantity *R* is defined as the ratio of the gravitational force to the electric force between the two objects. The separation is then increased slowly. What happens to *R*?

(A) *R* is constant for all separations.
(B) *R* increases as the separation increases.
(C) *R* decreases as the separation increases.
(D) *R* increases up to a distance *D*; then *R* decreases back to its original value.

31. An ideal gas is at a pressure *P* and a volume *V*. The gas is in a fixed volume, but is heated until the pressure doubles. What happens to the average speed of the molecules in the gas?

(A) The speed of the molecules on average remains unchanged.
(B) The speed of the molecules on average increases by a factor of $\sqrt{2}$.
(C) The speed of the molecules on average increases by a factor of 2.
(D) The speed of the molecules on average increases by a factor of 4.

32. Which of the following changes to a double-slit interference experiment would increase the widths of the fringes in the interference pattern that appears on the screen?

(A) Use light of a shorter wavelength.
(B) Move the screen closer to the slits.
(C) Move the slits closer together.
(D) Use light with a lower wave speed.

GO ON TO THE NEXT PAGE.

33. Tritium is an isotope of hydrogen consisting of one proton and two neutrons. The isotope has a mass of 5.008×10^{-27} kg. The mass of a proton is 1.673×10^{-27} kg and a neutron has a mass of 1.675×10^{-27} kg. What is the binding energy of tritium?

(A) 1.500×10^{-29} J
(B) 1.350×10^{-12} J
(C) 4.507×10^{-10} J
(D) 4.521×10^{-10} J

34. In an experiment designed to study the photoelectric effect, it is observed that low-intensity visible light of wavelength 550 nm produced no photoelectrons. Which of the following best describes what would occur if the intensity of this light were increased dramatically?

(A) Almost immediately, photoelectrons would be produced with a kinetic energy equal to the energy of the incident photons.
(B) Almost immediately, photoelectrons would be produced with a kinetic energy equal to the energy of the incident photons minus the work function of the metal.
(C) After several seconds, the electrons absorb sufficient energy from the incident light, and photoelectrons would be produced with a kinetic energy equal to the energy of the incident photons minus the work function of the metal.
(D) Nothing would happen.

35. Radioactive carbon-14 undergoes beta-minus decay. The atomic number of carbon is 6. The number of nucleons in the products after the beta decay is

(A) 6
(B) 7
(C) 13
(D) 14

36. A pipe carries water with a density of $\rho = 1{,}000$ kg/m^3. One end of the pipe has a diameter of 0.02 m. At that location, a pressure gauge says the pressure is $P = 1.25 \times 10^5$ Pa and the water is moving at a speed of 2.0 m/s. The other end of the pipe is 1.2 m higher than the end with the pressure gauge and has a diameter of 0.01 m. What will a pressure gauge installed in the upper end of the pipe read?

(A) 7.90×10^4 Pa
(B) 8.30×10^4 Pa
(C) 1.07×10^5 Pa
(D) 1.43×10^5 Pa

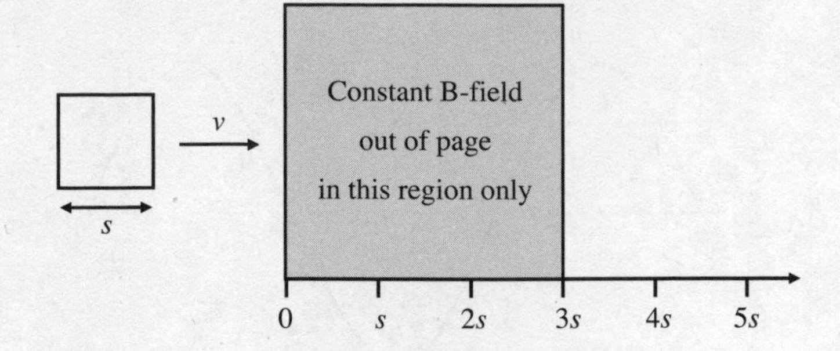

37. A square loop of conducting wire with side s is moved at a constant rate v to the right into a region where there is a constant magnetic field directed out of the page. Which of the following graphs shows the flux through the loop as a function of distance?

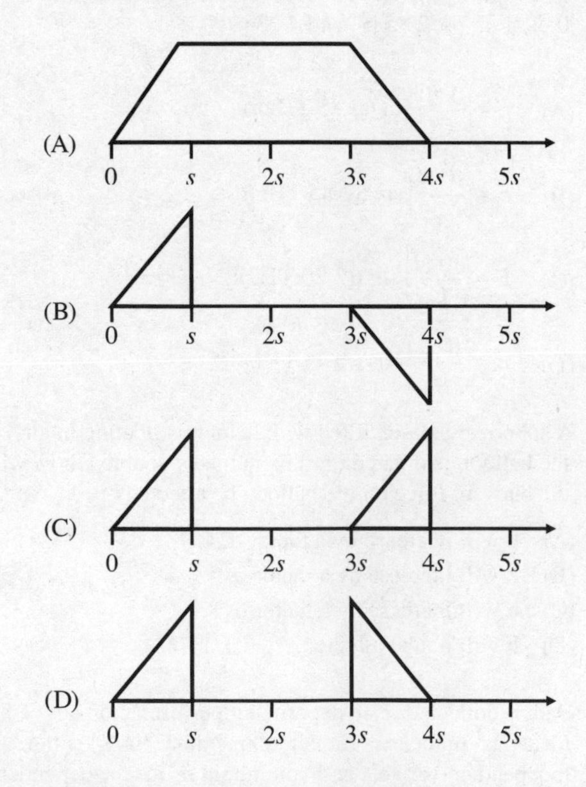

38. A hollow conducting sphere is placed around point P so that the center is at P. What happens to the electric field strength at point P?

(A) It doubles.
(B) It halves.
(C) It remains unchanged.
(D) It becomes zero.

GO ON TO THE NEXT PAGE.

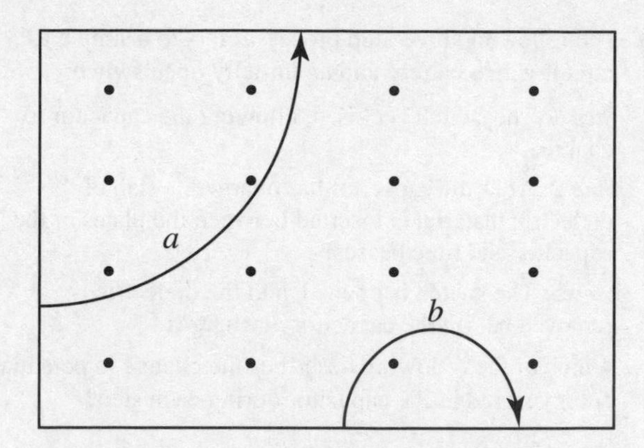

39. A machine shoots a proton, a neutron, or an electron into a magnetic field at various locations. The paths of two particles are shown above. Assume they are far enough apart so that they do not intersect and the magnetic field is going out of the page, as shown. What can you say about the paths that represent each particle?

    (A) *a* is the proton and *b* is the electron.
    (B) *b* is the proton and *a* is the electron.
    (C) Either may be a neutron.
    (D) You cannot make any conclusions without knowing the velocities.

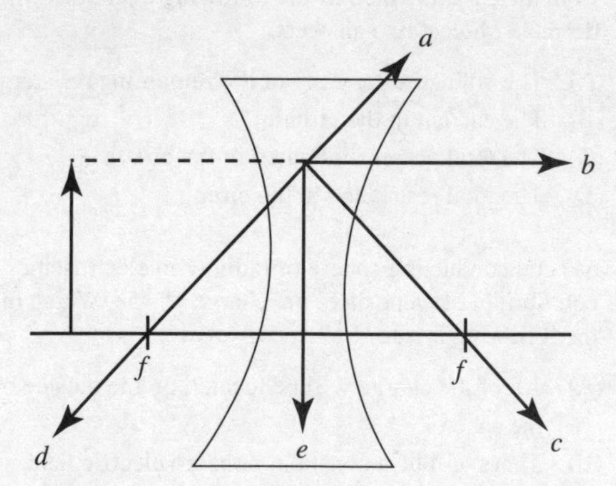

40. In the figure above, a ray of light hits an object and travels parallel to the principal axis as shown by the dotted line. Which line shows the correct continuation of the ray after it hits the concave lens?

    (A) *a*
    (B) *b*
    (C) *c*
    (D) *d*

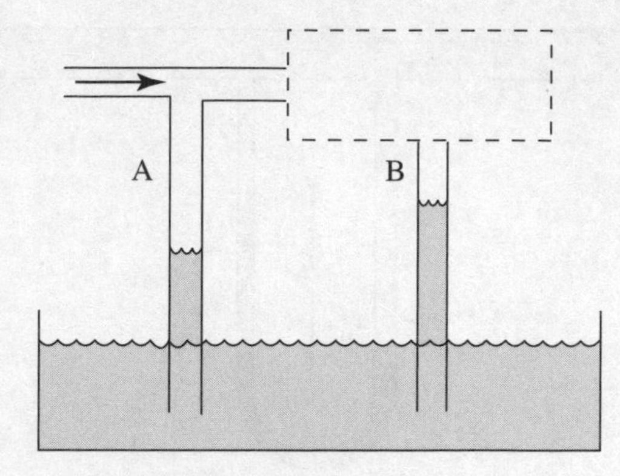

41. A tube with two T branches that has an open end is inserted in a liquid. However, the section of the tube above part B is hidden from view. The hidden section may be wider or narrower. Air is blown through the tube and the water levels change, as shown. You can conclude which of the following?

    (A) The picture as drawn is impossible—A and B must be at equal heights.
    (B) The tube is narrower and the air speed is greater above section B.
    (C) The tube is narrower and the air speed is less above section B.
    (D) The tube is wider and the air speed is greater above section B.

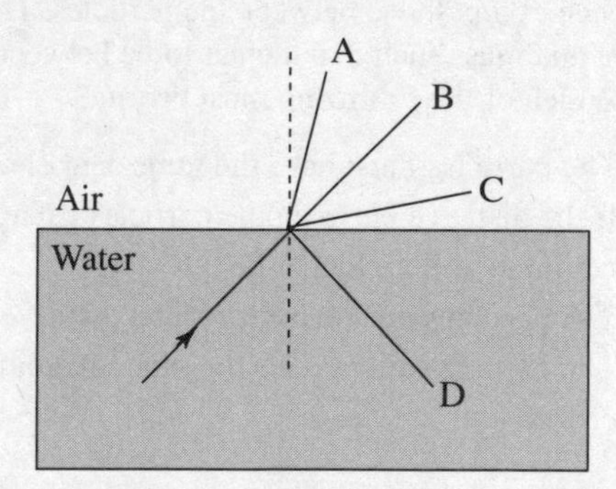

42. A beam of light goes from water to air. Depending on the actual angle that the light strikes the surface, which of the following is a possible outcome?

    (A) A only
    (B) B only
    (C) A or D
    (D) C or D

GO ON TO THE NEXT PAGE.

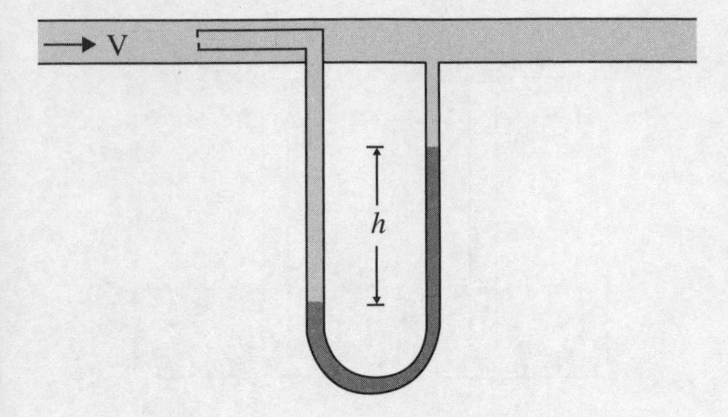

43. The instrument in an aircraft to measure airspeed is known as the pilot tube, shown in the figure above. The opening facing the incoming air (with the small aperture) is the part meant to capture the air at rest. The opening perpendicular to the flow of air (with the large aperture) is meant to capture air at speed. If $h = 1$ m and the liquid within the pilot tube is water, what is the airspeed? Take the density of air to be $\rho_{air} = 1.2$ kg/m^3.

(A) 27 m/s
(B) 68 m/s
(C) 95 m/s
(D) 128 m/s

44. Two charged, massive particles are isolated from all influence except those between the particles. They have charge and mass such that the net force between them is 0 N. Which of the following must be true?

(A) The particles must have the same sign of charge.
(B) If the distance between the particles changes, the net force will no longer be 0 N.
(C) The particles must have the same mass.
(D) The particle cannot have the same magnitude of charge.

45. The following three-step process refers to a simple RC circuit with a battery and an initially open switch.

Step 1: The switch is closed, allowing the capacitor to charge.

Step 2: After the capacitor has charged, a slab of dielectric material is inserted between the plates of the capacitor and time passes.

Step 3: The switch is opened, and the dielectric is removed before the capacitor discharges.

Which of the following describes the change in potential energy stored in the capacitor during each step?

(A) $\Delta U_1 < 0; \Delta U_2 > 0; \Delta U_3 > 0$
(B) $\Delta U_1 > 0; \Delta U_2 > 0; \Delta U_3 > 0$
(C) $\Delta U_1 < 0; \Delta U_2 > 0; \Delta U_3 < 0$
(D) $\Delta U_1 > 0; \Delta U_2 < 0; \Delta U_3 < 0$

Directions: For questions 46–50 below, two of the suggested answers will be correct. Select the two answers that are best in each case, and then fill in both of the corresponding circles on the answer sheet.

46. N resistors ($N > 2$) are connected in parallel with a battery of voltage V_0. If one of the resistors is removed from the circuit, which of the following quantities will decrease? Select two answers.

(A) The voltage across any of the remaining resistors
(B) The current in the circuit
(C) The total power dissipated in the circuit
(D) The total resistance in the circuit

47. A metal conducting sphere of radius r in electrostatic equilibrium has a positive net charge of $+5e$. Which of the following is true? Select two answers.

(A) All of the charge will be located on the outside of the sphere.
(B) There will be a constant, nonzero electric field within the sphere.
(C) There will be a constant, nonzero electric potential within the sphere.
(D) When the sphere is connected to a ground, the $+5e$ of charge on the sphere will flow into the ground to neutralize the sphere.

GO ON TO THE NEXT PAGE.

48. A circuit is created with a battery of negligible internal resistance and three identical resistors with resistance of 10 Ω. The resistors are originally arranged so that one is in series with the battery, and the other two are in parallel with each other. Which of the following changes to the circuit will result in an increase in the amount of current in the circuit? Select two answers.

(A) Rearranging the resistors so that all three are in parallel

(B) Replacing every resistor with a resistor of half the resistance

(C) Removing one of the two parallel branches entirely from the circuit

(D) Replacing the battery with a battery with half the voltage

49. An ideal gas is in state 1, with P_1, V_1, and T_1. The final volume will be the same as V_1 for which of the following processes? Select two answers.

(A) Triple P_1 and decrease T_1 by $\frac{1}{3}$

(B) Triple T_1 and decrease P_1 by $\frac{1}{3}$

(C) Quadruple P_1 and T_1

(D) Decrease P_1 by $\frac{1}{3}$ and decrease T_1 by $\frac{1}{3}$

50. In a double-slit experiment, students are attempting to increase the spacing of the fringes observed on the screen. Which modifications to the setup will result in increased fringe separation? Select two answers.

(A) Doubling the wavelength only

(B) Doubling the wavelength and doubling the slit separation

(C) Doubling the distance to the screen only

(D) Doubling the distance to the screen and doubling the slit separation

END OF SECTION I

PHYSICS 2
SECTION II
Time—90 minutes
4 Questions

Directions: Questions 1 and 2 are long free-response questions that require about 25 minutes each to answer. Questions 3 and 4 are short free-response questions that require about 20 minutes each to answer. On test day, you will be asked to show your work for each part in the space provided after that part. For this practice test, you may use scrap paper.

1. An unplugged freezer is at room temperature. The door is closed and the freezer is plugged in.

(A) Your friend observes, "When you open the freezer, the cold air comes out of the freezer. That didn't happen before the freezer was cooled down. It must therefore be the case that cold air is less dense than warm air, since the low density gas will expand more easily."

i. Is there any part of your friend's statement you agree with? Why?

ii. Is there any part of your friend's statement you disagree with? Why?

(B) You set out to perform an experimental investigation of the relationship between gas density and temperature. The following equipment is available. Place an X beside each item you will need to use:

	Chamber with sliding piston lid with a known area		Chamber with fixed lid with a known area
	Pressure monitor		Thermometer
	Meter stick		Stopwatch
	Large tub of boiling water		Large tub of ice water

(C) Write out a numbered procedure you will use to gather necessary data. Your description should be detailed enough that another student could reproduce your experiment.

(D) Your data analysis must include a graph. Explain what you would graph on the *x*-axis and the *y*-axis. Justify your decision and explain how your graph will help you understand the relationship between density of a gas and its temperature.

(E) What shape graph do you expect to see?

GO ON TO THE NEXT PAGE.

2. The figure below shows an electric circuit containing a source of emf, (ε), a variable resistor (r), and a resistor of fixed resistance (R). The resistor R is immersed in a sealed beaker containing a mass m of water, currently at temperature T_i. When the switch S is closed, current through the circuit causes the resistor in the water to dissipate heat, which is absorbed by the water. A stirrer at the bottom of the beaker simply ensures that the temperature is uniform throughout the water at any given moment. The apparatus is well-insulated (insulation not shown), and it may be assumed that no heat is lost to the walls or lid of the beaker or to the stirrer.

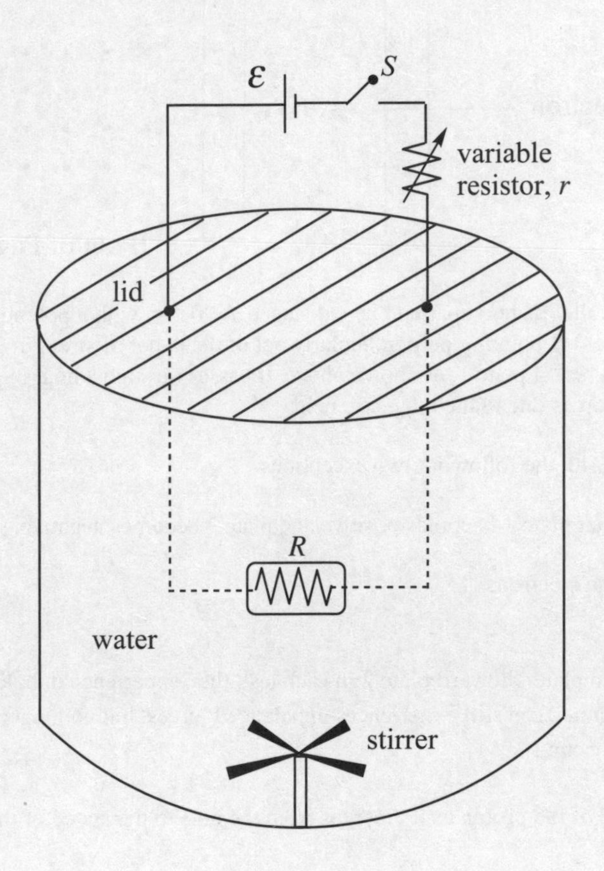

(A) Determine the current in the circuit once S is closed. Write your answer in terms of ε, r, and R.

(B) Determine the power dissipated by the resistor R in terms of ε, r, and R.

(C) Explain at the microscopic level why the water heats up when the switch is closed and how the stirrer helps ensure a constant temperature throughout.

(D) Assume the stirrer has a knob, which changes its speed. How can the temperature of the water be increased more rapidly by adjusting the rotation rate of the stirrer?

(E) As the temperature of the water increases, whether from the resistor or from the stirrer rate, explain the microscopic interactions responsible for the changing pressure in the container.

GO ON TO THE NEXT PAGE.

3. In an experiment, two tests are run. In both trials you may ignore the effect of gravity. The following is a diagram of the tests.

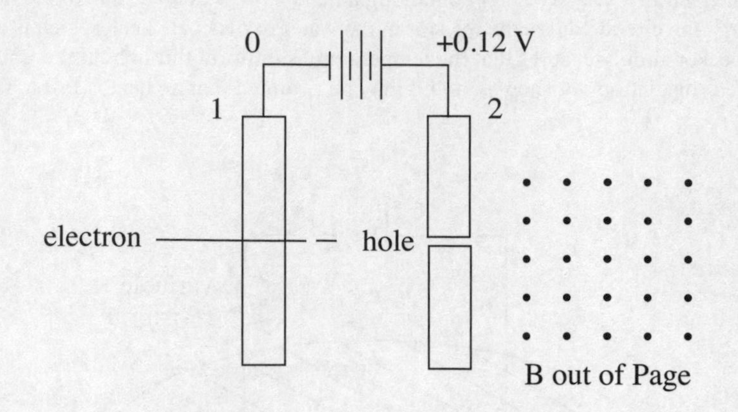

Test 1: There are two large parallel plates separated by a distance $d = 0.5$ m with a potential difference of 0.12 V across them. There is a uniform magnetic field **B** pointing perpendicularly out of the paper of strength 0.002 T starting to the right of plate 2. An electron is released from rest at plate 1, as shown above. It passes through a hole in plate 2 and enters the magnetic field, and it only experiences forces due to the magnetic field.

Test 2: The same setup is run with the following two exceptions.

1. The battery is switched so that plate 1 becomes positive and plate 2 becomes negative.

2. A proton is used instead of an electron.

(A) As the particles move from plate 1 toward plate 2 in each test, they experience unbalanced forces causing their speed to change. After plate 2, each particle still experiences unbalanced forces, but no longer changes speed. Use the concept of work to explain how this occurs.

(B) Find the ratio of the speed of the proton as it emerges from the hole to the speed of the electron as it emerges from the hole.

(C) Make a sketch of each path each particle will follow after emerging from the hole in plate 2.

(D) A third test is conducted similar to test 2, except an alpha particle is used instead of a proton. Explain how the path of the alpha particle after emerging from the hole will differ from the proton's path.

GO ON TO THE NEXT PAGE.

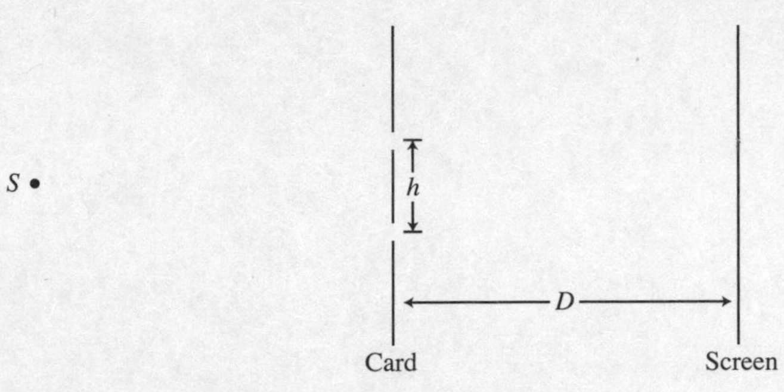

Note: Figure not drawn to scale.

4. In a double-slit interference experiment, a parallel beam of monochromatic light is needed to illuminate two narrow parallel slits of width w that are a distance h apart, where h >> w, in an opaque card as shown in the figure above. The interference pattern is formed on a screen a distance D from the slits, where D >> h.

   (A) Draw the first three wavefronts that emerge from each slit after the card.

   (B) In a clear, coherent, paragraph-length response, explain why a series of bright spots are seen on the screen after the light shines through the card.

   (C) In the interference patterns on the screen, the distance from the central bright fringe to the third bright fringe on one side is measured to be y_3. In terms of D, h, y_3, and S, what is the wavelength of the light emitted from the source S?

   (D) If the space between the slits and the screen was filled with a material having an index of refraction $n > 1$, would the distance between the bright fringes increase, decrease, or remain the same? Explain your reasoning.

STOP

END OF EXAM

Practice Test 2: Diagnostic Answer Key and Explanations

PRACTICE TEST 2: DIAGNOSTIC ANSWER KEY

Let's take a look at how you did on Practice Test 2. Follow the three-step process in the diagnostic answer key below and go read the explanations for any questions you got wrong, or you struggled with but got correct. Be sure to compare your scores on Practice Test 2 with your scores on Practice Test 1.

STEP 1 » Check your answers and mark any correct answers with a ✔ in the appropriate column.

Section I: Multiple Choice							
Q #	Ans.	✔	Chapter #, Section Title	Q #	Ans.	✔	Chapter #, Section Title
1	D		**5,** The Electric Field	22	C		**5,** Conductors and Insulators
2	A		**6,** Equipotential Curves and Equipotential Maps	23	B		**4,** The Ideal Gas Law
3	D		**3,** The Continuity Equation: Flow Rate and Conservation of Mass	24	D		**4,** The Ideal Gas Law
4	C		**3,** Pressure	25	C		**4,** The First Law of Thermodynamics
5	A		**7,** Resistors and Resistance	26	C		**5,** The Electric Field
6	A		**4,** The First Law of Thermodynamics	27	C		**3,** Buoyancy
7	B		**9,** Ray Tracing for Lenses	28	B		**4,** The Ideal Gas Law
8	B		**9,** Using Equations To Answer Questions About the Image	29	D		**4,** The Ideal Gas Law **6,** Electric Potential from a Point Charge **8,** Magnetic Field Created by Current-Carrying Wires
9	C		**9,** Using Equations To Answer Questions About the Image	30	A		**5,** Coulomb's Law
10	C		**10,** The Bohr Model of the Atom	31	B		**4,** The Ideal Gas Law
11	A		**10,** Nuclear Reactions	32	C		**9,** Young's Double-Slit Interference Experiment
12	D		**4,** The Ideal Gas Law	33	B		**10,** Binding Energy
13	B		**6,** The Energy Stored in a Capacitor	34	D		**10,** Photons and The Photoelectric Effect
14	A		**5,** Coulomb's Law	35	D		**10,** Beta Decay
15	A		**6,** Electric Potential	36	B		**3,** Bernoulli's Equation: Conservation of Energy in Fluids
16	A		**7,** RC Circuits with Capacitors in Steady State	37	A		**8,** Faraday's Law of Electromagnetic Induction
17	D		**7,** RC Circuits with Capacitors in Steady State	38	D		**5,** Conductors and Insulators
18	D		**7,** RC Circuits with Capacitors in Steady State	39	B		**8,** The Magnetic Force on a Moving Charge
19	D		**4,** The Ideal Gas Law	40	A		**9,** Ray Tracing for Lenses
20	B		**6,** Electric Field and Capacitors	41	B		**3,** The Bernoulli Effect
21	D		**8,** Magnetic Fields Created by Current-Carrying Wires	42	D		**9,** Total Internal Reflection

Section I: Multiple Choice—Continued							
Q #	Ans.	✔	Chapter #, Section Title	Q #	Ans.	✔	Chapter #, Section Title
43	**D**		**3,** Bernoulli's Equation: Conservation of Energy in Fluids	47	**A, C**		**5,** Conductors and Insulators
44	**A**		**5,** Coulomb's Law	48	**A, B**		**7,** Resistors in Parallel
45	**B**		**7,** Altering the Capacitance of a Capacitor	49	**C, D**		**4,** The Ideal Gas Law
46	**B, C**		**7,** Resistors in Parallel	50	**A, C**		**9,** Young's Double-Slit Interference Experiment

Section II: Free Response			
Q #	Ans.	✔	Chapter #, Section Title
1(A)	See Explanation*		**4,** The Second Law of Thermodynamics
1(B) 1(C) 1(D) 1(E)	See Explanation*		**4,** The Ideal Gas Law
2(A)	See Explanation*		**7,** Ohm's Law
2(B)	See Explanation*		**7,** Power Dissipation
2(C) 2(D)	See Explanation*		**4,** The Second Law of Thermodynamics
2(E)	See Explanation*		**4,** The Kinetic Theory of Gases
3	See Explanation*		**6,** The Electric Potential of a Uniform Field **8,** The Magnetic Force on a Moving Charge
4	See Explanation*		**9,** Young's Double-Slit Interference Experiment

**Explanations begin on page 345.*

Tally your correct answers from Step 1 by chapter. For each chapter, write the number of correct answers in the appropriate box. Then, divide your correct answers by the number of total questions (which we've provided) to get your percent correct.

CHAPTER 3 TEST SCORE SELF-EVALUATION

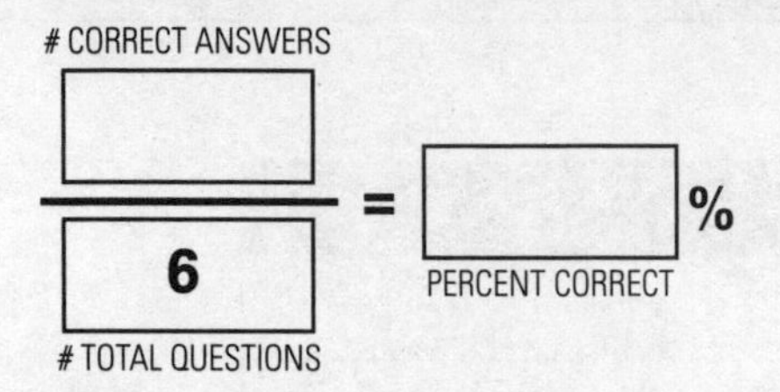

CHAPTER 4 TEST SCORE SELF-EVALUATION

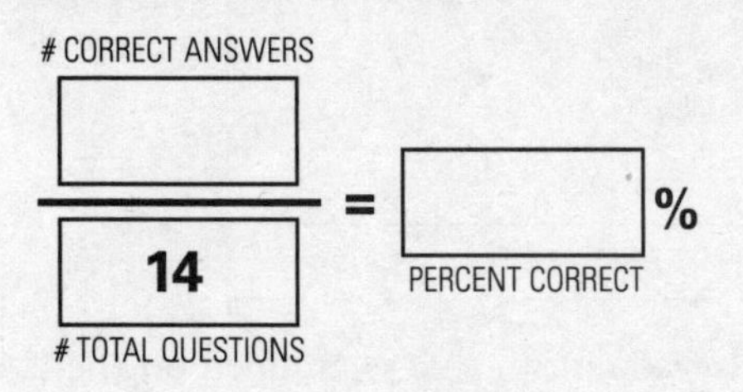

CHAPTER 5 TEST SCORE SELF-EVALUATION

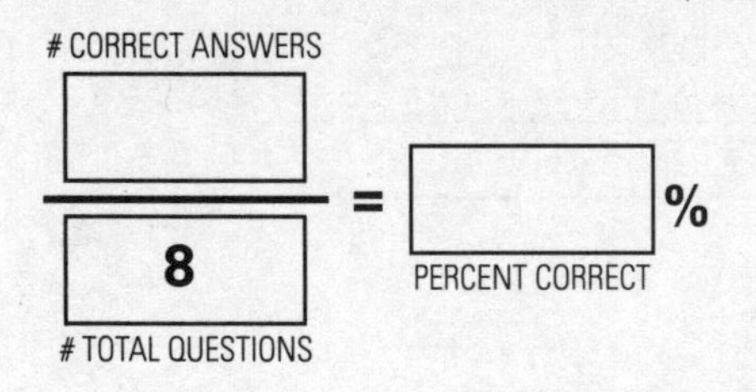

CHAPTER 6 TEST SCORE SELF-EVALUATION

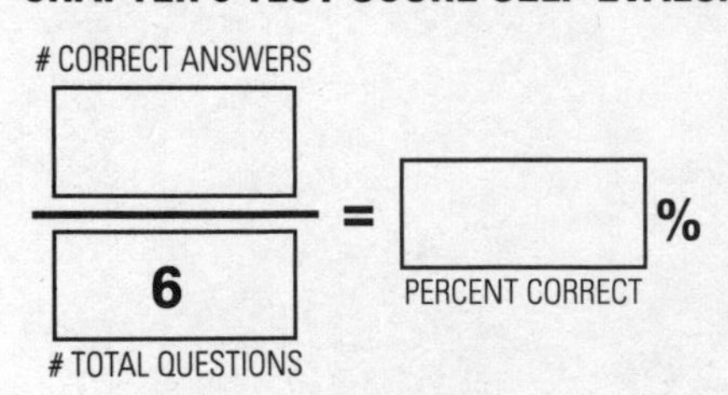

CHAPTER 7 TEST SCORE SELF-EVALUATION

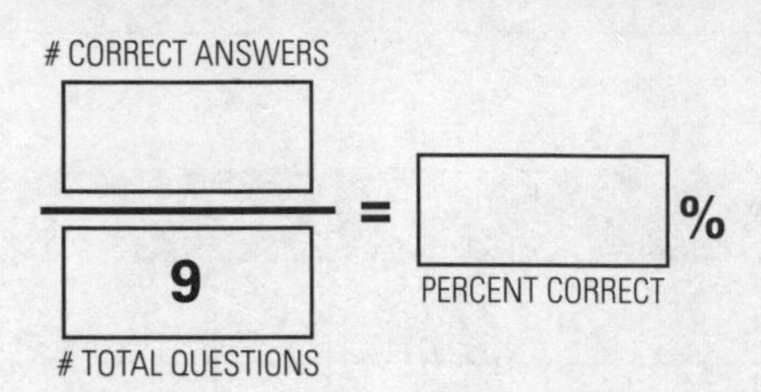

CHAPTER 8 TEST SCORE SELF-EVALUATION

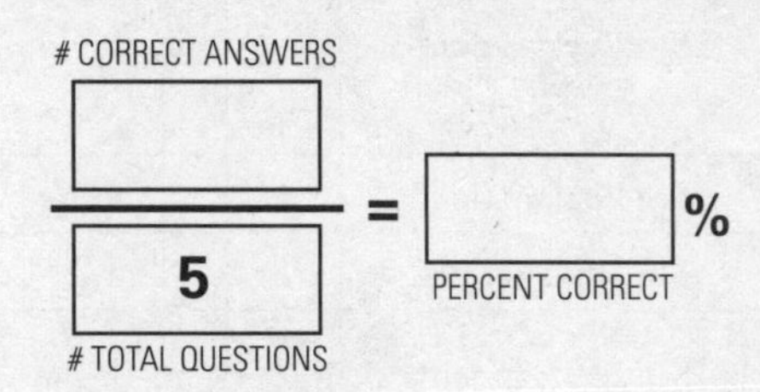

CHAPTER 9 TEST SCORE SELF-EVALUATION

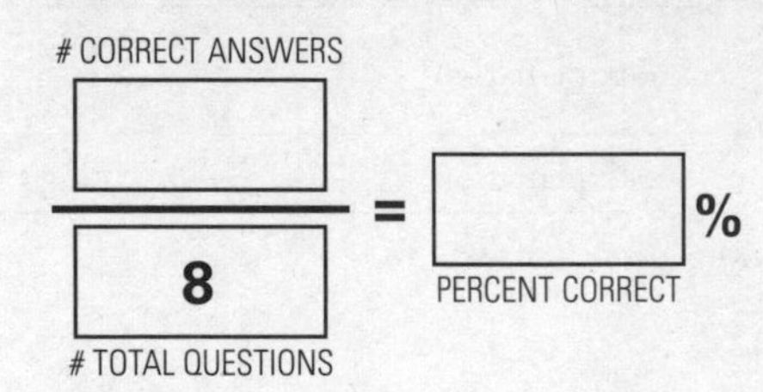

CHAPTER 10 TEST SCORE SELF-EVALUATION

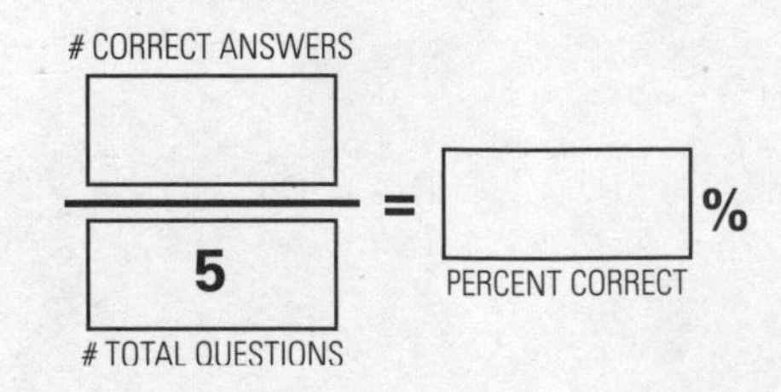

Use the results above to customize your study plan. You may want to start with, or give more attention to, the chapters with the lowest percents correct.

PRACTICE TEST 2: ANSWERS AND EXPLANATIONS

Section I: Multiple Choice

1. **D** The forces must balance in order for the proton to travel along a straight line. If the forces balance, $m(E_g) = q(E_{el})$. Since m is less than q for a proton, the gravitation field is stronger. Gravity points down, so the electric force must be up, and for a positive particle, the field must also be up.

2. **A** Equipotential lines will be perpendicular to the force lines. Gravitational force lines point vertically downward. The equipotential lines for the electric field must also be horizontal lines for the particle to move in a straight line at a constant speed.

3. **D** The Continuity Equation, $A_1v_1 = A_2v_2$ applies here as $A_1v_1 = 2A_2v_2$ where position 1 is in the large pipe. Solving for $v_1 = \dfrac{2\pi\left(\frac{d}{2}\right)^2 v}{\pi\left(\frac{D}{2}\right)^2} = \dfrac{2\pi\left(\frac{d^2}{4}\right)v}{\pi\left(\frac{D^2}{4}\right)} = \dfrac{2d^2v}{D^2}$

4. **C** The fluid pressure is the force divided by the area. The force is the same regardless of the container orientation, as the weight of the container is constant. To result in a minimum pressure, the largest face must be placed against the table.

5. **A** For any linear equation, $y = mx + b$, the slope is multiplied by the quantity that is graphed on the horizontal axis. In order to have the slope equal the resistivity, the horizontal axis must be the ratio of length to area.

   $R = \rho\dfrac{L}{A}$

6. **A** All processes that end at higher volumes will require removing work to be done on the system. When pressure is constant, the work is found from $W = P\Delta V$, so to minimize the work required, you want the change in volume to occur at the lowest pressure possible.

7. **B** A ray that travels into a converging lens along a line that goes through the focal point will refract to travel along a line parallel to the optical axis upon leaving the lens.

8. **B** The screen must be placed at the image location in order for the image to appear in focus on the screen. Use the equation $\dfrac{1}{f} = \dfrac{1}{s_o} + \dfrac{1}{s_i}$

   $$\frac{1}{s_i} = \frac{1}{25\text{ cm}} - \frac{1}{65\text{ cm}} \rightarrow s_i = 40.6\text{ cm}$$

9. **C** A concave mirror can make a real image only when the object is placed beyond the focus.

10. **C** Any photon that has energy that matches an energy difference in that atom can be absorbed. The 10 eV photon can be absorbed if there is an electron in the –10 eV state by ionizing the atom. A 5 eV photon can excite a photon from the –10 eV state to the –5 eV state or ionize an electron from the –5 eV state. The 1.5 eV photon can cause a transition from the –3.5 eV state to the –2 eV state.

11. **A** During nuclear decay processes, the total charge in the system before and after the decay must be the same.

12. **D** The internal energy is proportional to the temperature of the system, and the Ideal Gas Law explains that the product of pressure and volume is also proportional to the temperature. State A is at 5PV, State B is at 4PV, State C is at 16PV, and State D is at 20PV. The change from State D to State A requires the largest change in energy.

13. **B** The internal energy stored in a fully charged capacitor is $U_C = \frac{1}{2}CV^2$. The voltage of the battery will be the voltage across the capacitor when the capacitor is fully charged.

$$U_C = \frac{1}{2}(25 \times 10^{-6}\text{ F})(50\text{ V})^2 = 0.03125\text{ J}$$

14. **A** The acceleration of an object with charge $+q$ and mass m is given by $a = \frac{F_E}{m} = \frac{1}{4\pi\varepsilon_0}\left(\frac{q_1q_2}{mr^2}\right)$. Graph (A) shows an inverse square law.

15. **A** There is a change in electrical potential and no change in gravitational potential. The work is done because of the change in the potential. Therefore, the Work-Energy Theorem, $\Delta W_{net} = \Delta KE$, can be used to find the change in kinetic energy. Finally, using $\Delta KE = \frac{1}{2}mv_f^2 - \frac{1}{2}mv_i^2$ and the known initial speed and mass, the final speed can be determined.

16. **A** The capacitor will be fully charged a long time after the switch has been closed. So, no current will pass down the branch with the capacitor, and the circuit will behave like a battery connected to two resistors in series. Therefore, the voltage of the battery will be split between R_1 and R_2. Since $R_1 > R_2$, the voltage drop across R_1 will be greater than that across R_2. Finally, the capacitor is in parallel with R_2, so the capacitor will have the same voltage as R_2.

17. **D** You can eliminate (A) immediately. The current that flows into the capacitor changes over time, so the current in a parallel branch will also change over time. You can eliminate (B) as well, because before the switch is closed, no current flows in the circuit at all. Because the capacitor is discharged, as soon as the switch is closed, the parallel branch with the capacitor will receive all of the current, eliminating (C). In both the cases of (B) and (C), the current through R_2 is zero. Once the capacitor is fully charged, current flows around the loop creating the outside of the circuit, and whatever value it has at that time is greater than 0 A.

18. **D** In both scenarios, the capacitor will be fully charged a long time after the switch has been closed. So, no current will pass down the branch with the capacitor, and the circuit will behave like a battery connected to two resistors in series. Therefore, you can eliminate (A), (B), and (C). The charged stored in the capacitor is given by $Q = CV$. Since the voltage of the capacitor in both scenarios will be the same, matching that of resistor R_2, a greater capacitance will result in a greater stored charge in the capacitor.

19. **D** One way to find the pressure is with the Ideal Gas Law, which states that $PV = nRT$, so in addition to the volume, found from the length, one needs n and T to find the pressure, so I is correct. The impulse and time allow you to calculate the force applied against the gas in that time. Newton's Third Law states that this must be the same magnitude as the force applied to the wall. The force and area allow you to calculate the pressure. So II and III are also correct.

20. **B** A fully charged parallel-plate capacitor will have a uniform, nonzero electric field between its plates.

21. **D** Use the right-hand rule to find the magnetic field produced by the wire. The current is to the right, so at a position above the wire, the magnetic field points outward. Then use the right-hand rule again to find the force on the particle. The particle moves to the right in a magnetic field pointed outward. The force is toward the bottom.

22. **C** The positive charge at point P will draw negative charges toward the right side of the neutral sphere, resulting in an attraction between the two objects. This will cause the string to be pulled to the right. The vertical component of tension will still balance the weight of the neutral sphere, but now there will be a horizontal component of tension balancing the electrostatic attraction.

23. **B** The Ideal Gas Law is $PV = nRT$, or $P = (\frac{nR}{V})T$. Comparing this to $y = mx + b$, the intercept is 0 and the slope is related to $1/V$.

24. **D** Energy will flow from the hotter gas to the cooler gas. The speeds are the same, but the masses are different, so the more massive gas has more kinetic energy, which is proportional to temperature.

25. **C** Work is done when there is a change in volume. At constant pressure, the equation $W = -P\Delta V$ indicates that when P is higher (as it is at path A→B), a greater amount of negative work is done on the gas than at a lower P (as it is at C→D). Furthermore, no work is done along the constant volume paths. Thus, the work done by the gas during the entire cycle is positive.

26. **C** The field is a sine curve with an amplitude of 10 N/C. From the graph, the period, T, of the wave is 2.85 ps. The argument of sine needs to be $2\pi t/T = (2 \times 3.14/2.85 \text{ ps})t = (2.20\text{THz})t$.

27. **C** The buoyant force is found from $F_{buoy} = \rho_{fluid} V_{sub} g$. Because the entire balloon is surrounded by air, the entire volume is V_{sub}. The volume of the spherical balloon is $V = \frac{4}{3}\pi r^3$. Doubling the radius increases the volume by a factor of 8.

28. **B** The Ideal Gas Law equation is $PV = nRT$. Solving this for V (since volume is on the vertical axis of the graph described in the problem) yields $V = (\frac{nR}{P})T$. Comparing this to the equation for a line, $y = mx + b$, the slope of the line will be nR/P, which is inversely proportional to P, and $b = 0$ is the intercept.

29. **D** The Ideal Gas Law is $P = (nRT)\left(\frac{1}{V}\right)$.

The potential from a point particle is $V = \frac{q_1}{4\pi\varepsilon_0}\left(\frac{1}{r}\right)$.

The magnetic field from a wire is $B = \frac{\mu_0 I}{2\pi}\left(\frac{1}{r}\right)$.

All of these are inverse to the first power.

30. **A** The quantity R is defined as the ratio of the gravitational force to the electric force between the two objects, so $R = \left(\frac{GMm}{r^2}\right) / \left(\frac{1}{4\pi\varepsilon_0}\left(\frac{Qq}{r^2}\right)\right)$

$$R = GMm \,/ \left(\frac{1}{4\pi\varepsilon_0}Qq\right)$$

R is constant.

31. **B** Doubling pressure at constant volume causes temperature to double. Doubling temperature doubles the average kinetic energy of the molecules. The speed squared is proportional to the kinetic energy, so the speed increases by a factor of $\sqrt{2}$.

32. **C** Relative to the central maximum, the locations of the bright fringes on the screen are given by the expression $m\left(\frac{\lambda L}{d}\right)$, where λ is the wavelength, L is the distance to the screen, d is the slit separation, and m is any integer. The width of a fringe is therefore $(m + 1)\left(\frac{\lambda L}{d}\right) - m\left(\frac{\lambda L}{d}\right) = \left(\frac{\lambda L}{d}\right)$. The slit spacing will increase if there is a decrease in d.

33. **B** The binding energy is the mass defect, the difference in the mass of the constituent components and the mass of the nucleus, multiplied by c^2. That mass defect is

$m_p + 2m_n - m_{Tr} = 1.673 \times 10^{-27} + 2 * 1.675 \times 10^{-27} - 5.008 \times 10^{-27} = 1.500 \times 10^{-29}$ kg

$E = \Delta mc^2 = 1.500 \times 10^{-29} * (3 \times 10^8)^2 = 1.35 \times 10^{-12}$ J

34. **D** If the photons of the incident light have insufficient energy to liberate electrons from the metal's surface, then simply increasing the number of these weak photons (that is, increasing the intensity of the light) will do nothing. In order to produce photoelectrons, each photon of the incident light must have an energy at least as great as the work function of the metal.

35. **D** Beta-minus decay occurs when a neutron turns into a proton and a beta particle. The number of nucleons does not change in this process.

36. **B** Use the Continuity Equation to find the speeds.

$$A_1v_1 = A_2v_2$$

$$\pi\left(\frac{0.02 \text{ m}}{2}\right)^2(2.0 \text{ m/s}) = \pi\left(\frac{0.01 \text{ m}}{2}\right)^2 v_2$$

$$v_2 = 8.0 \text{ m/s}$$

Then use Bernoulli's Equation

$$P_A + \frac{1}{2}\rho_{water}v_A^2 + \rho_{water}gy_A = P_B + \frac{1}{2}\rho_{water}v_B^2 + \rho_{water}gy_B$$

$$1.25 \times 10^5 + \frac{1}{2}1{,}000(2.0)^2 + 1{,}000(10)(0) = P_B + \frac{1}{2}1{,}000(8.0)^2 + 1{,}000(10)(1.2)$$

$$P_B = 83{,}000 \text{ Pa}$$

37. **A** Flux is the product of the area enclosed by the loop of wire and the field strength. The flux will increase linearly up to a maximum value when the wire is completely within the field, and it will begin to decrease once the wire begins to leave the region within the field.

38. **D** Within a conducting sphere, the electric field is always zero. Therefore, in the region of P, as well as everywhere else within the conducting sphere, there is zero electric field.

39. **B** Path *b* follows the right-hand rule, so it must be positively charged. Path *a* is opposite the right-hand rule, so it must be negatively charged. Neutrons would follow a straight line path, so it is impossible for either *a* or *b* to be a neutron.

40. **A** The rules for ray-tracing diagrams for diverging lenses state that a line parallel to the principal axis bends away from the focal point, as shown.

41. **B** Because the fluid is higher in column B, the pressure above column B must be less than that above column A. According to the Bernoulli Effect, the drop in pressure is due to an increase in air flow speed above column B. This must be due to a narrower tube, according to Continuity Equation.

42. **D** Water has a higher index of refraction than air, so if light refracts through the water, the light will bend away from the normal—so C is possible. Another possibility is D, total internal reflection when the angle in water is greater than the critical angle.

43. **D** First, use Bernoulli's Equation. Call the point with the air at rest *B* and the point with the speeding air *A*.

$$P_A + \frac{1}{2}\rho_{air}v_A^2 + \rho_{air}gy_A = P_B + \frac{1}{2}\rho_{air}v_B^2 + \rho_{air}gy_B$$

You have $v_B = 0$ and $y_A = y_B$, so the equation reduces to

$$P_A + \frac{1}{2}\rho_{air}v_A^2 = P_B$$

The pressure difference, $P_B - P_A$, gives rise to a fluid depth $\rho_f gh$.

$$\rho_f gh = \frac{1}{2}\rho_{air}v_A^2$$

Solve for v_A and use the given values:

$$v_A = \sqrt{\frac{2\rho_f gh}{\rho_{air}}}$$

$$v_A = \sqrt{\frac{2 * 1{,}000\text{ kg/m}^3 * 9.8/\text{s}^2 * 1\text{ m}}{1.2\text{ kg/m}^3}}$$

$$v_A = 128\text{ m/s}$$

44. **A** It is possible for the gravitational force to balance the electrical force, making (D) false. In order for this to occur, since gravity is always an attractive force, the electrical force must be repulsive, and therefore the charges must have the same sign, making (A) true. Both the gravitational and electrical force are inverse square laws, so if the distance between the particles changes, both forces change by the same factor, making (B) false. Finally, (C) is false because both particles will experience the same gravitational force regardless of how the mass is distributed between them; the gravitational force depends on the product of the two masses.

45. **B** *Step 1:* Potential energy rises because the voltage across the plates, *V*, increases while the capacitance, *C*, is constant and $U = \frac{1}{2}CV^2$. Since $\Delta U_1 > 0$, eliminate (A) and (C).

Step 2: Inserting a dielectric increases potential energy for a given voltage, since inserting a dielectric always increases capacitance and $U = \frac{1}{2}CV^2$. $\Delta U_2 > 0$, so eliminate (D).

Step 3: At the moment the switch is opened, the voltage source is effectively removed, and charge on the capacitor will be constant. Removing the dielectric will decrease the capacitance and increase the potential energy stored in the capacitor according to $U = \frac{Q^2}{2C}$. A moment later, of course, the capacitor will begin to discharge, and the potential energy will decrease; however, *Step 3* only considers what happens before discharging occurs.

46. **B, C** The resistors are all in parallel, meaning their voltage is the same across each resistor and matches the voltage of the battery, which is constant. Eliminate (A). The resistors are in parallel, so eliminating one of the resistors will increase the equivalent resistance. Eliminate (D). Because $I = V/R$, this increase in resistance will decrease the current, making (B) true. Power's equation is $P = IV$. If the current decreases, the power decreases as well. If the current in the circuit decreases, the power in the circuit decreases as well, making (C) true.

47. **A, C** Conductors will always have all of their excess charge located at the surface, making (A) true. The entire volume of a conductor must be an equipotential in electrostatics because otherwise the charges in the conductor would move as a result of the potential difference, making (C) true. Choice (B) is false because within a conductor in electrostatic equilibrium, there is an electric field of 0 N/C. Choice (D) is false because, while the sphere will neutralize when connected to a ground, it is not the positive charges that flow out of the sphere, but negative charges that will flow from the ground into the sphere.

48. **A, B** From Ohm's Law, the current in the circuit will increase when the total resistance of the circuit decreases, or when the total voltage supplied to the circuit increases, which makes (D) false. Resistors in parallel have a lower equivalent resistance than those in series, so (A) is true and (C) is false. To see that (B) decreases the overall resistance, think of a simpler version of the problem. Replacing a resistor with a wire will have the same effect of either increasing or decreasing the overall resistance, but to a larger degree. If any of the three resistors is replaced with a wire, the overall resistance decreases. Therefore, decreasing the resistance of any of the resistors will decrease the equivalent resistance, and replacing all of them most certainly will decrease the equivalent resistance.

49. **C, D** The Ideal Gas Law states $PV = nRT$, so pressure is directly proportional to temperature. To keep volume constant, pressure and temperature must increase or decrease by the same amount.

50. **A, C** For the double-slit experiment, $m\lambda = dsin(\theta) \rightarrow m\lambda = dx/L$.

    The spacing of the fringes, x, increases for increasing λ and L and decreases for increasing d.

Section II: Free Response

1. (A) (i) I agree with the statement "When you open the freezer, the cold air comes out of the freezer. That didn't happen before the freezer was cooled down." The reason is that energy flows, according to the Second Law of Thermodynamics, due to a difference in temperature. When the freezer wasn't cooled down, there was no temperature difference.

(ii) I disagree with the statement "cold air is less dense than warm air, since the low-density gas will expand more easily." A low-density gas will have a lot of space between molecules, so a higher-density gas will expand more easily into the low-density gas.

(B) Using the chamber with the sliding lid allows for a changing volume at a constant amount of air, so that density can be the dependent variable in the experiment. The area of the lid is constant, so you can determine volume—using the meter stick—as the piston length changes. Using the large tubs of boiling water and ice water to heat and cool the chamber allows the independent variable to be the temperature, measured with the thermometer.

X	Chamber with sliding piston lid with a known area
	Pressure monitor
X	Meter stick
X	Large tub of boiling water

	Chamber with fixed lid with a known area
X	Thermometer
	Stopwatch
X	Large tub of ice water

(C) 1. Place the thermometer in the chamber with the sliding piston.
2. Place the chamber in the ice water and let it cool.
3. Record the temperature with the thermometer and the piston length with the meter stick.
4. Heat the chamber by placing it in the hot water.
5. Record the temperature and the piston length.
6. Repeat steps 4 and 5 to get temperature and volume data.

(D) The density is n/V, so since you have data for Δx and you know the area, plot $\frac{1}{\Delta x}$ versus temperature. The shape of the graph will tell you whether the increase in temperature causes the density to increase or decrease.

(E) Based on the Ideal Gas Law,

$$PV = nRT$$

$$P = (n/V)RT$$

$$(n/V)^{-1} = \left(\frac{R}{P}\right)T$$

you can expect to see an inverse relationship.

2. (A) The total resistance in the circuit is $r + R$. Using Ohm's Law, the current is then $I = \frac{\varepsilon}{r + R}$.

   (B) Power is $P = IV = \left(\frac{\varepsilon}{r + R}\right)\varepsilon = \frac{\varepsilon^2}{r + R}$.

   (C) Temperature increases when the average kinetic energy of a molecule increases. When the switch is closed, the electrons move through the circuit. As the electrons move within the metal of the circuit, they collide with the positive nuclei of the atoms. The nuclei are set into motion, so the resistor heats up. The Second Law of Thermodynamics states that energy will flow from high-energy to low-energy parts of a system. The molecules in the water are moving with a lower energy than the hot resistor, so as the water molecules collide with the resistor, energy transfers into the water molecules. The now heated water molecules transfer energy to one another through further collisions, with the colder, slower molecules heating up and speeding up. The stirrer helps to ensure that the average energy is equal throughout the water by dispersing the fast-moving water molecules so that they transfer energy throughout the container and not just locally near the resistor.

   (D) Energy can be transferred to the water molecules by the motion of the stirrer as well as by the high-temperature molecules in the resistor. If the stirrer spins at a rapid enough rate, on average, the water molecules will be sped up simply by its motion, thus increasing the temperature of the water.

   (E) The pressure in the container is a result of the collisions between the molecules within the container and its walls. As the molecules heat up and move more rapidly, they will collide more frequently with the walls of the container and cause the pressure against the walls to increase.

3. (A) Between the plates, there is an electric field that points from one of the plates toward the other plate. As the particles change their position within a potential difference, work is done on the particles, causing them to accelerate. After leaving plate 2, the particles are subjected to a magnetic field. Because the magnetic field is always perpendicular to the direction of motion, the magnetic field does no work, and the particles do not change their speed.

(B) Using the Work-Energy Theorem for each particle

$$W = q\Delta V = \frac{1}{2}mv_f^2 - \frac{1}{2}mv_i^2$$

v_i is zero for each particle, and each particle has the same ΔV and q, so

$$v_f = \sqrt{\frac{2q\Delta V}{m}}$$

The ratio for the speed of the proton to the electron is the square root reciprocal ratio of the masses:

$$\frac{v_{fp}}{v_{fe}} = \sqrt{\frac{m_e}{m_p}} = \sqrt{\frac{9.11 \times 10^{-31}\text{ kg}}{1.67 \times 10^{-27}\text{ kg}}} = 0.023$$

(C) Both charges will travel on circular paths; however, the radius of the proton's path will be greater due to its greater mass.

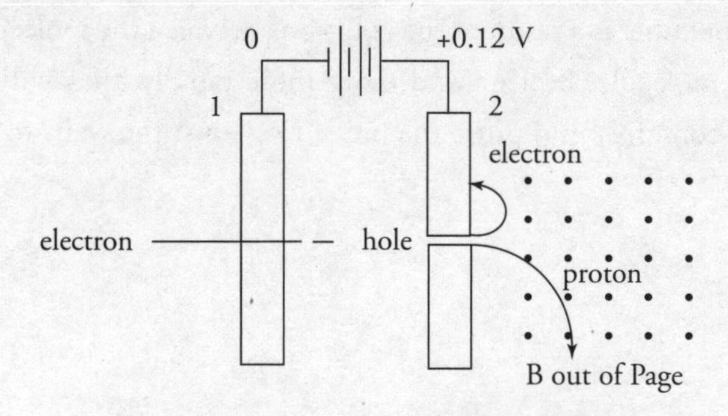

(D) The alpha particle will have twice the charge and four times the mass of the proton. Because

$$v_f = \sqrt{\frac{2q\Delta V}{m}}$$

the final speed coming out of the hole will decrease by $\sqrt{2}$. The path curves because the magnetic force is the centripetal force.

$$\frac{mv^2}{r} = qvB$$

$$r = \frac{mv}{qB} = \frac{4m_p\left(v_p/\sqrt{2}\right)}{2q_pB} = \sqrt{2}r_p$$

The radius of the circle will be $\sqrt{2}$ times as large, but the curve will be in the same direction.

4. (A)

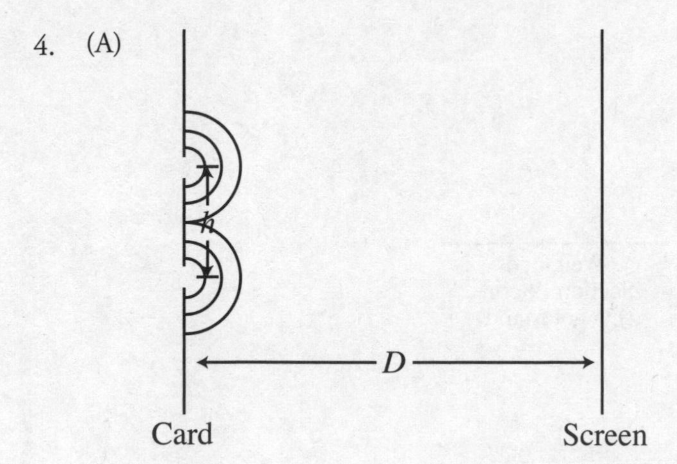

(B) The double-slit experiment works by producing coherent spherical wavefronts that interfere with each other. A spherical wavefront is produced when a plane wave impinges upon a small slit, such as either of the slits in the card, as shown in part (A). The two spherical wavefronts then propagate to the screen, and because there are two waves, they interfere with one another. Bright spots occur whenever there is constructive interference, while dark spots indicate that the interference is destructive. The interference will be constructive whenever the difference in the distance that the two waves travel is an integer multiple of the wavelength of the initial plane wave. There will be bright spots when the difference in traveled distance is 0 (in the middle), or ±1 wavelength, etc. There are a series of bright spots because there are many different possible locations on the card where constructive interference occurs.

(C) In the double-slit experiment, $m\lambda = dx/L$. Solving for λ,

$$\lambda = \frac{dx}{Lm}$$

d is the distance between the slits, x is the distance measured between fringes, L is the space from the slits to the screen, and m is the "order" or the number of dark spots between fringes. Using the given variables,

$$\lambda = \frac{hy_3}{3D}$$

(D) When light enters a medium with a different index of refraction, its speed changes. The light wave will have a constant frequency, so this change in speed results from a change in its wavelength. When light goes from an index of $n = 1$ to an index of $n > 1$, the wave speed decreases, resulting in a decreased wavelength.

$$\lambda = \frac{dx}{Lm}$$

So x is directly proportional to λ, and the distance between fringes will decrease.

HOW TO SCORE PRACTICE TEST 2

Section I: Multiple Choice

______ × 1.50 = ______

Number Correct (out of 50) | Weighted Section I Score (Do not round)

Section II: Free Response

Question 1: ______ (out of 12) × 1.7045 = ______ (Do not round)

Question 2: ______ (out of 12) × 1.7045 = ______ (Do not round)

Question 3: ______ (out of 10) × 1.7045 = ______ (Do not round)

Question 4: ______ (out of 10) × 1.7045 = ______ (Do not round)

Sum = ______

Weighted Section II Score (Do not round)

AP Score Conversion Chart Physics 2	
Composite Score Range	AP Score
107–150	5
90–106	4
73–89	3
56–72	2
0–55	1

Composite Score

______ + ______ = ______

Weighted Section I Score + Weighted Section II Score = Composite Score (Round to nearest whole number)